Geophysics of the Canary Islands

Results of Spain's Exclusive Economic Zone Program

Edited by

Peter Clift and Juan Acosta

Reprinted from *Marine Geophysical Researches*, Volume 24, Nos. 1-2, 2003.

Library of Congress Cataloging-in-Publication Data

A. C.I.P. Catalogue record for this book is available from the Library of Congress

ISBN 1-4020-3325-7

Published by Springer,
P.O. Box 17, 3300 AA Dordrecht, The Netherlands.

Sold and distributed in the U.S.A. and Canada
by Springer,
101 Philip Drive, Norwell, MA 02061, U.S.A.

In all other countries, sold and distributed
by Springer,
P.O. Box 17, 3300 AH Dordrecht, The Netherlands.

Printed on acid-free paper

Printed in the Netherlands

Table of Contents

Digital Terrain Model of the Canary Islands EEZ. Courtesy of Multibeam Maping Group, Instituto Español de Oceanografia, Madrid, Spain.

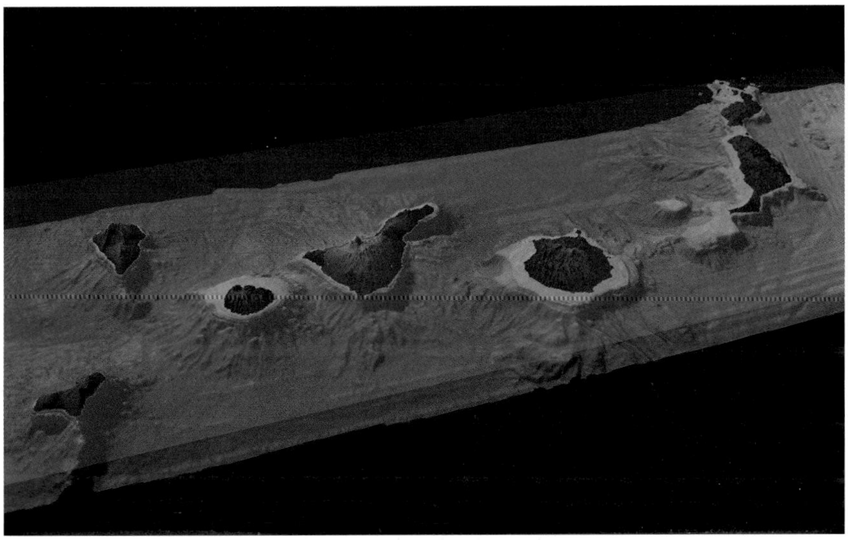

Foreword of the Director of the Oceanographic Spanish Institute

The eight papers contained in this special MGR issue reflect partly the successful collaboration between a group of institutions and researchers working on a major research program since 1995: *the Spanish EEZ Program* (Exclusive Economic Zone Program).

The Spanish EEZ program was set up in 1995 after a political mandate of the Spanish government to prepare a full study - comprising hydrography, geology, geophysics and oceanography - of the Spanish EEZ. Leaders of the study were the Instituto Espahol de Oceanografia (IEO) and the Instituto Hidrográfico de la Marina (IHM).

Since its beginning, the program disposed one month per year on the use of the R/V *Hespérides* and other Spanish oceanographic research ships like the R/V *Cornide de Saavedra,* the *Vizconde de Eza* and the *Tofiño*.

Other research groups, universities and private companies also have collaborated in the sea-cruises.

Present special issue shows the first scientific results of the EEZ Program in the Canary Islands area. Moreover, due to its scientific quality, the Program could be considered as an example of cooperation between different institutions, teams and individuals working on the same project.

Finally, the IEO, which supports part of the program, should express its deep gratitude to all captains, crews and technicians of the different oceanographic vessels that participated in the research campaigns. Their work made this issue possible.

Concepción Soto
Directora del IEO

Foreword of the Director of the Hydrographic Marine Institute

Article 132.2 of the Constitutional Law sets the grounds for public property of the natural resources of the Spanish Exclusive Economic Zone (ZEE for its Spanish acronym), implying the sovereign right to explore, exploit, conserve and administer all living and non-living resources coming from the sea-bed and the adjacent waters of the maritime area that goes from the end of the territorial sea to a distance of two hundred sea miles, counting from the base line from where its width is measured.

According to the agreement of the Council of Ministers of April 23rd, 1993, the Ministry of Defence is authorized to use the BIO "Hespérides" to carry-out research campaigns for data gathering during one month per year.

The Hydrographical and Oceanographic Research Plan of the Spanish Exclusive Economic Zone (ZEE) was approved by Ministerial Order 55/1994 of May 30th, amplified by the Ministerial Order 94/1993 of September 21st, in which the FAS Cartographic Plan was approved.

On May 25th of 1994, the framework cooperation agreement between the Ministry of Defence and the Spanish Oceanographic Institute (IEO for its Spanish acronym) concerning Hydrographical and Oceanographic Research on the ZEEE was approved. This Framework Agreement assigns the hydrography of the area to the Marine Hydrographic Institute, using the multi-beam sounding devices installed on the BIO "Hespérides"; and the data gathering that leads to a better knowledge of the physical structures of the seabeds to the IEO.

During the years 95, 96 and 97 data gathering has been carried-out on the Balearic Islands and during the years 98, 99 and 2000 on the Canary Islands with campaign heads coming alternatively from IHM and IEO personnel.

The results of the various campaigns were extraordinaire, especially because of the very close collaboration between the participating Institutions, more specific with the IEO, a collaboration of which the results clearly can be seen in this scientific work about the Canary Archipelago and in future joint works that surely will be carried-out

CN. D. Fernando Quirós Cebriá
Director del Instituto Hidrográfico de la Marina

Marine Geophysical Researches (2003) 24: 1–40
10.1007/s11001-004-1513-3

Geologic evolution of the Canarian Islands of Lanzarote, Fuerteventura, Gran Canaria and La Gomera and comparison of landslides at these islands with those at Tenerife, La Palma and El Hierro

J. Acosta[1,*], E. Uchupi[2], A. Muñoz[1], P. Herranz[1], C. Palomo[1], M. Ballesteros[1] & ZEE Working Group[3]

[1]*Instituto Español de Oceanografía. Grupo de Cartografía Multihaz. Corazón de María, 8, 28002 Madrid*

[2]*Woods Hole Oceanographic Institution, Woods Hole, MA 02543, USA*

[3]*A.Carbó, A. Muñoz-Martín, Univ.Complutense, Madrid; J.Martín-Dávila, M.Catalán and J.A. Marín, Real Observatorio de la Armada. S.Fernando, Cádiz; F.Pérez-Carrillo, C. Maté, Instituto Hidrográfico de la Marina. Cádiz.*

**Corresponding author (E-mail:juan.acosta@md.ieo.es)*

Key words: multibeam mapping, Canary Island, avalanches, geomorphology

Abstract

In this paper we discuss the results of a swath bathymetric investigation of the Canary archipelago offshore area. These new data indicate that volcanism is pervasive throughout the seafloor in the region, much more that would be suggested by the islands. We have mapped tens of volcanic edifices between Fuerteventura and Gran Canaria and offshore Tenerife, La Gomera, El Hierro and La Palma. Volcanic flows are present between Tenerife and La Gomera and salic necks dominate the eastern insular slope of La Gomera. This bathymetry also supports land geologic studies that indicate that the oceanic archipelago has acquired its present morphology in part by mass wasting, a consequence of the collapse of the volcanic edifices. In the younger islands, Tenerife, La Palma and El Hierro, the Quaternary (1.2 to 0.15 Ma) debris avalanches are readily recognizable and can be traced offshore for distances measured in tens of km. Off the older islands, Lanzarote, Fuerteventura, Gran Canaria and La Gomera (<20 to 3.5 Ma), the avalanches have been obscured by subsequent turbidity current deposition and erosion as well as hemipelagic processes. The failure offshore western Lanzarote is in the form of a ramp at the base of the insular slope bound on the seaward side by a scarp. Its size and the lack of evidence of rotation along its landwards side precludes the possibility that it is a slump. It probably represents a slide whose outer scarp is caused by break-up of the slide. Mounds on the ramp's surface may represent post-displacement volcanic structures or exotic blocks transported to their present locations by the slide. The failures offshore Fuerteventura are so large that, although they occurred in the Miocene-Pliocene, exotic blocks displaced from upslope are still recognizable in the insular margin morphology. The Canary Island insular margin appears to be a creation of Miocene-Pliocene mass wasting and more recent turbidity current deposition and erosion, and hemilepagic deposition. Failures offshore La Gomera are due to debris flows and/or turbidity currents. These events have obscured earlier mass wasting events.

Introduction

In this study we use multi-beam data acquired in the Canary Islands by the Instituto Español de Oceanografia to determine the role that landslides have played in the construction of the islands present morphology. In the course of our investigation we discovered that volcanic edifices are quite extensive, much more that expected from the extent of volcanism in the islands. Volcanic structures were imaged by multibeam recordings between Fuerteventura and Gran Canaria and offshore Tenerife, La Gomera, El Hierro and La Palma. Probable volcanic flows also were imaged between Tenerife and La Gomera as

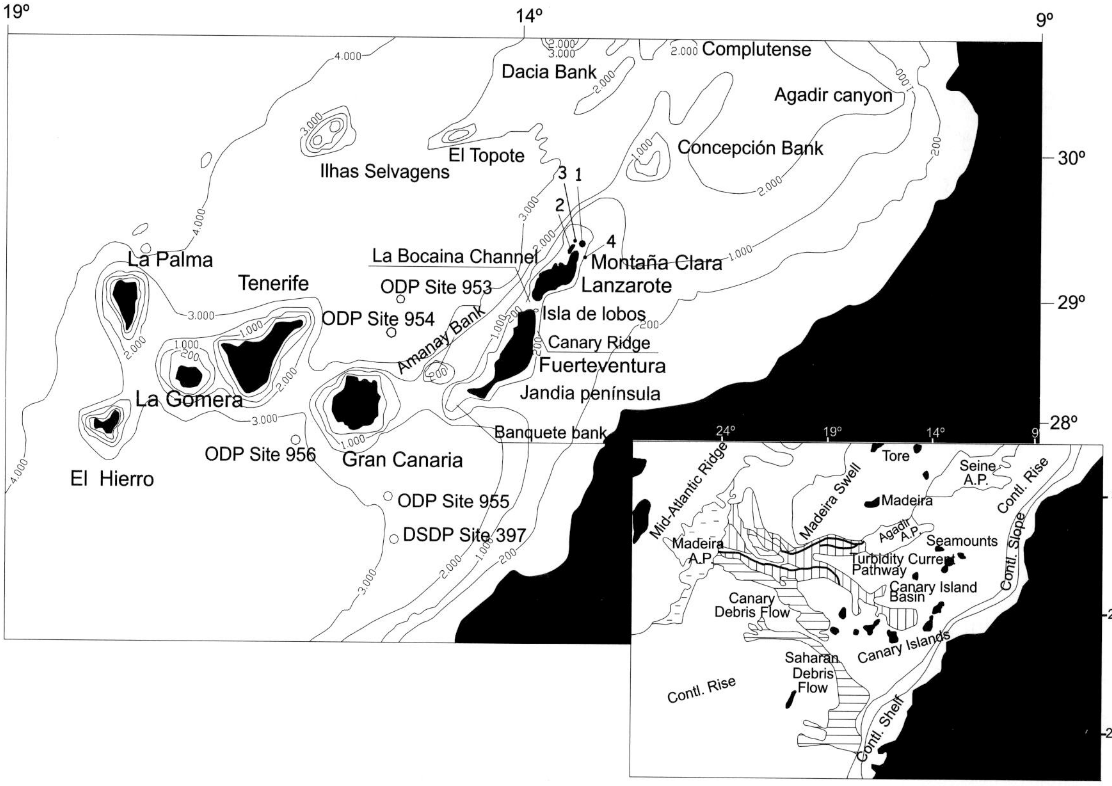

Figure 1. Submarine topography of the Canary Island archipelago region. Bathymetric contours are from GEBCO sheet 5.08 and base map from U.S. Naval Oceanographic Office 51017. Insert map is modified from Wynn et al. (2000). Contour in meters. 1= Alegranza; 2= Graciosa; 3= Roque del Oeste; 4= Roque del Este.

well as salic necks on the eastern insular slope of La Gomera. The most spectacular of these features are the mega-avalanches, particularly those off Tenerife, El Hierro and La Palma. As other studies have demonstrated, landslides are amongst the most significant processes in the creation of the morphology of mature oceanic volcanic islands. They have been reported from the Hawaiian Islands (Moore et al., 1989, 1994), Reunion Island (Labazuy, 1996) and the Canary Islands (Watts and Masson, 1995; Urgeles et al., 1997, 1999; Carracedo et al., 1999a, 1999b). Moore et al. (1989, 1994) recognized at least 68 such flows off Hawaii, some of which are 200 km long, incorporate as much as 5000 km^3 of volcanic material and cover an area of 100,000 km^2. Mass wasting facies off Hawaii take two forms: (1) slow moving slumps, up to 110 km wide and up to 10 km thick, characterized by transverse blocky ridges and steep toes that are up to 230 km long, and (2) 0.05–2 km thick, fast moving debris avalanches (Moore et al., 1989). To date eleven giant slides also have been mapped in the Canary Islands affecting the subaerial and submarine slopes of the islands of La Palma, El Hierro, Tenerife, Fuerteventura and Gran Canaria (Navarro and Coello, 1989; Holcomb and Searle, 1991; Carracedo, 1994, 1996; Masson and Watts, 1995; Watts and Masson, 1995; Masson, 1996; Masson et al., 1997; Guillou et al., 1998; Urgeles et al., 1998, 1999; Teide Group, 1997; Stillman, 1999). Masson et al. (2002) have summarized the results of these investigations.

Regional setting

The Canary archipelago is located on the continental rise off Cape Juby, northwest Africa (Figure 1). Fuerteventura and Lanzarote at the eastern end of the chain are 100 km from the African coast, and El Hierro and La Palma at its western end are 500 km from the coast. Lanzarote and Fuerteventura are along the crest of the northeast trending Canary Ridge, on the upper continental rise at a water depth of about 2000 m. Its northeast terminus is defined by the less than 200 m deep, flat-topped Conception Bank. This ridge may be

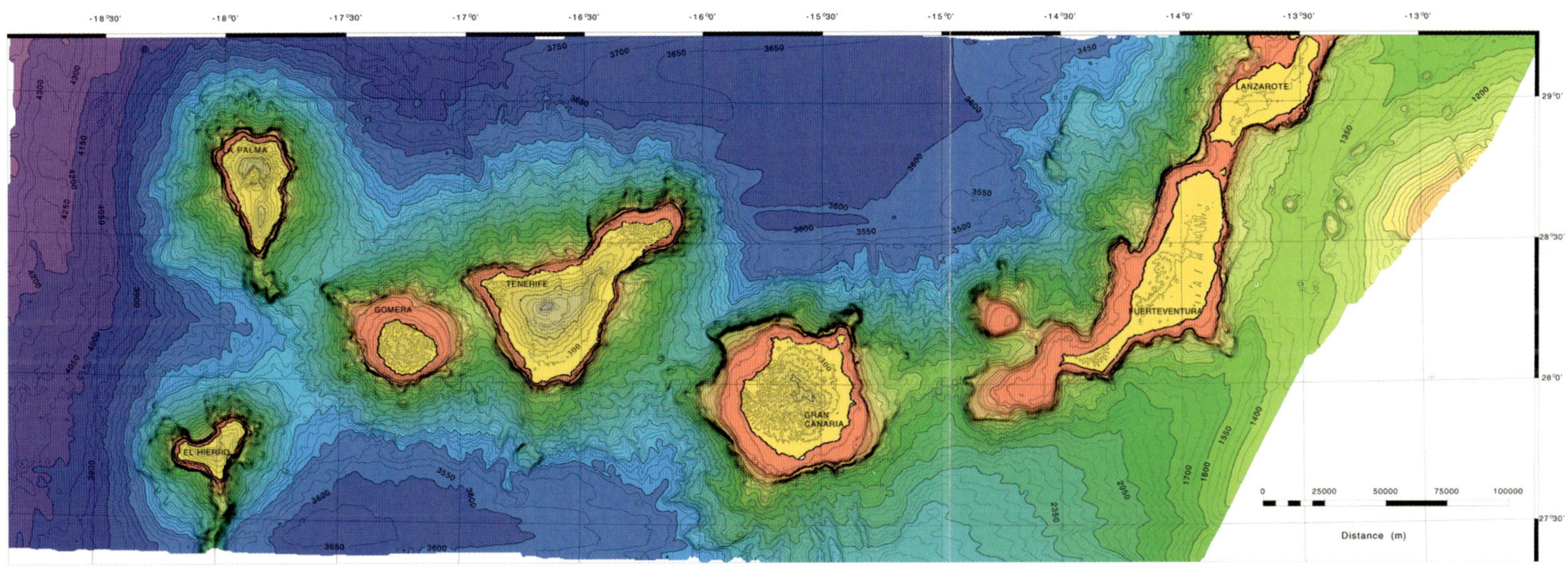

Figure 2. Topographic map of the Canary archipelago region. Offshore contours in meters are based on multi-beam echo sounding data collected during the present investigation and onshore ones also in meters from data provided by Servicio Geográfico del Ejercito, Cartografía Digital, 100 by 100 m grid, Madrid.

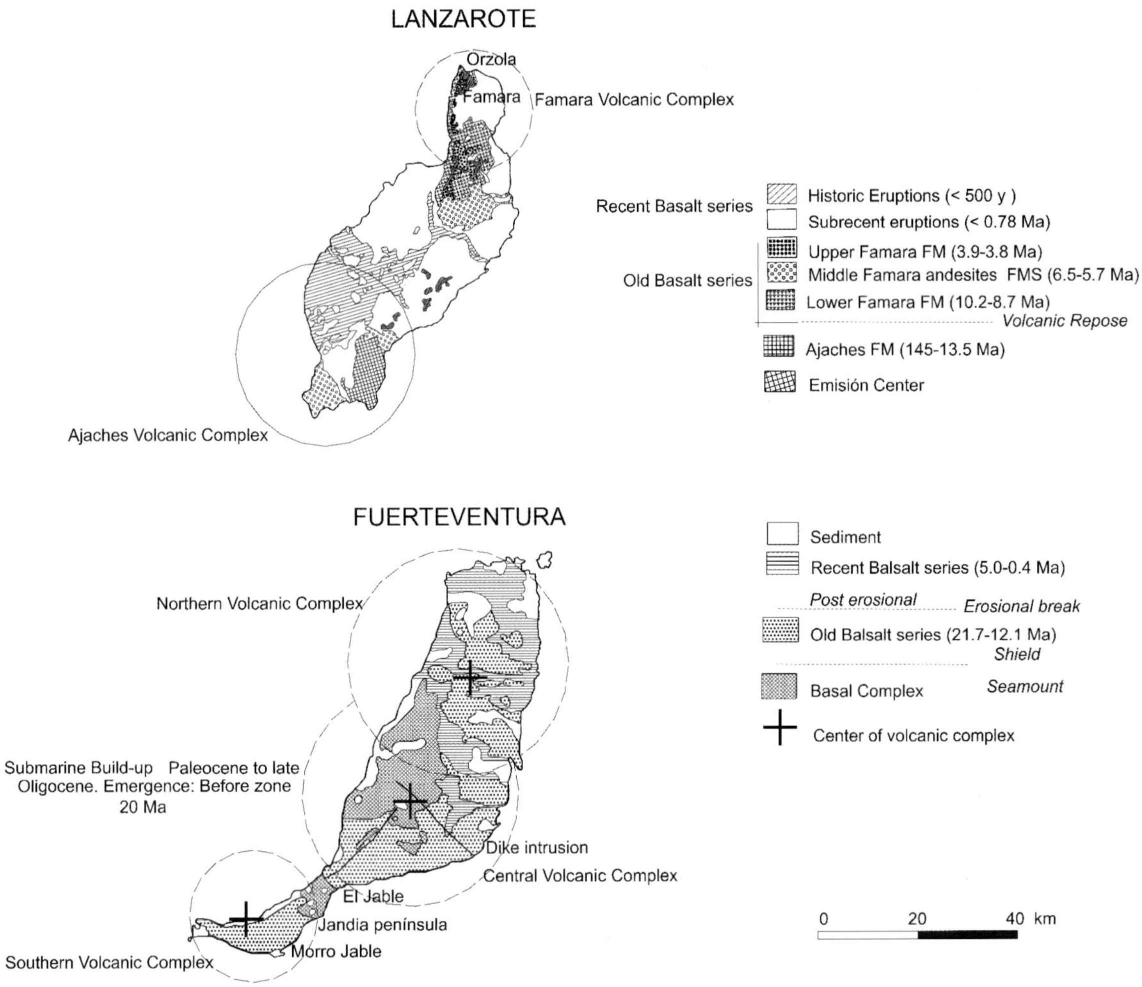

Figure 3. Geologic maps of Lanzarote and Fuerteventura. Compiled from Coello et al. (1992), Carracedo and Rodríguez-Badiola (1993), Ancochea et al. (1996) and Stillman (1999).

aligned along the contact between attenuated continental crust on the east and oceanic crust on the west (Emery and Uchupi, 1984). The rest of the archipelago, Gran Canaria, Tenerife, La Gomera, La Palma and El Hierro, at water depths of 3000 to 4000 m, are on oceanic crust of Jurassic age (Uchupi et al., 1976). Gran Canaria, Tenerife and La Gomera trend east-west parallel to the oceanic fracture zone trends in the region with Tenerife's long axis being oblique to this trend. La Palma and El Hierro, at the western end of the island chain, are offset to the north and south of this trend (Figure 1).

A subsurface oceanic basement high appears to link La Palma to Ilhas Selvagens (Figure 1) (Uchupi et al., 1976). From the Ilhas to the northeast end of the Canary Ridge are two northeast trending ridges, subparallel to the Canary Ridge, along whose crests are seamounts (Dañobeitia and Collette, 1989). One of these volcanic structures, Dacia Seamount, is flat-topped. Sediments in the lows between the ridges grade south into the Canary Island Basin west of the Canary Ridge and north into the Agadir Canyon with the drainage divide located near 30°30′ N (Figure 1). The seafloor of the Canary Islands Basin ranges from 3000 m on its eastern side to 4000 m along its western side. Here a slight shallowing of the basin floor to less than 4000 m reflects the subsurface high linking La Palma and Ilhas Selvagens. A gap in the high near 17°30′ W, 29°30′ N serves as passageway for turbidity currents into the Madeira Abyssal Plain to the west (insert, Figure 1). Gaps between the Canary Islands along the southern side of the basin serve as passageways for the southerly flowing North Atlantic Deep Water (NADW) at a depth of 2000-3800 m and northerly flowing Antarctic Bottom Water (AABW) below 3800 m.

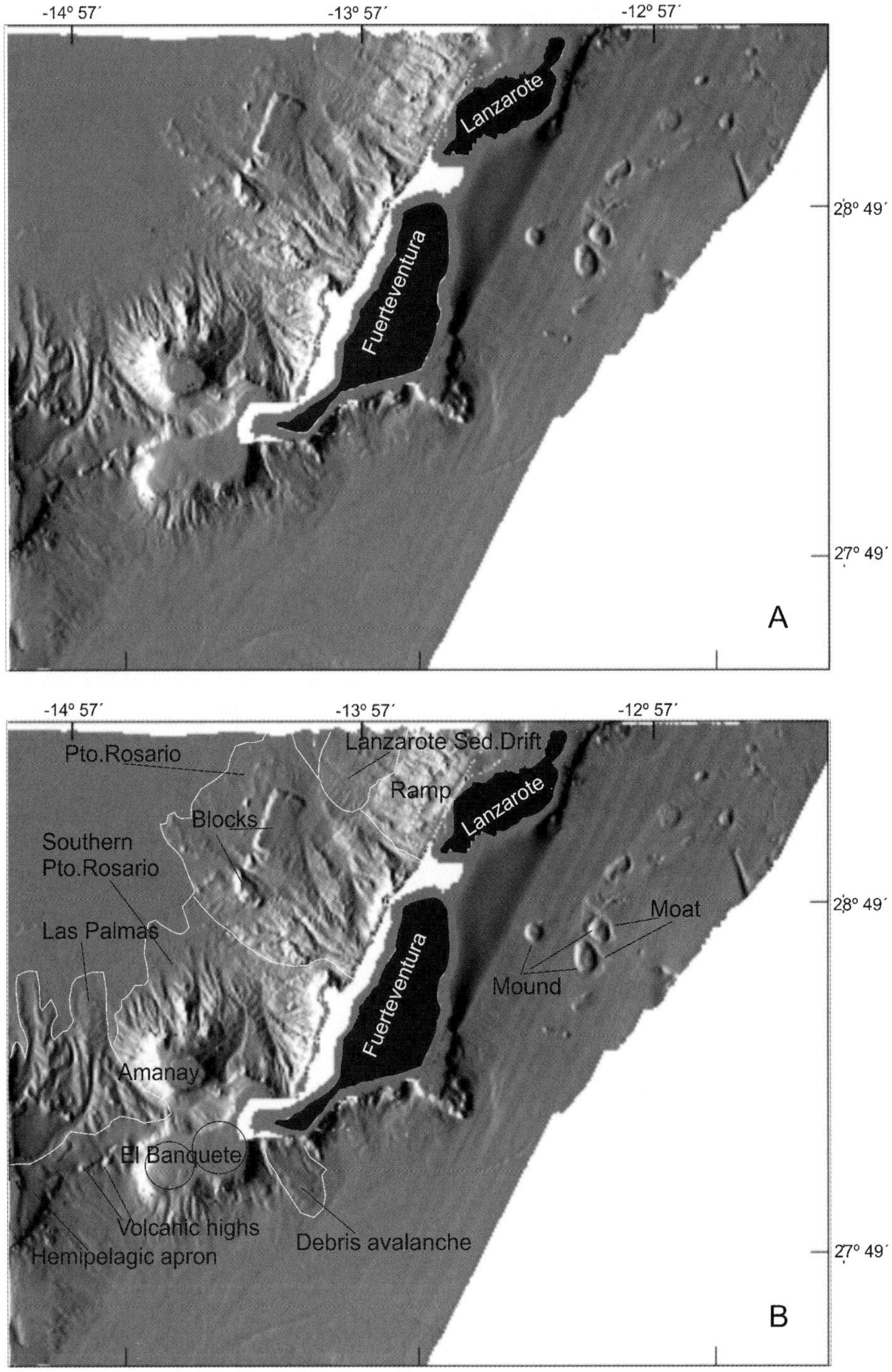

Figure 4. **A**. Shaded relief image of the western margins of Lanzarote and Fuerteventura and side slopes of Banquete and Amanay Banks . Illumination is from northwest. **B**. Shaded relief image of Lanzarote and Fuerteventura margins and Banquete and Amanay Banks showing the distribution of landslide areas west and south of Fuerteventura (Puerto Rosario, Southern Puerto Rosario and Las Palmas debris avalanches) and east of Gran Canaria. We infer that the highs along the northwest edge of the Puerto Rosario avalanche are exotic blocks. No such features appear to be present off Lanzarote where the margin consists of westerly dipping platform cut by northwest aligned channels and ridges some of which are capped by circular highs. This terrain probably is the creation of slumping and subrecent and historical lava lava flows. The seafloor beyond the platform is dominated by northeast trending highs and lows. We infer that this morphology is the creation of southerly flowing North Atlantic Deep water (NADW) and named the region the Lanzarote Sediment Drift. The mounds and moats east of Lanzarote and Fuerteventura are described and discussed in Acosta et al. (this issue).

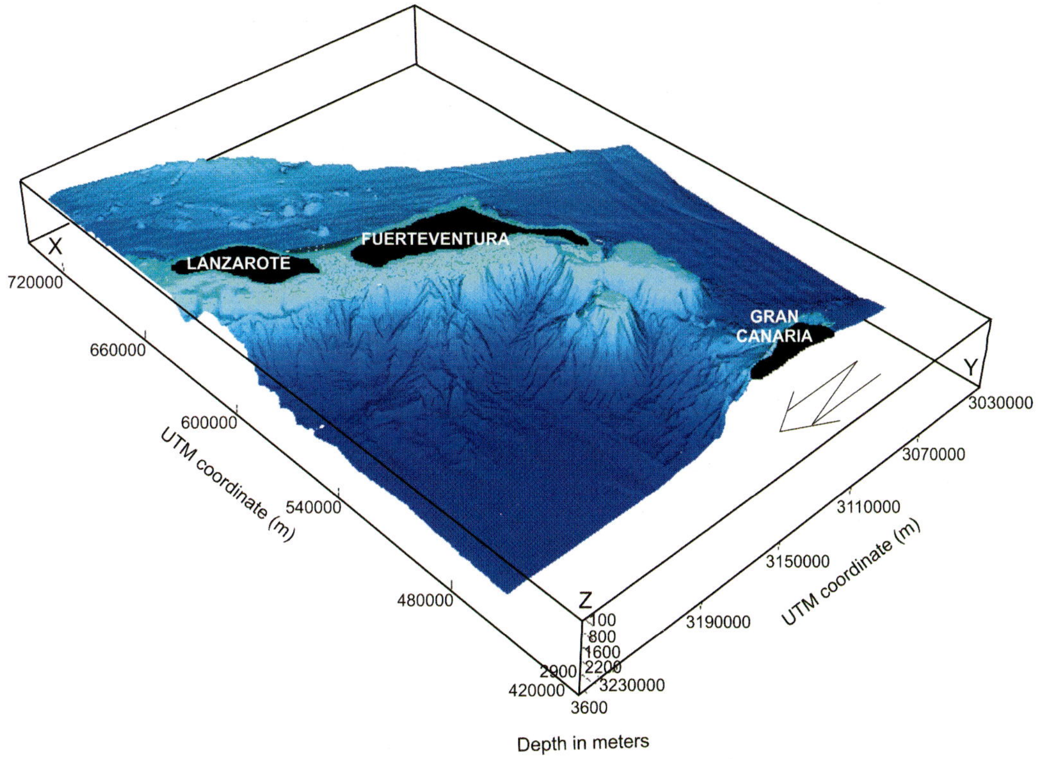

Figure 5. 3D diagram of the western margin of Lanzarote and Fuerteventura looking north. The two flat-topped highs south of Fuerteventura are Banquete (attached to Fuerteventura) and Amanay Banks. The high at the extreme right of the diagram is the tip of Gran Canaria.

Volcanism in the Canary Islands region has occurred at various times since the Late Cretaceous, the oldest represented on Fuerteventura, to the present in the western, La Palma and El Hierro (Le Bas et al., 1986). Based on their eruptive histories Carracedo (1994) divided the islands into three groups, those that have had eruptions in historic times (<500 years; Tenerife, La Palma; Lanzarote and probably El Hierro), those with a history of Quaternary volcanism (Fuerteventura and Gran Canaria) and those lacking evidence of Quaternary volcanism (La Gomera). Individual islands appear to have gone through four geomorphic phases: a seamount phase, a shield-building submarine and subaerial phase characterized by rapid growth and massive slope failures, a period of quiescence and deep erosion (erosional gap), and post-erosional stage of volcanic activity. In some islands the last phase also is characterized by renewed mass wasting. Lanzarote, Fuerteventura and Gran Canaria are in the post-erosional phase, Gomera in the repose stage (gap stage) and Tenerife, La Palma and El Hierro are in the shield stage of development (Carracedo, 1999). The tectonic setting of the Canary Islands consists of rift-type clusters of aligned eruptive vents (single or triple) and caldera-type depressions (Carracedo, 1994). Where the wide arcuate landslide depressions are related to triple rift geometry, they tend to be located at the junction between the two most active rifts with the third one acting as buttress for the landslide. It is the development of these triple rifts and the concentration of dikes leading to the destabilization of the flanks through magma overpressure and mechanical and thermal overpressure of pore fluids that lead to gravitationally unstable volcanic flanks (Carracedo, 1994, 1996; Elsworth and Day, 1999).

Methods

In 1995 the Instituto Español de Oceanografía and the Instituto Hidrográfico de la Marina began an investigation of the Spanish Exclusive Economic Zone. During this investigation the bathymetry of the region was mapped using a multibeam system during which 100 per cent coverage was obtained. At the same time high-resolution parametric seismic reflection profiles,

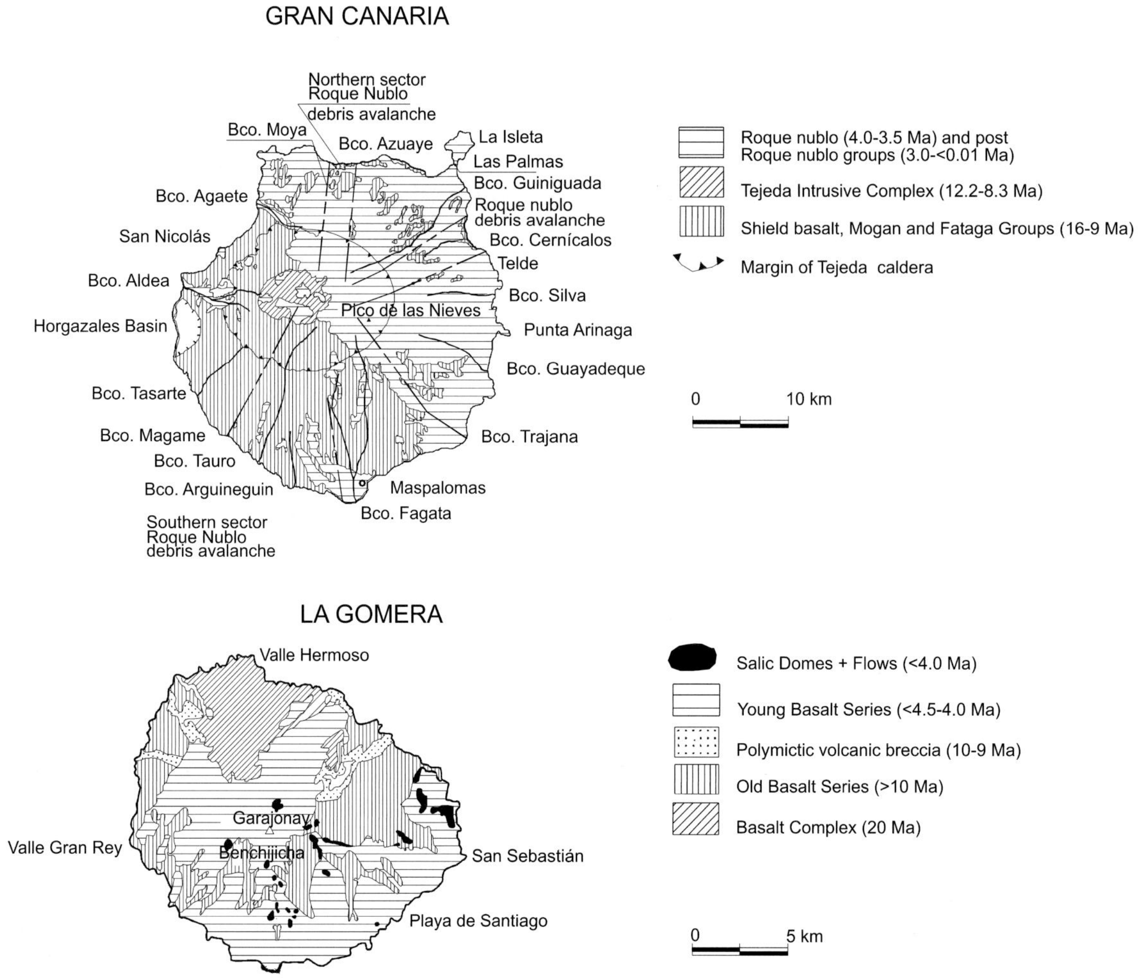

Figure 6. Geologic maps of Gran Canaria and La Gomera. Compiled from Hausen (1965), Cantagrel et al. (1984), Funck and Schmincke (1998), Mehl and Schmincke (1999), Van den Bogaard and Schmincke (1998) and Schmincke and Sumita (1998).

as well as gravity and magnetic, were also recorded in the surveyed areas (Muñoz et al., 1998). This paper describes the results of the multibeam bathymetric investigation of the Canary archipelago from 27° 10′ N to 29° 15′ N and from 13° 30′ W to 19° 15′ W.

The survey of the archipelago during four cruises from 1998 to 2001, aboard the R/V Hespérides and R/V Vizconde de Eza, was carried out using a variety of multibeam sounding systems that were used separately or in combination. Simrad EM1000, EM 1002 and EM 300, in conjunction with GPSD and inertial navigation systems, were used to survey shallow waters and EM12S in deep waters. Acoustic backscatter data were displayed in real time by means of sidescan sonar trace. A Microsystems SV-Plus Velocimeter was employed to correct depths for variation of sound velocity in the water column. The multibeam data were logged and post-processed in a Unix environment. Simrad Mermaid-Merlin software was used at sea for logging and real time quality assessment of the acquired data using color swaths bathymetric plots. As data collection per day in excess of 300 Mbytes is not uncommon, data validation is a major problem with multibeam systems. A Neptune package was used in post-cruise processes of the data. This post-cruise processing included the application of cleaning and editing tools for attitude, navigation and bathymetric data. The use of statistical analyses was of considerable help in detecting spurious data. The use of Geographic Information Systems (Cfloor from Roxar and IberGIS from ICI) allowed us to create, not only bathymetric maps, but also digital terrain models and 3D block diagrams of the surveyed area. In addition to computer mapping, the application of GIS also allowed complex spatial analyses of data. Thus, for the first time, the availability of such swath bathymetric and terrain maps of the whole Canary archipelago allows us to appreciate the extent of catastrophic slope failures. In the present study we describe for the first time the landslides that contributed to the present morphology of the seafloor surrounding the older islands of Lanzarote, Fuerteventura, Gran Canaria and La Gomera. We also incluce a brief discussion on the landslides of the younger islands, Tenerife, La Palma and El Hierro.

Recognition of gravity driven facies

Mass wasting is important in the geologic evolution of oceanic islands edifices during the shield-building phase. As described by Normark et al. (1993) landslides produced by the collapse of volcanic edifices take various forms ranging from slumps, debris avalanches to debris flows. Slump movement is an elastic-plastic flow involving rigid blocks that are internally undeformed and are displaced along curved slip surfaces. If not modified by concurrent or subsequent volcanism, slumps tend to have well-defined amphitheaters at their proximal ends that broadly correspond to the detachment surfaces of the slumps. The displaced material is generally found on the volcano flanks, reflecting the limited seaward transport of the displaced strata. The surface of the slump structures may be characterized by transverse ridges and scarps and on their seaward ends by scarps onlapping the undisturbed seafloor. These structures are rare in the Canary Islands and to date have only been described in El Hierro (Masson et al., 2002) and off the eastern side of Gran Canaria (Funck and Schmincke, 1998). Data from the present investigation suggest that such a structure may be present along the western margin of Lanzarote.

Debris avalanches cutting the flanks of volcanic edifices are elastic flows made up of internally rigid blocks that roll, slide and glide along shear planes. The avalanches are longer and thinner than slumps, with their inner side being marked by embayments. In Hawaii avalanches display a middle and distal train of hummocky debris (Normark et al., 1993) and scattered over their surfaces are blocks tens of km in horizontal dimension on their proximal ends to <1 km on their distal end. Jacobs (1995) proposed that avalanches might be the end result of slumping and suggested that as slumps travel across uneven slopes they break down and accelerate into avalanches. Why some terrains fail by slumping, while others having the same structure and petrology fail by formation of debris avalanches is yet to be resolved.

Debris flows are plastic flows that are characterized by shear throughout the flow. Such a flow may develop when a debris avalanche breaks apart. As defined by Masson et al. (2002) such flows only affect the sedimentary cover of the submarine island slopes, whereas debris avalanches and slumps cut into the extrusive and intrusive rocks of the island. Such a debris flow in the Canary Islands is the Canary Debris Flow off El Hierro that supposedly was triggered by the El Golfo Debris Avalanche. Both debris avalanches and debris flows in turn can evolve into turbidity currents, a viscous fluidal flow (Varnes, 1958).

Turbidity currents triggered by debris avalanches and debris flows are recorded in the Madeira Abyssal Plain. Volcanic turbidites on this plain go back 17 Ma with their occurrence increasing markedly at 7 Ma, a time when the volcanic edifice in Tenerife was growing (Weaver et al., 1998). According to Masson et al. (2002) turbidites during the last 7 my reached the Madeira Abyssal Plain every 100 ky. Thus Masson et al. postulated that if each turbidite event documents a debris avalanche in the Canary Island archipelago, then the 80 volcaniclastic turbidites at Ocean Drilling Program ODP Site 951 in the Madeira Abyssal Plain since 7 Ma is a minimum record of the volcanic collapses in the Canaries during that time.

In the present investigation we use data provided by the multi-beam bathymetric map and relief and 3D diagrams to identify the facies described above. Even though avalanche scars and valleys have been modified by later lava flows, sedimentation and by subaerial and submarine erosion, scars created by mass wasting can still be recognized in Gran Canaria and exotic blocks on the Fuerteventura insular margin. Slope segments not affected by such avalanches tend to terminate abruptly down slope, are more irregular and much steeper than scarps created as a result of an avalanche; they have gradients as high as 30°. In contrast, those slope segments affected by avalanches tend to be smoother than the original rough volcanic slope and have gradients ranging from 10° on mid slope to 5° on the lower slope/upper rise (Gee et al., 2001a; Masson et al., 2002). Slope segments created by avalanches also can be distinguished from those created by turbidity currents by their concave upward profile, their low relief and their linear flat-bottomed channels. In contrast, slopes created by turbidity currents are convex upward and are characterized by a high relief trunk tributary/distributary system (Masson et al., 2002). V-shaped channels in a slope created by an avalanche are restricted to the uppermost insular slope and merge down slope into a single flat-bottomed low, a channel that lacks distributaries and maintains its character to its distal end. These valleys also display features that have not been observed in the V-shaped turbidity current canyons, such as arcuate steps and longitudinal parallel ridges undulating down channel disrupting the valley floors. Such ridges, described from known volcanic avalanches, are either the creation of compression or scour. If compressional, they were created by velocity differences in adjacent flows such as in the ridges in the Mount St. Helens Avalanche, offshore British Columbia and offshore Norway (Voight et al., 1981; Prior et al., 1982; Bugge et al., 1988). The distal ends of avalanches consist of lobes onlapping each other and the undisturbed deep-sea sediments beyond the catastrophic flow. The surface of these lobes tends to be disrupted by linear hummocks and lows that Voight et al. (1981) have interpreted as grabens and horsts formed as a result of lateral spread within the apron.

The most unique feature displayed by 3D diagrams created from the multibeam surveys is the rough surfaces of the avalanches. Side-looking sonar and seismic reflection recordings demonstrate that some of these features represent post-avalanche volcanic cones and others represent exotic blocks scattered over the surface (Moore and Clague, 2002). The dimensions of the exotic blocks (hundreds of meters long and tens of meters thick) are clear evidence that they could only have been transported by a massive flow, not by turbidity currents. The lack of coherent deformation structures (e.g. transverse fault zones and scarps) in the avalanche deposits is another argument against slow and episodic slumping. Furthermore, the distances of the exotic blocks from probable sources also demonstrate that they could not have been transported to their present site by slumping.

Various processes may account for the present location of the blocks. They may have glided along the surface of the avalanche to their present sites with the excess pore water pressure in the avalanche acting as a lubricant. Possibly they were carried to their present location within the flow or along its surface where they were supported by a matrix of fine sediment (Bugge et al., 1988; Lee et al., 1993). As the blocks retained their identity during their transport it indicates that the blocks were more resistant to the internal shear of the flow than the other material making up the avalanche. Teide Group (1997) noted that some of the blocks had enough momentum to outrun the main part of the flow. Offshore Tenerife these blocks also controlled the locations of the post-avalanche turbidity current fans beyond the exotic blocks.

Ages assigned to the various debris avalanche deposits mapped in the Canary archipelago are generally (but not always) based on the ages of the collapse structures onshore from which the debris avalanche deposits appear to originate; alternative approaches include inferences from the thicknesses of later sediment cover as imaged by backscatter measurements, and indirect upslope correlations from major turbidite units in adjacent basins (Masson, 1996; Masson et al., 2002). However, as documented by the bathymetry map compiled by the present swath data, the onshore and offshore structures are not continuous, but are separated by a several hundred meter high scarp on the upper insular slope. Also differences in backscatter measurements between adjacent flows may not reflect different ages, but the presence of a volcanic cover or differences in rate of deposition.

Collapse of the Canarian Volcanoes

Islands in posterosional stage

Lanzarote-Fuerteventura: Inshore

Lanzarote and Fuerteventura, at the eastern end of the Canary Island archipelago, are separated by the less than 50 m deep La Bocaina Channel (Figure 1). They form a contiguous high, the Canary Ridge, with its northern end being defined by Conception Bank at 30–31°N. Ancochea et al. (1996) proposed that the Canary Ridge was constructed by a row of volcanoes aligned subparallel to the African coast, volcanic edifices that were built during several submarine and subaerial phases. The ridge and the islands along its crest appear to be the result of three igneous episodes and one erosional cycle. The earliest volcanic episode, the seamount phase, is represented by the Basal Complex along the west side of Fuerteventura (Figure 3). In Lanzarote this complex may be buried in the center of the island at depths of 900–2700 m below sea level (Coello et al.,1992). The tabular lavas and pyroclastic rocks resting unconformably or comformably on the Basal Complex represents the shield phase. This volcanic phase was followed by an erosional cycle that in turn was followed by another volcanic cycle represented by lava flows and volcanic cones oblique to the long axes of the islands.

Two Miocene volcanic edifices, Famara and Los Ajaches, constructed during the shield phase have been mapped over the Basal Complex in the northeastern and southwestern parts of Lanzarote (Figure 3). Famara, with an elevation of over 600 m, was built during late Miocene-Pliocene eruptive cycles 10.2-8.7, 6.5-5.7 and 4.9-3.9 Ma (Carracedo and Rodríguez Badiola, 1993; Coello et al., 1992; Ibarrola et al., 1988; Hausen, 1959). Los Ajaches edifice, with a maximum elevation of 560 m, was build up in the Miocene between 16 and 12 Ma (Ibarrola et al., 1988; Coello et al., 1992). The western part of this volcano is covered by younger flows (Abdel-Monen et al., 1971; Coello et al., 1992; Carracedo and Rodríguez Badiola, 1993). Ancochea et al. (1996) inferred that Famara was originally 1.0–1.3 km high and Los Ajaches 1.1–1.4 km high. The post erosional volcanic phase in Lanzarote that followed a 2 my long erosional phase following the construction of Famara (Miocene-Pliocene) and Los Ajaches (Miocene) consists of scattered volcanoes and associated lava fields of late Pliocene-Holocene age trending northeast and east-northeast (Coello et al., 1992). Some of these flows were extruded as recently as the 18th century. Raised marine terraces and beaches and Tertiary marine and lacustrine limestone at elevations of 55-60, 20 and 10 m document uplift of Lanzarote since the Pliocene (Hausen, 1959; Coello et al., 1992; Stillman, 1999). 12 Quaternary marine terraces from 0 to 70 m above sea level support such uplift (Zazo et al., 2002). The terraces document an uplift of 1.7 cm/1000 years for Lanzarote and Fuerteventura for the last million years. Present elevation of the last interglacial deposits further suggest that during the last 300,000 years Lanzarote has experienced subsidence of about 0.7 cm/1000 years whereas Fuerteventura has been stable during that time. Limestone and conglomeratic layers composed of rounded pebbles of colored limestone also occur in Graciosa and limestone ejecta in Roque del Este. Sediments similar to these, unsorted coral breccia-conglomerates, also have been described from Molakai and Lanai, Hawaiian Islands. Moore and Moore (1984, 1988), Moore et al. (1994) and Moore (2000) proposed that these sediments, occurring <60 m above sea level and nearly 2 km inland from the present shoreline in Molokai, were deposited by a giant tsunami wave triggered by one of the submarine slides mapped on the lower slopes of the Hawaiian Islands. However, the association of the deposits in Lanzarote and Fuerteventura with marine terraces and their occurrence at elevations of less than 10 m above sea level in Graciosa and Roque del Este suggest that such an origin is unlikely for these deposits. Some workers have even rejected a tsunami origin for the deposits in the Hawaiian Islands and have interpreted the elevation of these sediments in Molakai as a result of uplift due to lithospheric flexure rather than deposition by a giant tsunami wave (Grigg and Jones, 1997).

Fuerteventura can be divided into three topographic provinces parallel to the island's long axis (Figure 2). Along the west side is the Western Domed Area, in the center a Central Depression and on the east side the Eastern Rise. The Western Domed Area is an elongate oval area, with an elevation of 200 to 300 m, in which the seamount Basal Complex is exposed (Figure 3). Stillman (1999) inferred that the uplift, which occurred as a result of either isostatic rebound or thermal uplift, is recent. Prior to the uplift of the Western Domed Area the Central Depression extended westward to the coast. The Central Depression, with an elevation of 100 m to 200 m, probably first formed as much as 18 Ma (Stillman, 1999) and is now covered by a succession of recent sediment and Pliocene and Pleistocene volcanic rocks. The Eastern Rise dips eastward with a 20° gradient and is cut by 'barrancos' separated by sharp-crested divides (Cuchillos) draining radially away from the center of the island toward the west (Ancochea et al., 1996). This terrain is a remnant of the shield phase.

Volcanism in Fuerteventura was initiated sometime after the Cenomanian (Late Cretaceous) and before or during the Paleocene (Stillman, 1999). Submarine volcanic activity occurred, perhaps only episodically, between the Paleocene and the late Oligocene with initial emergence of parts of the island taking place before 20 Ma (Figure 3). Fragments of plutonic rocks indicate part of the island was undergoing erosion at that time (Robertson and Stillman, 1979). During the shield stage volcanic phase three volcanic structures were built in the northern (Northern Volcanic Complex, 14-12 Ma), central (Central Volcanic Complex, 20-18 Ma and from 17.5-13 Ma) and southern (Southern Volcanic Complex, 16-14 Ma) parts of the islands (Figure 3). Amanay Bank and El Banquete (Figures 1 and 2) offshore the southern end of Fuerteventura may correspond to other volcanic edifices. The summits of the northern central and southern volcanic complexes may have reached heights of 2300–3000 m, 2600–3300 m and 1600–2100 m above sea level (Stillman, 1999). The Southern Volcanic Complex at the southern tip of Fuerteventura in the Jandía peninsula is separated from the Central Volcanic Complex by El Jable, a narrow low covered by eolian sands and calcrete (Figures 1 and 3). Like the Central Volcanic Complex this volcano also was centered offshore of the present shoreline (Ancochea et al., 1996). The volcano remnant of the Southern Volcanic Complex consists of a northward facing convex scarp, La Pared, cut by radial 'arroyos' draining westward and separated by narrow divides (Cuchillos). North of La Pared is an arcuate low marked by prominent valley formed by the convergence of the gullies cut on La Pared. At least half or more of the South Volcanic Complex slid seaward in the Miocene producing the 3 to 7 km wide, 45 km long north-northwest trending low east of Amanay Bank. As in Lanzarote the shield phase in Fuerteventura was followed a major phase of erosion during which the Basal Complex was exposed and relief was decreased to around 200 m in less than 2 m.y. (Stillman, 1999). After the erosional phase Miocene basalts and Pliocene to Pleistocene volcanic rocks covered a small area of the Basal Complex and the Central Depression.

Stillman (1999) proposed that denudation of Lanzarote and Fuerteventura during the erosional cycle was by massive landslides, multiple slips that transported the volcanic structures northwestward into the sea. Stillman calculated that during this cycle as much as 3000 km^3 was removed from Fuerteventura in less than 2 m.y. The 17.6 to 16.5 Ma debris deposit cored at Deep Sea Drilling Project (DSDP) Site 397 (Figure 1) are evidence of such destruction in Fuerteventura or Banquete Bank southwest of the island. The flow has an average thickness of 20 to 30 m, forms a 5 to 25 km wide narrow tongue trending southwesterly along the base of the African continental slope, has an area of 2000 km^2 and a volume of 50 km^3 (Arthur et al., 1979). As the debris flows are mainly hyaloclastics with abundant palagonitized sideromelane shards they probably reflect a submarine shield-building stage of Fuerteventura or the bank (Schmincke and Von Rad, 1979). The microgabbro fragments in the flow suggest that part of Fuerteventura, or the bank, was already above sea level.

Lanzarote-Fuerteventura: Offshore

Our present discussion is limited to the west side of the Canary Ridge. In another paper (Acosta et al., this issue) we discuss the nature and origin of the morphology of sea floor of the Canary Channel east of the ridge. The characteristic topographic features of the western insular margins of Lanzarote and Fuerteventura are ambiguous. The presence of features that may be exotic blocks, however, has led us to infer that mass wasting has played a major role in the shaping of the margin. We argue that, the margin has been created by mass wasting, turbidity current activity and hemipelagic deposition.

Evidence of debris avalanches due to the collapse of the volcanic edifices is not apparent on the offshore swath bathymetry west of Lanzarote. The multibeam relief diagram of this slope does not display the morphologic characteristics of debris avalanches, such as coastal embayment and scars created by avalanches (Figure 4). Instead it indicates that the margin is dominated by a seaward-dipping platform bound by scarps on the landward and seaward sides. Seaward of the ramp is a northeast-oriented sediment wave, the Lanzarote Sediment Drift, that we infer to have been sculptured by the southerly flowing North Atlantic Deep Water (NADW) at a depth of 2000–3800 m. The ramp's surface is cut by narrow gullies or rills and above it rise 100 m high circular mounds (Figures 2, 4 and 5). These mounds are aligned at right angles to the slope's contours with the gullies between them terminating abruptly up slope. We infer that the highs are volcanic cones and the gullies were probably eroded out of a thin sediment cover. That the cones are volcanic structures is plausible as Lanzarote experienced extensive volcanism in the 18th century (1730-36). Contemporary accounts cited offshore explosions and discoloring of the water, and finding of unknown species of deep-sea fish killed and brought to the surface (Carracedo and Rodíguez Badiola, 1993), (Figure 3). The origin of the ramp is yet to be resolved. It could be a massive slump, but as no rotation appears to have taken place along the inner scarp it precludes such a possibility. The ramp probably represents an avalanche that broke apart during its displacement.

Like the Lanzarote the margin offshore Fuerteventura also lacks features characteristic of a mass wasting terrain. It too lacks coastal embayments and slopes scars produced by avalanches. However, the land geology with its history of catastrophic collapses of mass wasting indicates that landslides played a major role in the sculpturing the island's margin. A spur-like feature south of 28°30′ N divides the northwest insular margin of Fuerteventura in two (Figures 2, 4 and 5). This feature was inferred by Ancochea et al. (1996) to represent an erosional remnant of the Central Volcanic Complex whose center was located offshore of

Table 1. **Statistics of Landslides**

Dimensions of El Julan Debris Avalanche and Canary and Saharan debris flows are from Masson et al. (2002); thickness of units used to calculate volumes are from Masson et al. (2002) and Teide Group (1997); ages are from Cantagrel et al. (1999), Carracedo et al. (1999a), Urgeles et al. (1997; 1999), Masson et al. (2002) and extrapolation from onshore geology. Areas of units inshore and offshore were calculated by tracing their outlines; cutting and weighting them, and converting their weights into areas by dividing them by the weight of a known area. DA=Debris Avalanche; DF=Debris Flow; S=Slump.

Name	Type	Length (km)	Max. Width (km)	Area (km^2)	Volume (km^3)	Age
Lanzarote	S?	>40	>30	>800		18-16 Ma
Fuerteventura						
Puerto R.	DA?	70	50	3500		>17.5 Ma
S Puerto R.D.C	DA?	35	45	1200		>17.5 Ma
Gran Canaria						
Las Palmas	DA	45	25	1100		9 Ma; 4.0-3.5 Ma
Galdar	DA	30	10	300		4.0-3.5 Ma
Agaete	DA	30	7	200		12?/14? Ma
NW	S	50	7	400		15 Ma
SW	DA	30	10	250		4.0-3.5 Ma
R. Nublo	DA	12	11	150		4.03.5Ma
La Gomera						
I	DF?	15	10	80		4.0? Ma
II	DF?	10	15	80		4.0? Ma
III	DF?	45	15	340		4.0? Ma
IV	DF?	45	8	160		4.0? Ma
V	DF?	40	16	300		4.0? Ma
VI	DF?	20	5	40		4.0? Ma
VII	DF?	24	7	50		4.0? Ma
VIII	DF?	32	25	300		4.0? Ma
Tenerife						
Teno	DA	35	15	400		6 Ma
R. Garcia	DA	95	30	2200		0.6-0.7? Ma
Icod	DA	95	18	1500		<0.15 Ma
Tigaiga	DA	30	10	200		>2.3 Ma
Orotova	DA	75	40	2200		0.69/0.54 Ma
Anaga	DA	33	15	500		
Güimar	DA	85	45	2600		<0.84 Ma
A	DA	22	4	80		
B	DA	17	5	80		
C	DA	7	5	30		
La Palma						
PV	DA	50	35	1600	520	1.0-0.8 Ma
CN	DA	43	30	700	80	<536->125 Ka
W-PN	DA	>40	15	>300		
E-PN	DA	40	11	400		
SC	DA	50	35	1700		>1.0 Ma
El Hierro						
Golfo	DA	60	50	1700	170	9/15-10/17;17-9 Ka
PN	DA	35	25	1300		
LPI	S	30	15	1300		545-261/176 Ka
LPII	DA	45	340	350	<50	545-261/176 Ka
Julan	Da	48		1800	130	500-300;130 Ka
Debris Flows						
Canary	DF			40,000	400	13–17 Ka
Saharan	DF			48,000		60 Ka

the present shoreline. The insular slope/rise northeast of the spur is characterized by northwest-trending lobes separated by V-shaped to box-shaped valleys. Along the crest of one of the lobes is a meandering valley with short distributaries. This lobe and valley morphology terminates rather abruptly at a depth of 3000 m. Along this seaward edge are a 22 × 11 km rectangular flat-topped high, a 4 × 4 km high and an 11 × 6 km elliptical shaped high (Figure 4). Short box-shaped valleys separate these highs from one another. We infer that these highs represent very large exotic blocks carried to their present sites by the debris avalanches that originated during the collapse of the Northern and Central Volcanic Complexes. We have named this unit, with an area of 3500 km^2, the Puerto Rosario Debris Avalanche (Figure 4; Table 1). Such an interpretation of the highs may not be unrealistic because a seismic reflection profile recorded by Uchupi et al. (1976) shows that basement immediately west of Banquete Bank is 1.5 seconds (1500 m; assuming a sediment velocity of 2 km/sec) below the seafloor. However, in the absence of seismic reflection profiles directly over the features themselves the possibility that they represent peaks in oceanic basement cannot be dismissed.

The insular slope/rise south of the spur at 28°30′ N consists of sediment lobes cut by a complex of valleys. Although the terrain has the appearance of being formed by turbidity current processes, we propose that this process only contributed to the modification of a pre-existing terrain. We infer that debris avalanches triggered by the collapse of the Central and South Volcanic Complexes contributed to the original construction of this terrain. At least half or more of the South Volcanic Complex slid seaward in the Miocene (Ancochea et al., 1996) producing part of the 3 to 7 km wide, 45 km long north-northwest trending low east of the Amanay and northeast of Banquete Bank (Figures 4 and 5). Banquete Bank, southwest of Fuerteventura, is a flat-topped high with a minimum depth of 60 m that is attached to the island. It appears to consist of two circular highs whose geometry is suggestive of volcanic construction. If so then the geometry of the South Volcanic Complex differs from that described by Ancochea et al. (1996). It may have consisted of three overlapping volcanoes, rather than a single edifice (Figures 2, 4 and 5).

The west and southwest sides of the Banquete Bank are scalloped suggestive of mass wasting. Northwest of the bank is an apron cut by gullies draining into a valley between El Banquete and Amanay Banks. Northeast of Banquete Bank is a triangular shaped smoothed top sediment lobe whose northern tip is cut by low relief gullies radiating outward. Its southern tip, next to the bank, terminates on a northeast-trending scarp whose surface is gullied. The feature may define a detachment surface suggesting that the triangular sediment body is a debris avalanche or slump. East of this slump is an amphitheater-like indentation on the Fuerteventura insular slope whose surface is characterized by small gullies and intervening ridges. This feature also may document a mass wasting episode.

Immediately north of Banquete Bank is Amanay Bank that Ancochea et al. (1996) interpreted as a submarine volcano. The apron on the southwest side of the bank is narrow and terminates against a valley that originates in the low between Banquete and Amanay Banks. On the bank's west side is a sedimentary apron cut by gullies that fan out down slope. The bank's northeast side is dominated by two box-shaped valleys whose distal ends are bordered by levee-like ridges. Southeast of this valley system is a small circular high that is either an exotic block or a volcanic cone and a grid system of narrow highs and lows. These highs and lows are flanked on their east side by a valley that separates them from a sediment lobe off the northern tip of Banquete Bank (Figure 4). Overall the morphology of Banquete and Amanay Banks appears to be the creation of mass wasting followed by a more recent cycle of turbidity current deposition. We have named this multiple event deposit feature, covering an area of 1200 km^2, the South Rosario Debris Avalanche Deposit Complex (Table 1). If this morphology does reflect mass wasting it was created by the collapse of the South Volcanic Complex in Fuerteventura and the volcanic edifices in Banquete and Amanay Banks.

Gran Canaria: Inshore

Gran Canaria, a nearly circular island with a diameter of 45 km and an area of 1532 km^2, is separated from the southwest end of the Canary Ridge by a 50 km wide north-south trending low (Figures 1 and 2). The highest point in the island, Pico de Las Nieves, has an elevation of 1949 m (Figure 6). The peak is located in the center of the island from which a radial system of canyons, known locally as barrancos, extends to the coast (Funck and Schmincke, 1998). Several valley-embayments dissect the island coastal region. The largest of these, Caldera de Tejada, extends from the western edge of the island to nearly its center (Hausen, 1962). Another large embayment, Caldera de Tirajana occurs on the southeastern sector of Gran Canaria. Young volcanic cones occur over the northern half of the island down to the coast forming a promontory, La Isleta (Figure 6). Many of the lava streams from these structures have partly filled the barrancos, subduing the topography. The island's coast is generally smooth being disrupted by a few promontories such as La Isleta in the northeast, Punta de Arinaga in the east and Maspalomas, a broad headland, in the south.

The major subaerial magmatic episodes, which build the island, took place in the Miocene between 16 and 9 Ma (Figure 6) (Funck and Schmincke, 1998; Schmincke and Sumita, 1998). Volcanic activity was absent from 9 to 5 Ma, and the island, which had an elevation of about 2000 m, was deeply eroded and debris flow fans were deposited over the eroded Miocene igneous rocks. Above these sediments are 5 Ma nephelinites and basanites and the Roque Nublo Group, a post-erosional feature, emplaced between 4.0 and 3.5 Ma. A stratocone, possibly as high as 3000 m, was constructed above the eastern rim of the Miocene caldera during the Roque Nublo cycle (Funck and Schmincke, 1998; Mehl and Schmincke, 1999). A possible brief gap in volcanism was followed by another period of volcanism 3.2 to 1.8 Ma. Quaternary volcanism has been restricted to the northern half of the island. Such activity may have constructed the volcanic edifice off La Isleta, whose radial lobes may represent submarine volcanic flows (Funck and Schmincke, 1998), and the circular highs in the divide between Gran Canaria and Fuerteventura. The occurrence of 3000 year BP basanite scoria cones and lavas flows indicates that Gran Canaria continues to be volcanically active.

Landslides and debris avalanches took place at the end of the Miocene basaltic shield phase 13.3-9.0 Ma and during the Pliocene Roque Nublo phase 4.0 to 3.5 Ma (Funck and Schmincke, 1998; Mehl and Schmincke, 1999). The Roque Nublo Debris Avalanche resulting from the collapse of the Pliocene Roque Nubo stratocone that crops out along the south, northeast and northwest present coast of Gran Canaria, a distribution that demonstrate that the flow must have entered the sea. Included within and on top of the avalanche are at least seven megablocks with diameters of >100 m, blocks 0.25 to 100 m in diameter, and blocks up to 100 m in diameter that were highly brecciated during the transport. The largest of the megablocks blocks (Roque Nublo Plateau) is 1200 m long, 800 m wide and about 300 m thick (Mehl and Schmincke, 1999).

Gran Canaria: Offshore

Like the insular margin offshore Lanzarote and Fuerteventura, the margin of Gran Canaria may be mainly the creation of turbidity currents and hemipelagic deposition. However, segments of the margin appear to reflect mass wasting, particularly in the divide between Gran Canaria and Fuerteventura and offshore the north and west sides of the island. The region where the Gran Canaria edifice overlaps the Fuerteventura structure is one of the most complex topographic regions in the Canary Island archipelago. It is here that Amanay and Banquete Banks offshore Fuerteventura are found. Funck and Lykke-Andersen, (1998) and Funck and Schmincke (1998) inferred that the a 200-m-deep and 3-km-long erosional channel was formed in the region by bottom currents, but the new swath data demonstrate that the topography is more complex (Figures 7 and 8). Linking Banquete Bank and a northeast-trending swell offshore Gran Canaria are two small highs separated by south and north draining channels. Along the southeast side of the swell and highs are numerous circular highs whose topography is suggestive of volcanism (Figures 4 and 7). Southeast of the swell is a sediment apron whose surface is cut by low relief gullies. This apron, flanked on the east by Banquete Bank and on the west by a deep-sea fan offshore the south coast of Gran Canaria, was formed from sediments washing over the divide between Fuerteventura and Gran Canaria (Figure 7). The contact between Banquete Bank and the overlying apron to the north is sharp with the apron gullies curving along the edge of the bank. The contact between the apron and the deep-sea fan to the west is one of overlap with the apron prograding over the fan.

Two curved north-draining valleys originating on the east side of Gran Canaria dominate the sea floor between Amanay Bank and Gran Canaria. The northern end of the valley divide is irregular and shallower, whereas its southern end is deeper and smoother. The rough topography of the northern end of the divide appears to be due to mass wasting and the smooth topography of the southern end of the divide is the result of either recent slumping or sediment being trapped behind the higher northern end of the divide. If due to sedimentation, then the deposits were dammed behind the detachment surface of debris avalanche. At the northern end of the seaward extension of Barranco Silva (Figure 6), is a rough surfaced debris

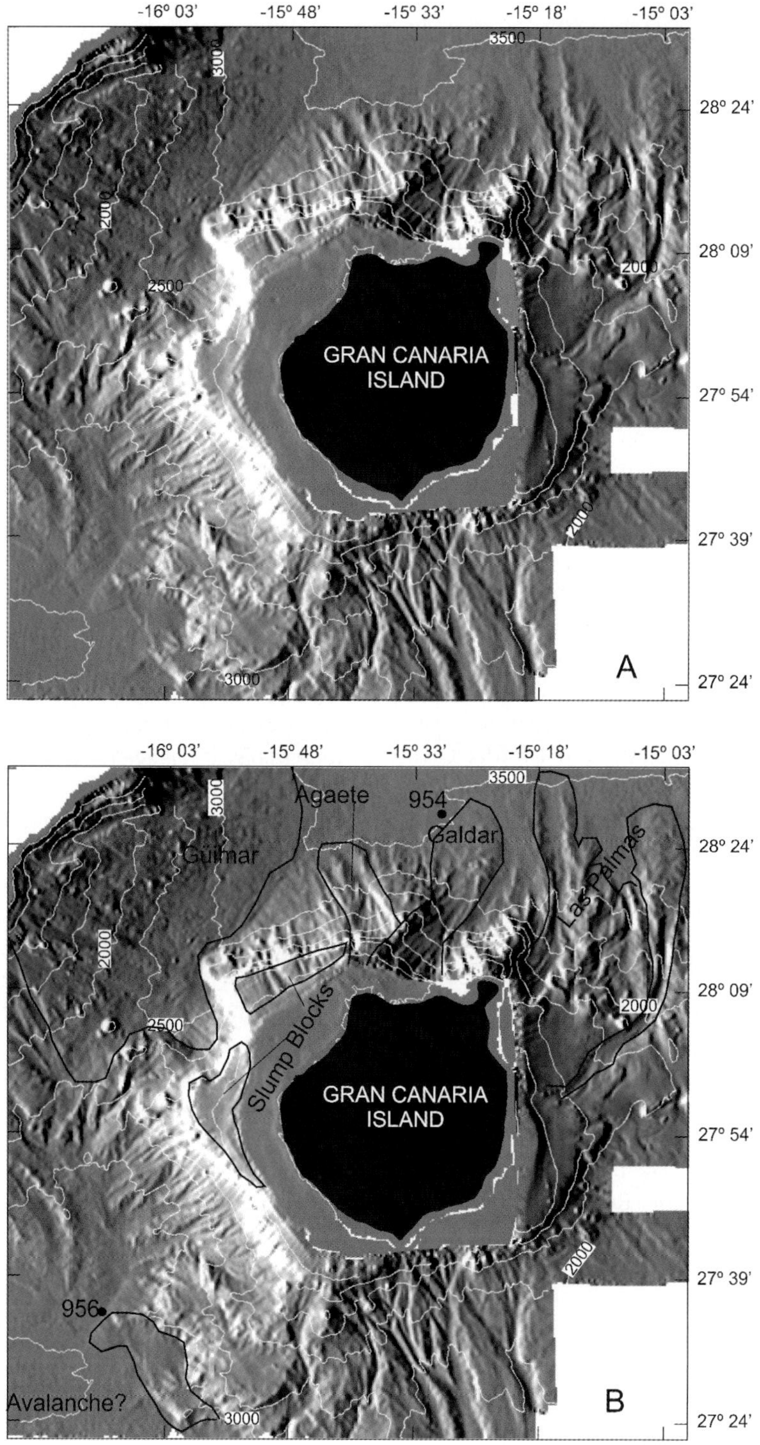

Figure 7. **A**. Shaded relief image of Gran Canaria margin. Illumination is from northwest. **B**. Shaded relief image of Gran Canaria displaying the extent of possible mass movements off Gran Canaria and eastern Tenerife.

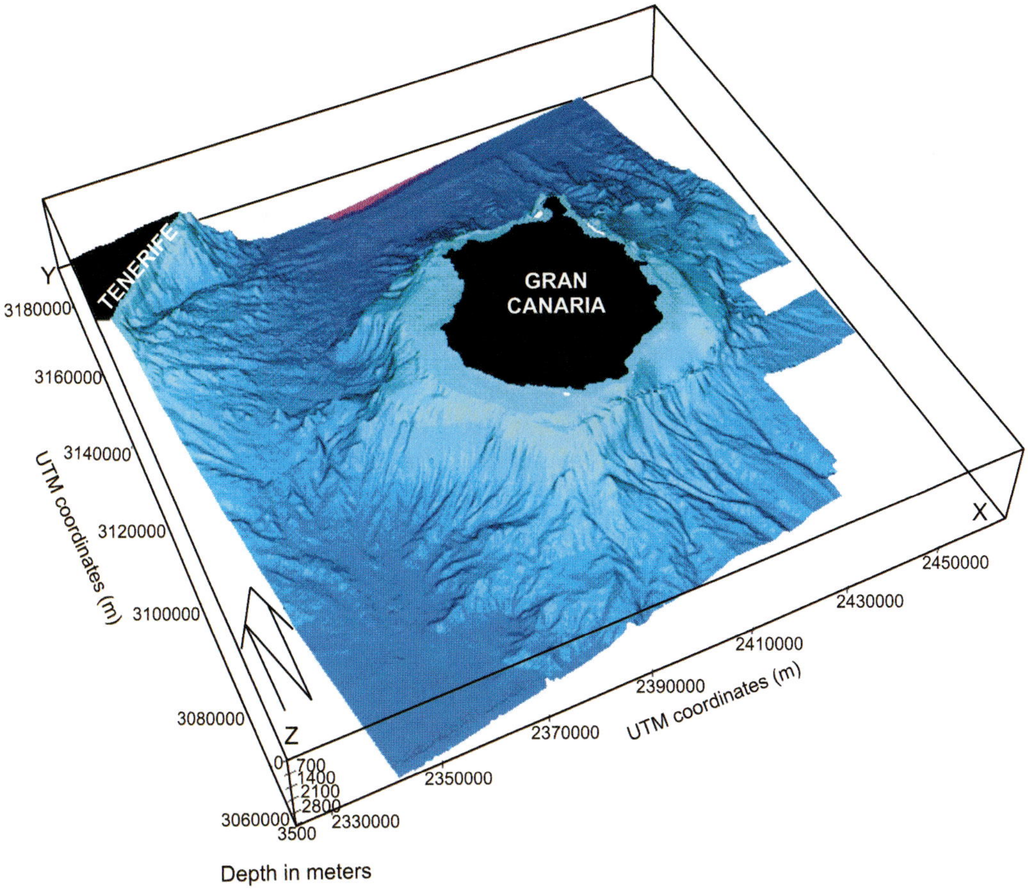

Figure 8. 3D image of Gran Canaria margin and tip of the Tenerife margin looking north. Note high linking Gran Canaria and Tenerife. It consists of basement highs and sediment ridges created by the southern flowing North Atlantic Deep Water. This high forced the Güimar Debris Avalanche off Tenerife northeast. The fan off southern Gran Canaria is separated from the apron off western Gran Canaria by a triangular shaped high, which we infer was created by intrusive/extrusive igneous activity along a rift zone. The two terraces on the upper slope separated by an embayment may be due to slumping. High on east side links Gran Canaria with Banquete Bank off Fuerteventura.

flow or avalanche. Roughness is due to northeast and northwest-trending grooves segmenting the flow. A western valley, the seaward extensions of Barranco Guiniguada, displays low relief grooves aligned parallel to its axis. They represent scour marks and the ridges flanking the distal end of this valley represent levees. Seaward of this valley is a sediment lobe that represents either a debris flow, on avalanche or a turbidite fan. We infer that much of this topography is due to a combination of mass wasting and subsequent turbidity current activity and have named the feature the Las Palmas Debris Avalanche (Figure 7; Table 1). The avalanche has an area of 1100 km^2 and is probably the consequence of three events, a Miocene (9 Ma) event that led to the construction of the bulk of the unit, the Pliocene Roque Nublo collapse at 3.5 Ma that led to the avalanches associated with the seaward extensions of barrancos Silva and Guiniguada, and post-Pliocene erosion and deposition by turbidity currents.

Between La Isleta and Galdar, at the northeast tip of Gran Canaria, is a wide valley, the seaward extension of Barranco Azuaye (Figures 6, 7 and 8). At its mouth is a rough surface sediment lobe that can be traced to a water depth of 3400 m. This valley may have been formed by an avalanche, debris flow or turbidity current with the low grooves paralleling its axis, representing scour marks. The lobe at the mouth of the low represents sediments deposited by the avalanche or debris flow or turbidity currents. We have named the event that eroded this scar the Galdar Debris Avalanche (Table 1).

A debris avalanche origin for this valley is supported by the lithology at ODP Sites 953 and 954. (Figures 1 and 7; Shipboard Scientific Party, 1995a,

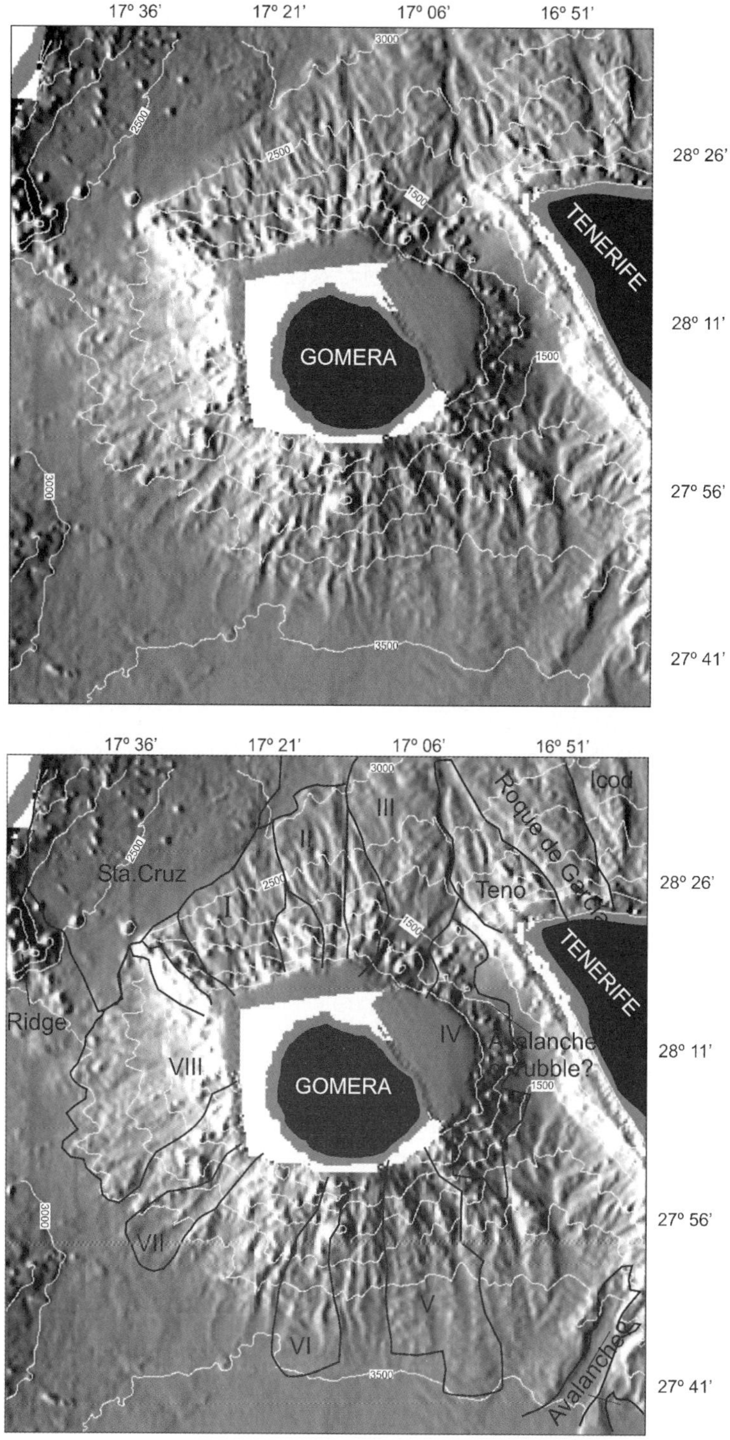

Figure 9. **A**. Shaded relief image of La Gomera margin. **B**. Shaded relief image of La Gomera region displaying possible debris flows on the margin. Note that those on the northern slope merge down slope.

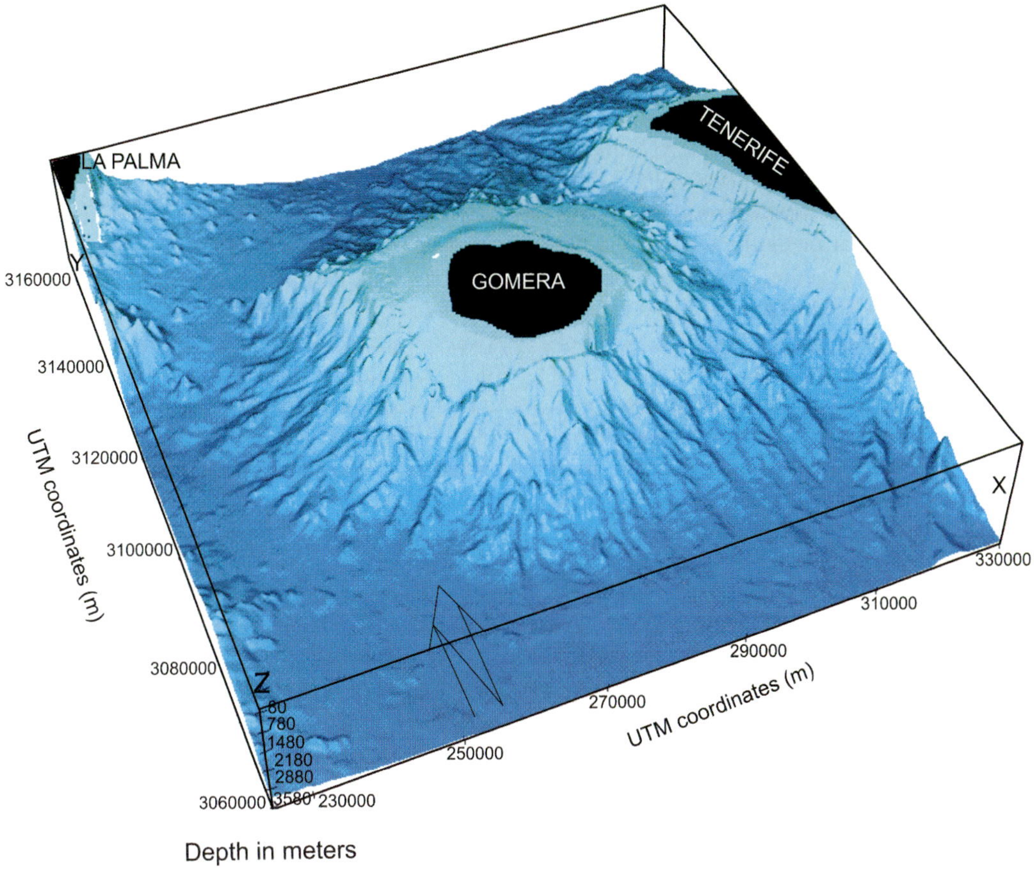

Figure 10. 3D image of La Gomera (G) margin looking north. We infer that the chutes on the right and left may represent scars eroded by debris flows. Another debris unit may be represented by the low on the southeastern slope of the island.

b). The Miocene sediment section at ODP Site 953, the more distal site, consists volcanic debris flow and turbidites, nannofossil sediment, volcaniclastic material and distal submarine ash flows correlative with the Fataga and Mogan phases of volcanism. Also included in the section is late Miocene to early Pliocene nannofossil ooze correlative with the hiatus in volcanic activity and early to late Pliocene volcaniclastics and nannofossil ooze correlative with the Roque Nublo phase. The section at ODP Site 954, the more proximal site, consists of 15-14 Ma breccia composed of basaltic clasts with minor green hyaloclastic tuffs, lapillistone, calcareous sediment and clay and late Pliocene lapilli stone with volcanic clasts (Shipboard Scientific Party, 1995b).

Funck and Lykke-Andersen (1998) inferred that the breccia unit with an area of >60 km^2 was a debris deposit emplaced >14 Ma and transported 70 km from its source. In contrast Schmincke and Segschneider (1998) suggested that the breccia unit is a consequence of the collapse of part of the shield volcano at the end of its shield-building phase. The swath bathymetry is not conclusive in regards to the origin of the breccia unit, but the location of the breccia at the mouth of valley appears to support Funck and Lykke-Andersen (1998) speculation that the breccia is a debris flow that originated 70 km away. We suggest that the valley was formed or exhumed during the Roque Nublo phase 4.0-3.5 Ma. We further propose that the path of the Miocene avalanche responsible for the deposition of the breccia lay farther west, probably the low offshore Galdar. We infer that the valley was created by a flow that we named the Agaete Debris Avalanche (Figure 7; Table 1). It was this Miocene flow that deposited the lobe offshore Galdar, a lobe that was cut during the Roque Nublo event leading to the formation of the Galdar Scar.

The northwest coast of Gran Canaria, between Galdar and San Nicolas, is dominated by a broad 19 km wide and 4.5 km deep coastal re-entrant. Its landward side is characterized by a >1000 m high cliff displaying a gradient of 45°. Funck and Schmincke (1998) proposed that this feature is the head of a landslide comparable to those found in El Hierro. They suggested that the collapse is related to the formation of the Tejada Caldera during the Miocene about 15 Ma. Offshore the feature is characterized by protrusion on the insular shelf at the north and a re-entrant at the south. A seismic reflection profile recorded by Funck and Schmincke (1998) shows that two wedge-shaped sedimentary units, pinching out at a depth of 1000 m, mantle the protrusion. The top of the wedge forms the insular slope and its outer surface the upper insular slope. The 180 m thick upper unit consists of volcaniclastics and biogenic sediments emplaced after the collapse of the coastal low. The over 200 m thick lower unit may represent debris flows emplaced during the collapse of the low with the shelf protrusion representing a massive slump block. The displacement of the slump supposedly formed the shelf's re-entrant south of the protrusion. The seafloor seaward of the protrusion ranges from mounds to northeast-trending furrows on the surficial sediments blanketing the Tenerife volcanic edifice. The mound topography may be related to sediment recently displaced from up the Tenerife insular slope. The furrows are probably of erosional origin and were created by southerly flowing NADW (Teide Group, 1997).

West of Barranco Aldea (Figures 6 and 8) is a northeast-southwest trending irregular surfaced ridge linking the Gran Canaria and Tenerife margins. Seismic reflection profiles recorded by Teide Group (1997) indicate that the high consists of oceanic basement highs and sediment waves formed by the south-flowing NADW. The protrusion of Gran Canaria shelf's edge south of San Nicolás is the seaward edge of a scar in the Miocene shield volcano (Figure 6). The low defined by this scar, the Horgazales Basin of Schmincke (1968), is filled on land with late shield basalts of the Mogan Group. If so this suggests that the scar is >14.1 Ma, Funck and Schmincke (1998) suggested that the offshore extension of this low is filled with volcanic debris comparable to those onshore. They interpreted a 650 m thick lower sediment unit imaged by seismic reflection profiles as possibly representing this volcanic debris flow. Schmincke and Segschneider (1998) speculated that this debris flow was created by the collapse of the subaerial and submarine flank of Gran Canaria to form Horgazales Basin. Funck and Schmincke (1998) further speculated that the hummocky topography seaward of Horgazales Basin represents slump blocks formed during the creation of the low. An 80 m thick Miocene debris flows cored at ODP Site 956 support such a collapse model (Shipboard Scientific Party 1995c).

The Gran Canaria southwest slope is characterized by a V-shaped northeast-trending valley system, draining into a northwest-trending trough at a depth of 3000 m. A narrow ridge, possibly generated by volcanism along a rift, dominates the margin off the southwest tip of Gran Canaria. At the end of this ridge is a smooth northwest-trending platform partially fronting a sediment apron (Figure 7). The northeast side of this platform is gullied and its southwest side is scalloped suggesting that the high has experienced mass wasting. Whether the high is a volcanic construction or represents an exotic block displaced from Gran Canaria cannot be determined from our data set. That it may represent an exotic block is not unrealistic as mass wasting deposits have been documented at ODP Site 956 just west of the high (Figures 1 and 8). The early Pliocene to late Miocene nannofossil ooze at this site is characterized by slump structures and slump debris that were emplaced during the collapse of the shield volcano that led to the formation of Horgazales Basin >14.1 Ma (Shipboard Scientific Party, 1995d).

The Plio-Quaternary section at Site ODP 956 consists of nannofossil ooze interbedded with coarse debris rich in pumice whose source is Tenerife. Schmincke and Segschneider (1998) speculated that the 75 to 100 m thick sandstone interbedded with turbidites of mixed volcaniclastics and biogenic sediments above the debris flow is a tsunami deposit. They pointed out that its composition indicates that the sandstone was derived from a mafic/ultramafic source and proposed that it was derived from La Gomera. They also point out that Gran Canaria itself could have been the source of this sediment if Jurassic oceanic crustal material was transported to the island surface by phreatic explosions.

On the southern Gran Canaria margin is a fan cut by a system of branching canyons (Figures 7 and 8). This fan, centered on Barranco de Arguineguin, was probably constructed from material discharged offshore by the radial canyons between Barranco de Mogan and Barranco de Fagata (Funck and Schmincke, 1998). Mehl and Schmincke (1999) inferred that the Pliocene Roque Nublo debris avalanche traveled more than 50 km beyond the present south coast and that a major part of the avalanche debris was deposited as a debris flow on the apex of the fan. From seismic reflection data Funck and Schmincke (1998) proposed that the lobe had an approximate volume of 20 km^3. They also proposed that canyon cutting in the fan took place during the Roque Nublo phase 4.5-3.5 Ma, an erosive phase that was aided by the wet climate that prevailed at that time. If the sediment lobe along the crest of the fan is the seaward extension of the Pliocene Roque Nublo and canyon cutting took place at 4.5-3.5 Ma, then the fan beneath the debris lobe must have been constructed during the Miocene. If so then a significant volume of debris produced by the collapse of the Miocene shield volcano at 9 Ma was transported southward to construct the fan offshore the southern coast of Gran Canaria.

Island in Repose Stage (Gap Stage)

La Gomera: Inshore

La Gomera is semi-circular in outline with an east-west diameter of 24 km, a north-south length of 20 km and an area of 378 km^2 (Figures 1 and 2). The center of the island is occupied by an undulating upland with a maximum elevation of 1375 m at Alto de Garajonay (Figure 6). Barrancos radiate in all directions from the upland to the coast and steep and rocky cliffs, particularly in the north and south, surround the island (Hausen, 1965). Hausen (1971) inferred that the barrancos are the products of two cycles of erosion with some of the older features being cut off by precipices in the vicinity of the coast and others being deepened by younger narrow canyons when reaching the coast. With the exception of a small indentation on the west coast, La Gomera lacks the amphitheaters and buttresses characteristic of the other islands suggesting that La Gomera either has not experienced massive catastrophic failures of its volcanic edifices or these features have been erased by erosion. This denudation has not yet been followed by post-erosional volcanic activity as the island is devoid of younger volcanoes with its morphology, giving an impression of old age.

Hausen (1971) proposed that faulting has uplifted the island 900 m at the north and tilted it to the south. It is this uplift that supposedly was responsible for the creation of mountain blocks on the north and west coasts and the initiation of the younger erosional cycle that rejuvenated the older forms. Evidence for post-tectonic erosion includes the deep barrancos radiating from the island's center, the landslides in the barrancos and along the coast, the truncation of flat-lying basalts by the steep sea cliffs at the north and west coasts and the presence of a phonolitic neck at the water's edge at the northwest coast. Hausen (1971) proposed that this neck was emplaced at the time when the coast lay far to the north.

As noted by Abdel-Monem et al. (1971) the geology of La Gomera is quite simple. At the base is a basal complex comparable to the complex in Fuerteventura and La Palma (Bravo, 1964a, b; Hausen, 1965, 1971; Cendrero, 1967; Cantagrel et al., 1984). Resting unconformably on basement are the older tilted basalts (Basaltos Antiguos of Bravo, 1964a) exposed on the northeast coast. Resting unconformably on the basement and the older tilted basalts is the shield phase of volcanism consisting of basalt lavas, tuffs and agglomerates. Pliocene salic lava sheets, sills, dikes and laccoliths, represent the concluding phase of this volcanic episode.

Cantagrel et al. (1984) inferred the following history for La Gomera: 1. The age of the basement complex is not well constrained, an age of 20 Ma reported by Abdel-Monem et al. (1971) may be conservative; 2. The age of the unconformity truncating the basement complex also is unknown; 3. The Older Basalts are older than 10 Ma (Cantagrel et al., 1984); 4. Polymictic breccias above these basalts are 10 to 9 Ma and the upper old basalts 9 to 6 Ma (Cantagrel et al., 1984); 5. The hiatus truncating these deposits is 6 to 5 Ma; 6. The radial dikes were intruded at 5 Ma; 7. The Younger Basalts were extruded 4.5 to 4.0 Ma with local igneous activity taking place 2.8 Ma. Prior to the extrusion of the younger basalts the island subsided southward at 8 to 5 Ma, accounting for the greater thickness of the basalt in that direction. In time the whole island was covered by basalts. This basaltic extrusion was not continuous, but was disrupted by periods of erosion and barranco formation that were later filled by younger lavas. The volcanic phase was terminated with the emplacement of salic lavas of phonolitic composition in the form of flows, cones, dikes, sills and laccoliths. The salic volcanic phase was followed by a long period of erosion during the Pliocene-Quaternary. How extensive this erosion was is attested by the Roques. These columnar structures, that rise hundreds of meters above their surroundings, were formed by erosion of the cone leaving the conduit fill of salic lavas as erosional remnants.

La Gomera: Offshore

Much of the morphology of the insular margin of La Gomera has the appearance of a terrain created by turbidity current processes (segment II) modified to some degree by hemipelagic sedimentation. Not all

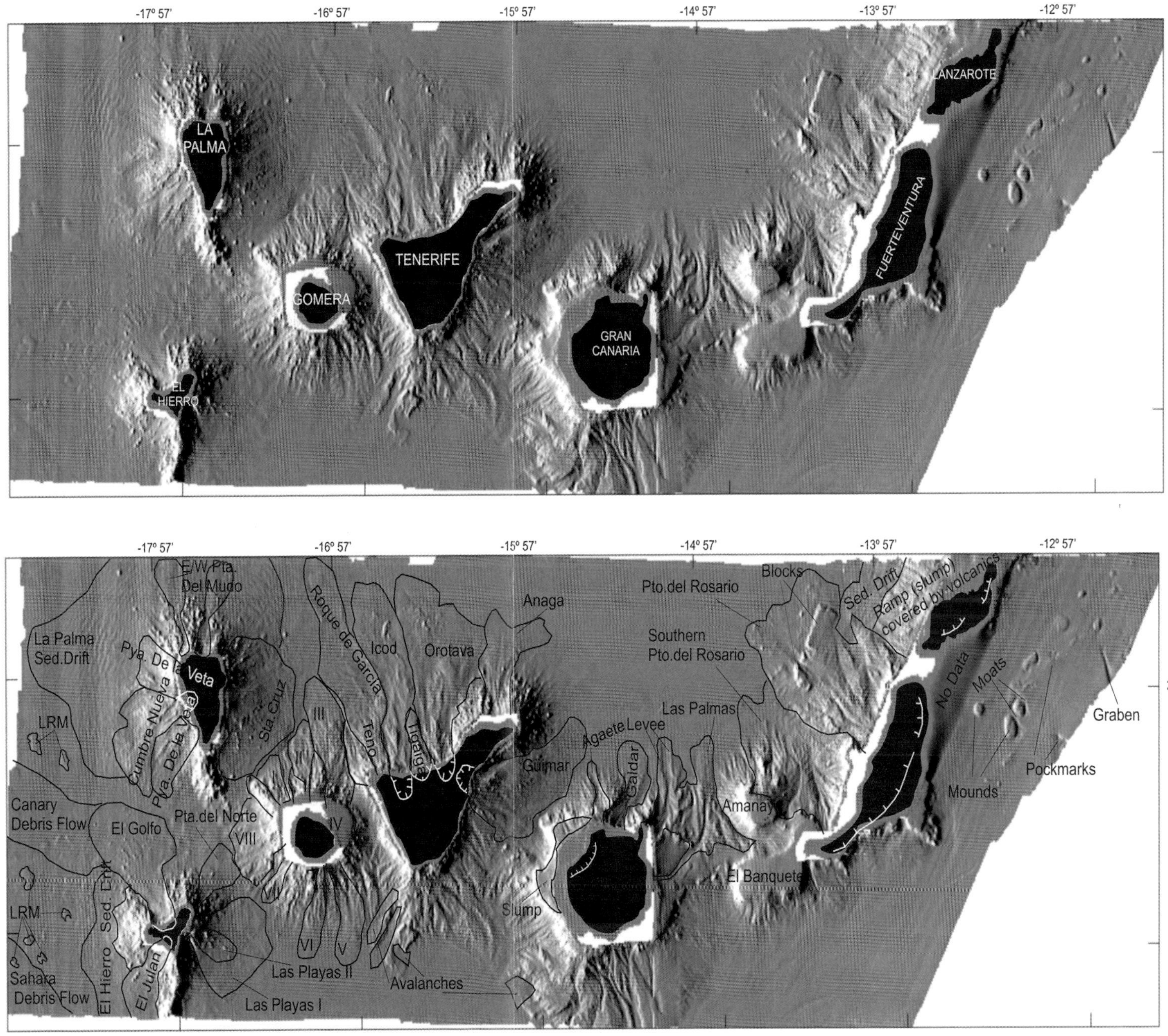

Figure 11. **A**. Shaded relief map of the Canary archipelago. **B**. Relief map showing the distribution of landslides on the flanks and tops of the Canary Islands. LRM= Low Relief Mounds.

the morphology, however, appears to be the creation of these two processes. For example, segment III, has the appearance of a debris flow, but again it could be interpreted as a fan. Whatever mass wasting events took place that led to the creation of the present morphology of La Gomera they are no longer recognizable offshore. What few features on the slope that can be attributed to gravitational processes originated on the insular slope and may have been triggered by turbidity currents. For example, the slope on the east side of La Gomera, segment IV, is rough and may consist of exotic blocks and represent an avalanche deposit (Figures 9 and 10). However, in the absence of sampling or bottom photographs the possibility that the highs may represent salic necks (comparable to the Roques on land) or volcanic cones or even simply valley divides cannot be eliminated. On La Gomera's northwest sector, segment I, the margin is in part in the form of a northwest-trending ridge over 100 m high which may represent a rift, with a chain of volcanic cones along its crest. The slope/rise on the island's south side consists of wide flat-bottomed valleys (segments V-VII) entrained in lobes having the appearance of deep-sea turbidite systems. Segment VIII on the southwest side of the island also displays the same morphology. All in all the morphology of La Gomera, both inshore and offshore differs from the other islands. Nowhere does the morphology display features diagnostic of mass wasting.

Younger Islands

Tenerife, La Palma and El Hierro

Whereas in the older islands landslides triggered by collapse of volcanic edifices are obscured by subsequent events, those in the younger islands, Tenerife, La Palma and El Hierro, are readily recognizable and can be traced tens of kilometers offshore. On land such catastrophic episodes are documented by breached calderas, impressive coastal embayments and massive breccia flows. It is for these reasons that all investigations of mass wasting to date have been concentrated in these islands. Although our survey has provided us with an extensive data base on these events (Figures 2, 11 and 12), our description of them is brief. The reader is referred to a recent summary by Masson et al. (2002) for a discussion of the younger islands. We should mention, however, that our survey revealed avalanches north of La Palma and Hierro that were not known before and that the Santa Cruz Debris Avalanche east of La Palma is more extensive than proposed by Masson et al. (2002). Our survey also demonstrates that although the offshore landslides are aligned with the failures on land, they are not topographically continuous, but are separated by a several hundred meter high scarp on the insular slope.

Tenerife: Inshore

The pyramid-shaped Tenerife, with the double peak stratovolcano, Pico Viejo (3100 m) and Teide (3710 m), with the Cañadas Caldera at its apex, began to form subaerially 11.6 Ma (Miocene) (Ancochea et al., 1990). Resting on the older volcanic units, the Anaga, Teno and Roque del Conde massifs, are the Cordillera Dorsal and Las Cañadas volcanic edifice built between 3.5 and <0.15 Ma. The island acquired part of its present morphology as a result of volcanic collapses that took place along north draining valleys (>2.3 Ma Tigaiga Debris Avalanche; the Roques de García Debris Avalanche 0.7?-0.6 Ma; the Cordillera Dorsal and Cañadas III failure <0.84 Ma to <0.15 Ma; Orotova 0.69-0.54? Ma and Icod de los Vinos <0.15 Ma) north of the Cordillera Dorsal and a south draining valley (Güimar Valley <0.84 Ma) south of the cordillera (Cantagrel et al., 1999). The many small individual basaltic volcanoes scattered throughout the islands and the Teide-PicoViejo which erupted in historical times have added to Tenerife's morphology (Ancochea et al., 1990).

Tenerife: Offshore

On top of the offshore extension of the Anaga Massif is a valley/ridge system that can be traced to a depth of 3000 m. They define the pathways of the flows making up the Anaga Debris Avalanche lows (Masson et al., 2002). Seaward of the Teno Massif is another debris avalanche, represented on land by a breccia overlain by 6.7 to 6.3 Ma and 5.6 to 5.0 Ma volcanic rocks. Cantagrel et al. (1999) dated this avalanche as being approximately 6 Ma. East of the Teno Avalanche is another unit that Masson et al. (2002) correlated with the Roques de García Debris Avalanche on land. Seaward of the Tigaiga Massif is a triangular shaped rough surfaced high that extends to a water depth of 1200 m (Figure 11). Topographically this insular slope spur appears to be an expression of the Tigaiga Avalanche on land, but such a supposition needs to be verified by sampling. The high could simply be part of the flank of the volcanic edifice that survived the avalanches.

Between the Anaga and Teno Massifs offshore the northwest coast are two avalanche deposits that can

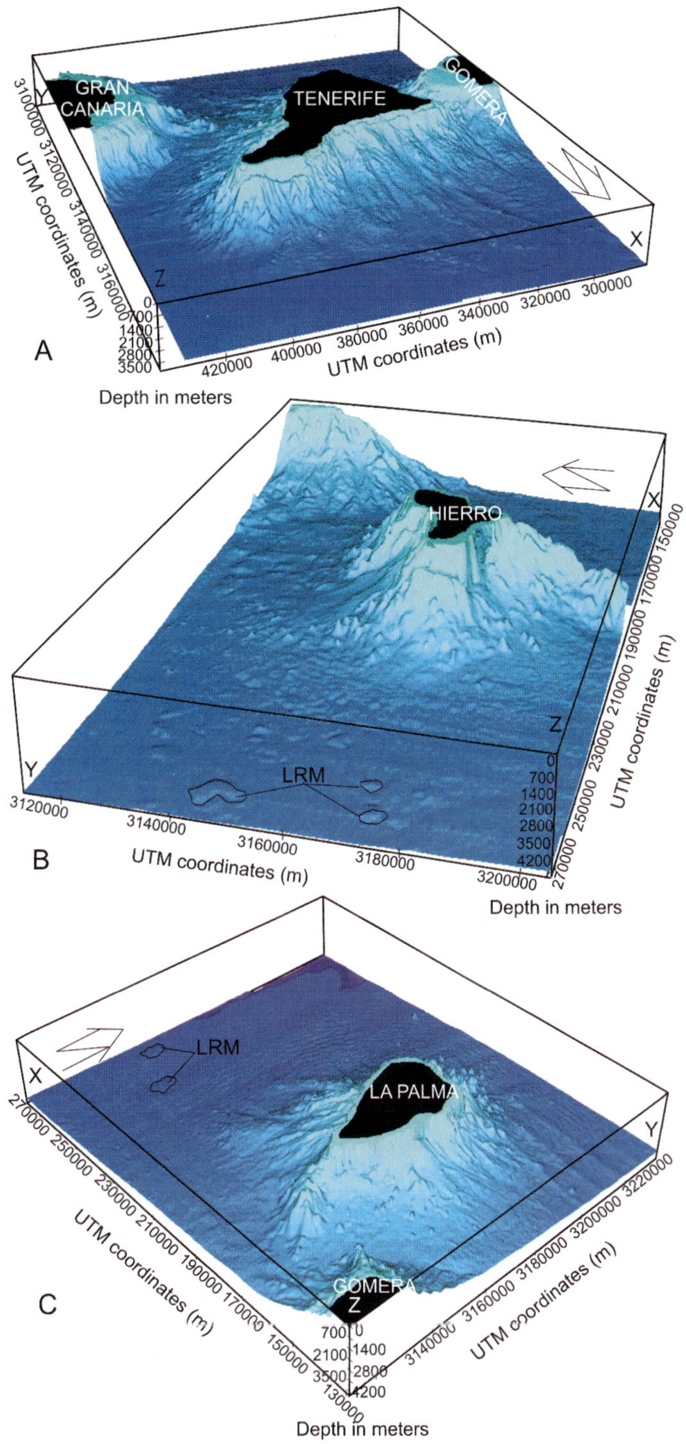

Figure 12. 3D images of the Younger Canary Islands: A) Tenerife, B) La Palma and C) El Hierro. The smooth low relief slope segments represent avalanche scars. Note the sediment drift off the distal ends of the avalanches offshore La Palma and El Hierro. We infer that these sediment waves were formed by currents of AABW. LRM= Low Relief Mounds

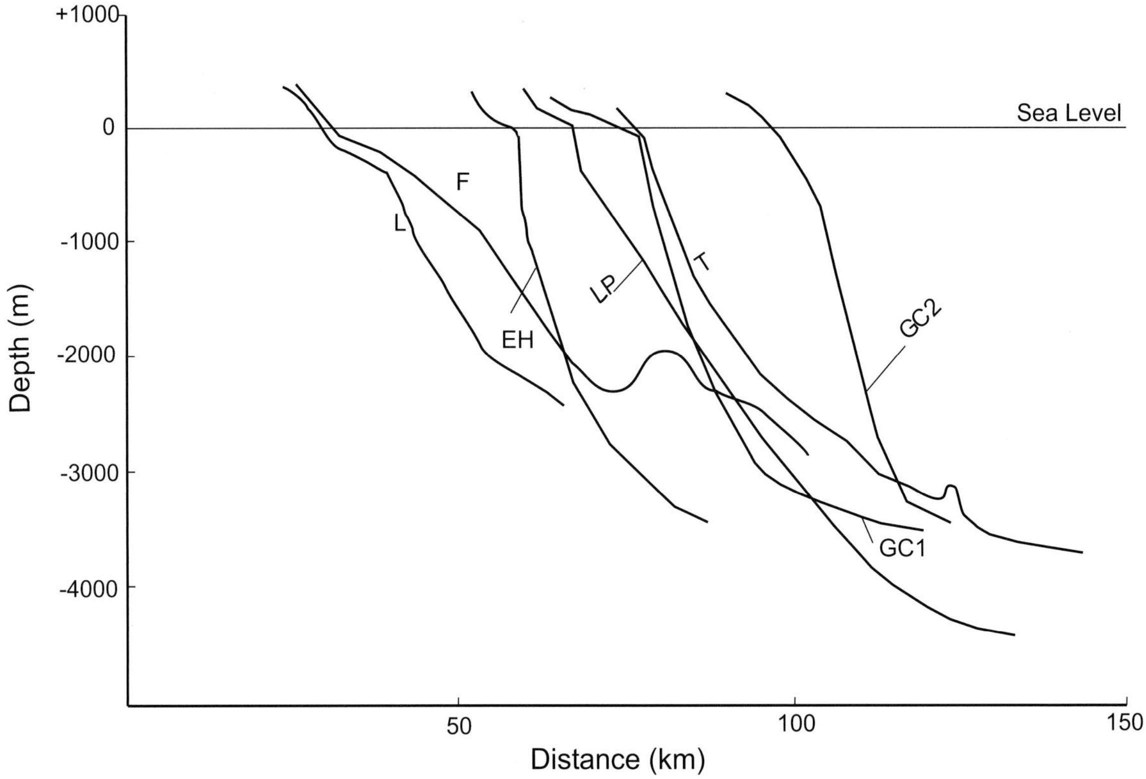

Figure 13. Depth profiles of the Canary Islands compiled from Figure 2. Note that whereas the younger islands margins display concave upwards profiles characteristic of mass wasting, Fuerteventura's and La Gomera's margins are convex upwards characteristic of a turbidity current regime. The Gran Canaria margin is both concave and convex upwards suggestive of mass wasting and turbidity currents. The platform outline of the Lanzarote margin is suggestive of slumping. EH= El Hierro; F= Fuerteventura; GC= Gran Canaria; L= Lanzarote; LP= La Palma; T= Tenerife.

be traced for tens of kilometers offshore, the Icod and Orotova Avalanches (Figure 11). Associated with them are small highs that Watts and Masson (1995) and Teide Group (1997) interpreted as exotic blocks transported by the avalanches. Not all the highs within the flows are of this origin. At least one high at the head of Orovota flow, however, displays a circular morphology more suggestive of volcanic construction (Teide Group, 1997). The Orotova/Icod failures supposedly triggered a turbidity current, which deposited a volcaniclastic layer in the Madeira Abyssal Plain west of the Canary Islands (Figure 1) (Rothwell et al., 1998) and Agadir Basin north of the Canary Islands (Figure 1) (Masson et al., 2002). Teide Group (1997) noted that, although the offshore slope failures are aligned with the Orotova and Icod Valleys, no topographic continuity exists between them; a north-facing 100–600 m high scarps separates them.

The southwest margin of Tenerife is rectilinear and has the appearance of a fault scarp whose movement has truncated the Teno and the Cañadas volcanic series (Figures 11 and 12). At its steepest the scarp, whose base is at depths of 1500 to 1125 m, descends at a rate of 330 m per km. The scarp is divided in two by a transverse spur oriented at right angles to the coast (Figure 9). At the base of the scarp, southeast of this transverse structure, is a rise whose channels drain southwestward terminating against the rise south of La Gomera. Most of this terrain appears to be the creation turbidity current processes (Figure 9). The apron is missing northwest of the transverse high and the insular slope is bordered by a smoothed floored low. Apparently deposition on this low was primarily by pelagic processes, with little or no slope degradation from either Tenerife or La Gomera. Alternatively the lack of relief is due to lava flows.

Morphologically Tenerife's southeast insular margin, buttressed by Gran Canaria, can be divided into two provinces (Figure 12). From the Bandas del Sur to the Anaga Massif the Tenerife coast displays two embayments separated by a buttress at the mouth of Güimar Valley. The buttress is due to lava flows ex-

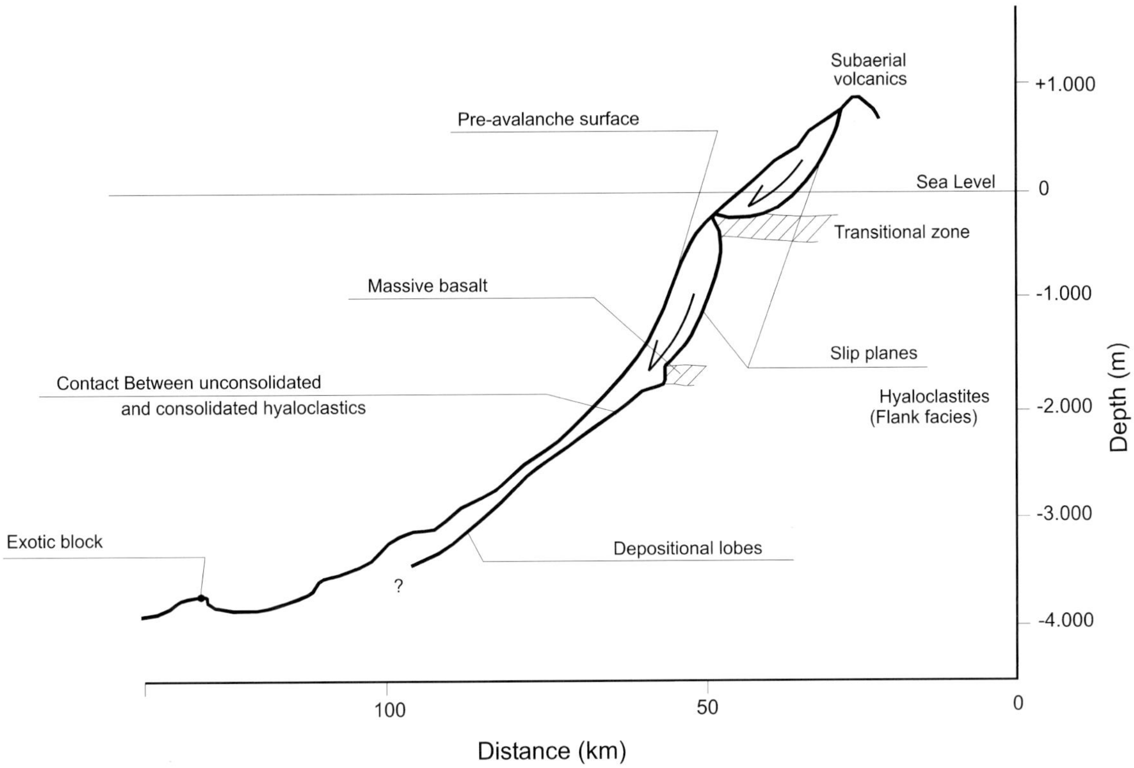

Figure 14. Schematic cross-section of the slip surfaces of the inshore and offshore debris avalanches. The inner scarp represents the detachment surface of the inshore avalanche with the base of the avalanche being along the top of the transitional zone between the subaerial and submarine lava flows. The slip surface of the offshore deposit is along the outer surface of the transitional zone. This model is modified from Jacobs (1995) using data from DePaolo et al. (2001).

truded by several volcanoes at the mouth of the valley. The embayments are the result of a catastrophic failure extending southeastward to the rise northwest of Gran Canaria where it fills a re-entrant on the Gran Canaria margin with the distal end being deflected southeastward by Gran Canaria (Masson et al., 2002). Scattered over the surface of this failure, the Güimar Debris Avalanche, are numerous highly reflective small objects, which we infer to be exotic blocks. A ridge extending from Gran Canaria to Tenerife near 28°05′N also prevented the avalanche from flowing southwestward (Figures 11 and 12). Like the debris avalanches offshore Orotova and Icod Valleys on the northwest coast, this debris flow is separated from the unit in Güimar Valley by a 700 m high scarp that is slightly convex to the east. This slope is cut by V-shaped canyons at the mouths of which are fans resting on debris avalanche. Except for one V-shaped canyon on its northeast side due to post-failure local erosion by turbidity currents (Krastel et al., 2001), the offshore avalanche surface is smooth. Two volcanic highs within the debris avalanche deposit, indicative of post failure volcanic construction, may be correlative with the latest Pleistocene-Holocene volcanoes in the Güimar Valley. At the distal end of the debris avalanche deposit, next to Gran Canaria, is an erosional trough that Teide Group (1997) inferred to be cut by the southwest flowing NADW.

At the southwest corner of Tenerife is the Bandas del Sur Fan, a thick pyroclastic apron that correlates with the upper group felsic formations exposed in the Las Cañadas Caldera (Ucanca Formation 1.59-1.18 Ma; Guajara Formation 0.85-0.57 Ma; Diego Hernández Formation 0.53-0.18 Ma; Martí et al., 1994). It is composed of pumice, ignimbrites, basaltic scoria cones, lavas and mudflows (lahars) with steep-sided erosional remnants of the older basaltic series rising above its general surface. Offshore the Bandas is a northwest-trending fan that can be traced to a depth of more than 3000 m (Figures 11 and 12). Its surface is cut by a system of canyons that is slightly convex toward the east. The fan is bordered on its west side by a sediment-covered ridge that may have been constructed by igneous activity along a rift. Toward the northeast the fan terminates against an oceanic basement high offshore Gran Canaria that, preliminary results suggest may represent a rift related to the volcanic construction of Gran Canaria. Further south the fan is separated from an apron offshore Gran Canaria by a northwest trending low.

La Palma: Inshore

Pear-shaped La Palma is the second youngest island of the Canary Island archipelago. Its surface is cut by many barrancos, the largest of which is the westward draining Barranco de las Angustias that originates in the Caldera de Taburiente (type caldera) whose vertical walls are up to 800 m high (Hausen, 1969) and whose floor has a maximum elevation of 1500 m (Ancochea et al., 1994). Extending southward from the rim of the Caldera de Taburiente to Punta de Fuencaliente at the southern tip of La Palma is a narrow ridge, Cumbre Vieja, whose crest is over 1300 m above sea level. Along the crest and slopes of Cumbre Vieja, at Punta Fuencaliente, and on the outer slopes at the northern end of the island are many young cinder cones. Lava flows originating from the cones along the crest of the Cumbre Vieja extend along the east and west coasts of La Palma filling many of the barrancos in the slopes. The island's coasts are generally steep and rocky with noticeable re-entrants along the north, east and west coasts.

The earliest volcanic collapse in La Palma took place at 1.2 Ma as a result of which a south-southwest trending paleocaldera was formed at the site of the Taburiente Caldera. This relief was subsequently subdued by the construction of the southern ridge (Cumbre Nueva high) by three overlapping edifices to form the northern shield of La Palma (Carracedo et al., 2001). This structure in turn was destroyed about 0.4 Ma to form the 'Cumbre Nueva collapse scar' and the 'Caldera Taburiente'. Ancochea et al. (1994) inferred that large landslides formed both the Cumbre Nueva collapse scar and the Caldera de Taburiente 700-710 Ka, whereas Carracedo et al. (1999b) and Carracedo et al. (2001) inferred that only the Cumbre Nueva collapse scar was formed by a collapse 560 Ka. Carracedo et al. (1999b) suggested that the Caldera de Taburiente is erosional in origin and formed by a drainage system trapped between the growing Bejenado Volcano and the 558 Ka Cumbre Nueva collapse scar. More recently Carracedo et al. (2001) inferred that the geology of the caldera is due to gravitational collapse, the emplacement of the Bejenado Volcano inside the collapse structure and the incision of the Barranco de las Angustias along the western boundary of the collapse to form the Caldera de Taburiente. The Cumbre Vieja Volcano at the southern end of Cumbre Vieja Ridge began to form on the flank of the Taburiente-Cumbre Nueva Caldera about 123 Ka, a volcanism that has continued until the present (Guillou et al., 1998; Carracedo et al., 2001).

La Palma: Offshore

Like Urgeles et al. (1999) we infer that the aprons on the east and west insular margins of La Palma are the creations of mass wasting (Figures 11 and 12). They recognized two avalanche deposits on the west side of the island consisting of three lobes. They inferred that the central lobe is Cumbre Nueva and the lobe north this lobe as Playa de Veta (Urgeles et al., 1999). From its sediment cover they suggested that the southern lobe is older than the Cumbre Nueva Debris Avalanche and correlated it with the Playa de la Veta Debris Avalanche north of Cumbre Nueva. According to Urgeles et al. the onshore headwall of the Cumbre Nueva Debris Avalanche is defined by the Caldera de Taburiente and Cumbre Nueva Ridge and is 558 Ka. They correlated the Playa de la Veta Debris Avalanche Deposit with the collapse of Garafía Volcano, beneath south-southwest trending beneath Caldera de Taburiente dates from 1.2 Ma (Figure 11). According to Carracedo et al. (2001) this volcano developed steep walls closely centered over the seamount forming the foundation of La Palma. This volcano was at least 2000 m high with a diamater of 25 km and covered much of the northern end of La Palma. On land the Playa de la Veta flow may be represented by a several hundred meter thick breccia atop the seamount series in the east and southeast of the Caldera de Taburiente (Carracedo et al., 2001). We question the age assigned to the southern lobe by Urgeles et al. (1999). Instead we propose that this is part of the Cumbre Nueva Debris Avalanche. We ascribe differences in sediment cover between the northern and southern sides of the Cumbre Nueva Debris Avalanche, not to different ages, but to differences in rates of deposition, a difference documented by the thick sediment apron on the west flank of the Cumbre Vieja Ridge. Thus the avalanches off the west consist of two lobes, the Playa de la Veta to the north overlapped to the south by the Cumbre Nueva Avalanche. How far south the Playa de la Veta Avalanche extends beneath the Cumbre Nueva cannot be determined from our data.

Several small debris of limited extent were imaged by the new multi-channel data on the north insular slope (Figure 11). A third large debris avalanche off the east coast was identified by Masson et al. (2002) and named the Santa Cruz Debris Avalanche deposit. They inferred it to be older than 1 Ma and that the plane of separation of the avalanche might be located in the embayment in Santa Cruz. According to Carracedo et al. (2001), however, there is no geological evidence onshore for this slide. Data collected during the present survey indicate that the deposit is much more extensive than inferred by Masson et al. (2002). In plan view the Santa Cruz Debris Avalanche deposit displays a fan shape with its apex located in the vicinity of Santa Cruz. From its morphology the Santa Cruz Debris Avalanche appears to consist of three units, a northern one entrained by northeast-trending lows and ridges with an extensive coverage of exotic blocks, a central one also cut by similar lows bordered by ridges but with sparce coverage of exotic blocks on its surface, and a southern one lacking the lows and ridges, but having a noticeable coverage of highs that are circular suggesting that they represent volcanic cones (Figure 11). The contact between the northern and central units is gradational, with the flows being distinguished from one another by block concentration, whereas the boundary between the central and southern flow is sharp, being defined by a chain of blocks. Along the southern side of the Santa Cruz Debris Avalanche is a line of three volcanic cones extending from the offshore ridge south of La Palma (Figures 11 and 12). Along the southwest border of the Santa Cruz Debris Avalanche is a submarine ridge on strike with the Cumbre Vieja Ridge on land. This ridge plunges offshore to a depth of 3000 m, is convex in plan view and curves toward the east in the direction of the northwest trending fracture zone off La Gomera.

As there is no obvious geological evidence onshore for this slide (Carracedo et al., 2001), we suggest that its path to a potential source was buried by subsequent volcanic activity. Consequently we infer that the Santa Cruz Debris Avalanche was the creation of either the collapse of the southeast flank of the Garafía Volcano or the east flank of the Cumbre Nueva Rift. If the failure is due to the collapse of the Garafía Volcano then the Santa Cruz Debris Avalanche is 1.2 Ma and if due to the collapse of the Cumbre Nueva Ridge then the avalanche is 560 Ka. If the flow originated as a result of the collapse of the Garafía Volcano, then it probably was continuous with the Playa de Veta west of La Palma and was cut in two by the construction of Cumbre Nueva and Cumbre Vieja Ridges. If due to the collapse of the Cumbre Nueva Ridge then it may have been continuous with the Cumbre Nueva Flow Avalanche west of La Palma and was cut in two with the construction of the Cumbre Vieja Ridge.

On the distal ends of the western and eastern avalanche deposits and the rough terrain off the northern end of La Palma, at a water depth of 3600 to 4500 m, is a sediment drift. Northwest-trending channels originating on the northwest insular slope of La Palma segment the field. South of the drift are the El Golfo Debris Avalanche and the Canary Island Debris Flow (Figure 11). According to Masson et al. (1992) the Canary Debris flow is younger than the sediment drift, implying that the sediment drift is older than 15 Ka. The sediment drift is absent in the gap between the avalanche deposits offshore southern La Palma and northern El Hierro. Wynn et al. (2000) inferred that this is the result of a predominance of mass wasting processes in the region.

The crests of the sediment waves in the sediment drift are parallel or subparallel to the insular slope, with their crests being sinuous and bifurcate. According to Wynn et al. (2000) the sediment waves are asymmetrical with a steeper slope facing down slope, become smaller down slope, have average wave heights of 21 m and wavelengths of 1.2 km. Core samples indicate that the sediment waves were formed from volcaniclastic turbidites interbedded with pelagic/hemipelagic layers, and apparently have migrated upslope through time.

Two origins have been proposed with regard to the sediment drift. As the features lie within the northeast flowing Antarctic Bottom Water (AABW) Jacobi and Hayes (1992) and Masson et al. (1992) proposed that the sediment waves were or are formed and maintained by this current. Wynn et al. (2000) rejected such an origin on the basis of the following observations: (1) AABW flow in the region is weak and bottom photographs display no evidence of significant bottom current activity; (2) waves formed in this manner should be oriented at an angle to the current and will migrate up current and to the right of flow direction; those in the La Palma field, however, are aligned approximately parallel to the current and display an upslope migration; (3) the waves are located within a zone of mixing of northeast flowing AABW and southwest flowing NADW, a zone of 'no motion'; and (4) the waves display no change in direction and orientation across the zone of mixing. Although rejecting such an origin they did point out that features may be relict and may have formed during glacial periods when the AABW current was stronger.

Wynn et al. (2000) favored formation of waves by down slope turbidity currents. They point out that such an origin is supported by the following: (1) the waves are at right angles to such flow; (2) the waves display evidence of migration up slope (up current); (3) the waves decrease in size down slope, reflecting a reduction in the velocity of the turbidity currents with a decrease in slope gradient, and (4) the features occur in an area of numerous fine grained volcaniclastic turbidites. They proposed that the features were formed at the distal ends of the avalanches where the turbidity currents escape the confines of channels and are able to spread laterally as sheet flows.

We question the validity of a turbidity current origin for the sediment waves off La Palma and El Hierro for the following reasons. Where the waves are the creation of turbidity currents, they tend to be found in areas where turbidity currents are of low-speed and have low sediment concentration, such as the backside of levees (Normark et al., 1980). They can also be found within the channels themselves where currents may die out. This is not the setting of the waves seen offshore La Palma. Channels cutting the waves indicate that the region is not one of weak turbidity currents. The wave field also is too extensive to be diagnostic of low-density, low-velocity turbidity currents. Wave orientation and migration patterns also cannot be used to support a turbidity current origin. As summarized by Normark et al. (1980) such waves can be oriented not only transverse, but also parallel to the currents and they can migrate up current, possibly down current and may even be stationary relative to the current. Another observation that makes us question a turbidity current origin for the waves is their relation to the volcanic peaks northeast of La Palma. The waves tend to be warped along the edges of the volcano, a geometry that is not supportive of a turbidity current. Such currents would erode on the up-current side of the high and deposit on the down-current side of the volcanic features. Such a geometry is not present. It is for these reasons that we support the concept that the waves are relict and was formed at the time when AABW flow was stronger than now.

El Hierro: Inshore

In plan view El Hierro has an irregular outline, being essentially three lobes whose intersection forms a tableland in its center. This trilobate morphology is typical of three-armed rifts (Carracedo, 1994). The tableland with a few erosional barrancos displays the greatest concentration of cinder cones in the archipelago (Hausen, 1973; Fuster et al., 1993). In contrast to La Palma that has had seven eruptions in the last 500 years, only one of these cones has been active in historical time (Hernández Pacheco, 1982; Carracedo et al., 2001). The island's periphery consists of steep cliffs indented by three embayments, Las Playas, El Julan and El Golfo. The semicircular El Golfo, opened to the northwest, has an arc of 25 km with its steep side descending abruptly about 1100 m to a 5 km wide platform filling the embayment. The inner side of this platform has an elevation of about 300 m and its irregular outer side is at sea level. Las Playas has an arc of 9 km, is opened to the southeast and is bordered by 900 m high cliffs and El Julan has an arc of 8 km, is opened to the southwest and bordered by 600 m high cliffs. These cliff bound embayments have been ascribed to the collapse of volcanic edifices (Ridley, 1971; Hausen, 1973; Bravo, 1982; Holcomb and Searle, 1991; Fuster et al., 1993; Guillou et al., 1996; Carracedo, 1994, 1996).

The main bulk of El Hierro was built during the Tiñor and El Golfo volcanic phases. According to Carracedo et al. (2001) the volcanic edifices in El Hierro, like those in La Gomera and Gran Canaria, overlap concentrically. In contrast those of Fuerteventura-Lanzarote and La Palma are aligned in prominent directions. We think, however, that the edifices in La Palma overlap more than those in Lanzarote and Fuerteventura. The Tiñor Volcano at the eastern part of the island was built between about 1.12 and 0.882 Ma and El Golfo from about 550 to 130 Ka (Guillou et al., 1996). A break in volcanic activity between these two volcanic edifices coincides with the regional rapid growth of Cumbre Nueva volcanic activity in La Palma (Carracedo et al., 2001). El Hierro acquired much of its morphology as a consequence of volcanic collapses. The westerly-directed Tiñor collapse took place about 882 Ka during which more than half of the northwest flank of the edifice was removed (Carracedo et al., 1999a, 2001). This displacement was followed by the construction of the El Golfo Volcano between 545 and 130 Ka that filled the depression formed by the collapse of El Tiñor Volcano (Carracedo et al., 2001). The failure of the El Golfo structure formed the El Golfo Embayment (Carracedo et al., 1999a). Correlation of the El Golfo Debris Avalanche with the Canary Debris Flow offshore and a turbidite in the Madeira Abyssal Plain (turbidite b of Weaver et al., 1992) led Masson (1996) and Urgeles et al. (1998) to suggest that the El Golfo collapse took place between 9 and 15 Ka or 10 and 17 Ka. Geologic observations along galleries excavated for ground water exploration led Carracedo et al. (1999a) to infer that the collapse took place earlier, about 130 Ka, that a marine terrace was eroded in El Golfo 130 to 25 Ka and that lavas 20 to 15 Ka covered the terrace. It was the collapse of the seaward edge of this lava cover and the submarine slope beyond 17 to 9 Ka that supposedly produced the offshore avalanche debris, the Canary Debris Flow and the turbidite sequence b at its distal end, not the collapse of the volcanic edifice.

El Hierro: Offshore

The El Hierro insular margin is divided into three compartments by a system of ridges trending northeast, northwest, west and southwest (Figures 11 and 12). The NE, W and SW (South Ridge) ridges, continuations of those found onshore, extend as much as 20 km offshore and can be traced to a water depth of about 3200 m. The NE and W ridges are lobate and have the appearance of volcanic aprons, whereas the SW high displays a ridge morphology. This high, the Southern Ridge, is 38 km long, 2000 m high, is 20 km wide on its proximal end, 10 km wide at its distal end and curves gently southwestward (Gee et al., 2001a). One minor and three major avalanches are present within and between these ridges (Figure 11). The minor one occurs within the northwest rift whose head is at the upper end of the insular slope (Figure 11). Two major landslides, Las Playas I and II, occur off Las Playas Embayment (Gee et al., 2001a; Masson et al., 2002). Within these avalanches are less than 200 m high and 2 to 4 km wide pinnacles that Gee et al. (2001b) inferred to represent dikes. At the head of this collapse is a narrow, steep-sided embayment parallel to the southern end of the San Andrés Fault System. Day et al. (1997) inferred this fault system to be the head of an aborted landslide whose displacement took place between 545 and 261-176 Ka and gave rise to Las Playas I and II Avalanches. Carracedo et al. (1997) inferred that the slumping was due to one major movement along the fault, a movement that took place 250 to 150 Ka.

Carracedo et al. (1997) proposed that the landslide off El Julan Embayment took place prior to 158 ± 4 Ka and Masson (1996) suggested that the failure took place about 500 to 300 Ka. Carracedo et al. (1999a) inferred that the avalanche probably took place when the El Golfo Volcano was well-developed, before 130 Ka. More recently Carracedo et al. (2001) stated that that lavas belonging to rift volcanism in the water galleries in the El Julan Collapse Embayment constraints the minimum age of the collapse to 150 Ka.

Sidescan sonar recordings described by Masson et al. (2002) and swath bathymetry obtained during the present study display a 124 km^2 wave field within an 8 km long and 2 km wide channel in the El Julan debris (Figure 11). The waves have lengths of 0.4 to 1.2 km and heights of 6 m and are aligned parallel to the contours. These features are part of the El Hierro Sediment Drift. Sidescan sonar recordings also indicate that the surface of the El Julan slide is cut by a series of sediment draped steps of fault origin with a few exotic blocks (Masson et al., 2002). Displacement along the steps may have been triggered by the erosion of the toe of the avalanche deposit by the Saharan Debris Flow (Masson et al., 2002). This flow originated in the African margin about 60 Ka and flowed westward along 24° N (Insert, Figure 1). According to Gee et al. (1999) 40 km^3 of the El Julan Debris Avalanche deposit was eroded and incorporated into the Saharan Debris Flow.

Offshore El Golfo Embayment is the El Golfo Debris Avalanche deposit whose landward termination is a 400 high scarp on the upper insular slope (Masson, 1996). Like other landslides in the region, sidescan sonar images of the surface of the El Golfo Avalanche show typical exotic blocks up to 1.2 km wide and as much as 200 m high. These blocks are particularly abundant at a depth of 3500 m with a few blocks extending to a depth of nearly 4000 m. As previously described correlation of the debris avalanche with the Canary Debris Flow and the flow with turbidite b in the Madeira Abyssal Plain (Weaver and Rothwell, 1987; Weaver et al., 1992; Masson, 1996) implies that the avalanche must be 15 Ka and that the collapse of the El Golfo volcanic edifice that created the avalanche must have occurred at 15 Ka. Onshore geology, however, suggests that the collapse of the El Golfo volcanic edifice took place 130 to 100 Ka. This variance in age led Carracedo et al. (1997) to propose that the failure in El Golfo Embayment occurred in two phases, the collapse of the El Golfo volcanic edifice 130 to 100 Ka and the collapse of the volcanic platform filling the embayment and the submarine slope beyond the platform 17 to 9 Ka. It was the collapse of the platform, not the volcanic edifice that produced the Canary Debris Flow. More recently Carracedo et al. (2001) inferred that the geology of the marine abrasion platform beneath the lavas filling the El Golfo Embayment is incompatible with features associated with a single event collapse. From the geologic data they suggested that the subaerial embayment might have formed after the extrusion of lavas over the cliff sequence at

134 Ka. They also point out that the magnetic studies of Szérémeta et al. (1999) of the El Golfo and Rift sequences at the El Golfo escarpment suggests that the section has experienced continuous clockwise rotation.

The El Golfo Avalanche grades westward into the Canary Debris Flow. Immediately west of El Golfo Debris Avalanche the Canary Debris Flow is characterized by an opaque facies partially overlapped by the avalanche. At the distal west end of the debris flow is a transparent facies onlapping the stratified pelagic/hemipelagic sediments of the Madeira Abyssal Plain. The debris flow is lobate in form, is 60 to 100 km wide and flowed 600 km westward from the El Hierro Rise (Urgeles et al., 1997; Masson et al., 1997). The flow is a mixture of clasts and matrix with the clasts ranging in size from cm up to 300 m. The larger sediment clasts were derived locally rather than up-slope indicating that the debris was capable of eroding the seabed in the middle part of its course.

Unusual features imaged by the relief diagrams constructed from the swath bathymetry off El Hierro are low relief mounds surrounded by moats (Figures 11 and 12). They have reliefs of 25 to 50 m and are 10 to 2 km long and 1 to 5 km wide. A few are circular and others are kidney shaped. These features occur within the El Hierro Sediment Drift, within the Canary Island and Saharan Debris Flows and between the flows. They also appear to be present west of the West La Palma Sediment Drift (Figure 11). They tend to occur at water depths of 4000 m. Within the debris flows they could represent blocks eroded locally by the currents. Outside the flows they could represent exotic blocks that outran the main part of the debris avalanches that originated in the insular slopes. Such origins do not explain the moats around them. That they surround the highs also suggests that they are not due to erosion by northeasterly-flowing AABW. If they were formed in this manner the lows would to be eroded only up current of the mounds with their down-current sides being sites of deposition. Morphologically they resemble the mounds described by Hovland et al. (1994) offshore western and northwest Australia, that were created by seepage of hydrocarbons along faults that first eroded the seafloor to form the moats with local nutrient enrichment of the overlying waters and enhancement of carbonate accretion in the moats. If they occurred in this manner then the gas must of be hydrothermal origin. In the absence of any information, besides that of their surficial shape, any attempt to determine their origin is premature.

Discussion

Sedimentary processes in the Canary Island Basin during the Neogene were influenced by the stages of volcanic development of the islands. Offshore deposition around those islands in the shield phase, Tenerife, La Palma and El Hierro, is dominated by debris avalanches that extend tens of km offshore to a water depth of 3000-4000 m (Figure 11). Construction of post-avalanche volcanic cones, salic necks and extrusion of lava flows offshore, reworking by bottom currents, turbidity current processes and hemipelagic deposition has modified the surface morphology of these deposits. The margins of those islands in the erosive cycle, Lanzarote, Fuerteventura, Gran Canaria and La Gomera reflect mainly turbidity current activity and hemipelagic processes and secondarily avalanche deposition and volcanic activity. Those in the shield phase, Tenerife, El Hierro and La Palma are dominated by mass wasting processes and widespread volcanic activity.

That turbidity currents were a major contributor to the construction of the insular margins of Fuerteventura and La Gomera is indicated by their depth profiles. Whereas the younger islands display concave upwards profiles these islands display profiles that are convex upward (Figure 13). Both islands, including Lanzarote, lack calderas, coastal embayments and scars produced by avalanches cascading down the slope. However all three islands have undergone massive erosion and uplift that may have erased these features. In addition geologic data have shown that Fuerteventura's volcanic edifices were eroded by mass wasting. As the centers of two of these cones were located off the present west coast, construction of the island's western margin must have been primarily by mass wasting processes. In contrast to the other islands the Gran Canaria margin displays profiles that are both convex and concave upwards indicating that mass wasting and turbidity currents have played roles in the formation of its margin. The profile of Lanzarote's margin is in the form of a ramp, suggesting that mass wasting played a role in creating its morphology (Figure 13). La Gomera is unusual in that the island has been so eroded that it does not display any evidence of mass wasting, such as collapse scars or coastal embayments. Landslides on this island are young and are limited to the 'barrancos', the base of the seacliffs and the insular slope.

Mass wasting

Massive failures in the younger Canaries are related to rift geometry with some of the volcanism occurring along or fanning out from rifts. Other volcanic structures trend at an angle to the rifts, or occur along troughs created by avalanches, or at the junction of two islands, such as Fuerteventura and Gran Canaria, where volcanic trends change abruptly. Associated with some of the rift volcanic edifices are what appear to be lava fronts and topographic features resembling lava falls and lobes spreading out over the sea floor adjacent to the rifts. Massive failures in the Canary Islands occur at the junction between the two most active rifts, with the third one acting as buttress for the landslide. It is these triple rifts, the concentration of dikes leading to the destabilization of the flanks through magma overpressure, mechanical and thermal overpressure of pore fluids, oversteepening due to rapid volcanic deposition, and seismic activity that leads to formation of gravitational unstable volcanic flanks (Carracedo, 1994, 1996; Elsworth and Day, 1999; Masson et al., 2002). Masson et al. (2002) proposed that failures tend to be concentrated on that flank of the rift having the greatest gravitational potential for lateral movement due to its steeper gradient.

Seismicity and sea level fluctuations affecting the stress regime in a volcanic island also may contribute to slope instability. McGuire et al. (1997) proposed glacially induced sea level regressions, which reduce the water volume on the flanks of the islands by about 100 m, to decrease radial compressive stresses by as much as 1 Ma. Such a reduction would tend to enhance extrusion on the volcanic edifice. Such lowering in sea level also reduces shear-stress on the volcanic flanks that in turn would tend to enhance slope instability. This model, however, is too simplistic. As documented by Guillou et al. (1998) volcanism along the Cumbre Vieja in La Palma was pronounced during the glacially induced regressions and transgressions 125 to 80 Ka. Volcanism was sporadic from 80 to 20 Ka when sea level reached its lowest level and Cumbre Vieja Volcano underwent massive erosion forming seacliffs as high as 700 m. There was intense rift volcanism during the subsequent rise in sea level from 15 Ka to the present. Masson et al. (2002) also suggested that climatic changes, affecting rainfall and groundwater levels within the volcanic edifices, also contribute to its failure. Preferential deposition and accumulation of weak layers within a previous failure would tend to encourage failures in the same area.

Most workers have proposed that the onshore and offshore avalanches in the Canary Islands are continuous. Mapping of the offshore deposits is based mainly on swath methodology that has a minimum depth of 200 m. Thus these marine data do not provide information on the critical link between the onshore and offshore avalanches. Although the swath bathymetric map compiled during the present investigation (Figure 2) is limited to depths greater than 200 m it does show that the onshore and offshore avalanches are not continuous, but are separated by a several hundred meter high scarp on the upper insular slope. Such a scarp also was reported off El Hierro by Masson (1996) and off Piton de Gran Brulé, Reunion Island by Lénat et al. (1990). It is the existence of this topographic divide between the onshore and offshore deposits that led Teide Group (1997) to infer that failures off Orotova, Icod and Güimar Valleys in Tenerife may have taken place in two stages. The failure on land created an avalanche or slump that extended some distance offshore. It was the failure of the distal end of the onshore structure along the scarp of the upper slope that Teide Group (1997) proposed to have created the offshore avalanche. They further inferred that the area of rough topography at the base of the scarp offshore Tenerife represented a matrix-rich, poorly sorted, chaotic-rich debris avalanche that originated on the 100–600 m slope. Teide Group (1997) further suggested that blocks on the lower slope represent earlier failures of the front and that these blocks obstructed the more recent flows. Carracedo et al. (1999a) suggested such a model for the offshore segment offshore El Golfo Embayment in El Hierro.

Failure of the upper slope avalanche may be the consequence of tectonic uplift, seismic activity or sediment loading on the slope. The intertidal fossils found by Bravo (1952) 8 m above sea level on the seacliff northwest of Tenerife may support such recent tectonic uplift. Other evidence for uplift comes from the gravity field of the island. Bosshard and MacFarlane (1970) inferred that a large positive gravity anomaly in the region indicates that a magma body may be beneath the Las Cañadas Caldera and that basement in Tenerife has been uplifted 1 km above its surroundings. The time interval between these two events is yet to be determined, although in the chronology described by Carracedo et al. (1999a) for El Hierro indicates that they occurred about 100 ky apart. We, however, do not infer that the offshore deposits are not due to uplift or faulting on the upper slope. The intertidal fossils found 8 m above sea level in Tenerife reflect, not tectonic uplift, but glacially induced changes in sea level during the Pleistocene. Funck and Schmincke (1998) and Carracedo (1999) have demonstrated that, except for Lanzarote and Fuerteventura, the Canary Islands have remained relative stable with respect to sea level for some time.

Masson et al. (2002) rejected the model that suggests that the offshore avalanches, at least in Tenerife, were the result of the collapse of the wall of the crater. According to them such a model is not supported by the volume assessment of the offshore avalanche, or by offshore mapping evidence. However, offshore volume measurements are too uncertain as to be more than educated guesses. The avalanche deposits are acoustically opaque, so that low energy single channel reflection profiles used in most investigations only provide information on their surface configuration. Changes in the landslide scars by post-avalanche erosion and volcanism, show that the scars may have served as chutes for more than one avalanche. Some of the avalanche detritus may come to rest within the collapse structures or within the chute itself, and that material along the path transversed by the avalanches is probably incorporated into the avalanche, making use of volume estimates to determine the extent of insular degradation above sea level unrealistic. It is, however, a relative estimate of the degree of degradation of the volcanic edifice above and below sea level. Although our estimates of the volumes of the avalanches, except for Las Playas II, in Table 1 are comparable with those reported by Masson et al. (2002) we still tend to put little reliance on them. It's even possible that the features we identified as debris flows off La Gomera may be turbidite systems that originated on the uppermost insular slope. That little reliance can be placed on the offshore avalanche volumes is shown by a comparison of the offshore and inshore data in Orotava Valley. As a result of the numerous galleries that have been dug for ground water in the valley (Coello, 1973), the geologic parameters in the valley are reasonably constrained. The valley has an area of 100 km^2 and is filled with 50 km^3 of avalanche debris, 35 km^3 of post avalanche volcanic rock and has a space volume from the top of the volcanic rocks to its rim of the valley of 30 km^3. Thus the valley could only have provided a volume of about 70 km^3 of avalanche debris offshore yet Masson et al. (2002) and the Teide Group (1997) reported that the volume of the offshore Orotova Debris Avalanche was about 400 and about 500 km^3 (minus the volume in Orotova Valley) respectively. Apparently the path traversed by the avalanche provided most of its volume.

Not only is the volume of the offshore avalanche not an index of the extent of erosion above sea level, but also the detritus contributed by the volcanic collapse above sea level may not be as voluminous as postulated. Were the calderas (in the topographic sense of the word) formed by mass wasting, or are they due to magma collapse as it has been proposed for the Caldera de las Cañadas (Martí et al., 1997) or fluvial erosion as proposed by Carracedo et al. (1999a, b), or a combination of collapse and fluvial erosion as inferred by Carracedo et al. (2001) for the Caldera de Taburiente in La Palma. Masson et al. (2002) stated that geologic mapping on land and the huge volume of the debris avalanche offshore Tenerife do not support a magma chamber collapse as a major factor in the formation of the Caldera de las Cañadas. However, if the offshore avalanche contains a large volume of sediments derived from the insular slope, then collapse of the north wall may have been sufficient to account for the amount of offshore deposit derived from above sea level. In contrast to Masson et al. (2002), Martí et al. (1994) believe that surface geologic data on land does support a magma chamber collapse origin for Las Cañadas Caldera. Martí et al. (1997), Martí and Gudmundsson (2000) and Soriano et al. (2002) inferred that the Las Cañadas Caldera is the creation, not of one, but of three vertical collapses from 1.6 Ma to 179 Ka, that triggered the Orotova, Güimar and Icod Valleys by lateral collapses. Data from the galleries excavated for groundwater (Coello, 1973) and wells drilled in the caldera (Martí et al., 1997), however, are too limited to verify or reject the validity of the existence of a north wall.

Present model

We propose instead that the distinct difference in volume between the inshore and the offshore debris avalanches is a reflection of the manner that failures take place along the flanks of volcanoes. We infer that failures in the Canary Islands took the form described by Jacobs (1995) for the Nuuanu Debris Avalanche in Oahu, Hawaii. Here failure was in two synchronous or near synchronous avalanches with the slip-surface on the onshore delineated by the steep cliffs parallel to the coast, the Nuuanu Pali, and the offshore one by a steep scarp on the upper slope. We farther propose that the location of the surface separating these two slip surfaces was controlled by the geology of the flank of the volcanoes. The stratigraphic succession on the slope of the Hawaiian Volcano was documented by a 3098 m boring on the flank of Mauna Kea, Hawaii (DePaolo et al., 2001). It is true that the Canary Islands differ from the Hawaiian chain. They differ in magma production, eruptive frequency and the Canary Islands have not subsided significantly during their construction. Differences in the rate of subsidence probably are related to differences in the age of the oceanic crust on which they rest, Jurassic for the Canaries and Cretaceous for Hawaii. However, the Canaries do display similar features with the Hawaii chain in having a two-stage subaerial evolution (shield stage volcanism separated from the post-erosional volcanism by an erosional gap), as well as common structural features such as rift zones, and multiple gravitational collapses (Carracedo, 1999). Thus the stratigraphic section revealed at the drill site at Mauna Kea may be applicable to the slopes of the Canary Islands volcanoes.

From this stratigraphic section we infer that the scarp on the upper slope in the Canary Islands archipelago defines the transition from subaerial lavas above to the clastic facies below, a transition marked by volcaniclastic rocks, glassy lavas and massive basalt layers. It is this contact that defines the surface separating the slip-planes of the subaerial and submarine avalanches (Figure 14). We further speculate that these two avalanches were probably near or synchronous with the subaerial failure occurring first followed soon after by the submarine one. As in Jacobs (1995) description of Nuuanu Debris Avalanche, the avalanches offshore the Canary Islands were the creation of these two events. As the subaerial debris avalanche entered the sea it eroded small V-shaped canyons out of the harder units along the transition between the subaerial and submarine series on the upper slope and triggered the offshore avalanche, possibly by sediment loading, and became mixed with it. The slip surface of the offshore unit was within the units on the transition and the décollement along which the offshore avalanche was displaced down slope is the contact between the unconsolidated and consolidated hyaloclastics. It was the mass wasting of the slope facies that created the flat-bottomed channels and irregular crested divides. The ridges in the valleys are due to scour compression between different flows. Secondary debris flows may be responsible for the discontinuous bedded insular units described by Schmincke and Sumita (1998). The hummocks and lows at the end of the debris flows represent grabens and horsts constructed by the lateral spreading of the distal ends of the avalanches. We further speculate that some of the exotic blocks found on the surface of the avalanche may have originated on the transitional zone on the slope rather than inshore and that the steps noted in some of the offshore avalanches may represent either local exposure of massive basalt beds that are intrusives, large lithic clasts or subaerial flows that may continued offshore. All in all the morphology of the offshore avalanche appears to be the result of several events of varying intensity. These distinct flows may reflect displacement along bedding planes of the different volcanic flows dipping away from the center of the volcanic edifice. As the failure continued mass wasting may have gradually changed to debris flows that in turn led to the formation of turbidity currents. It is these currents that were responsible for the fans seaward of the avalanches offshore Tenerife. According to the Teide Group (1997) the sediments in these fans are made up of at least six turbidite sequences, further verification that catastrophic failure of the slope was not a single event, but by repetitive landslides. However, in the absence of detailed seismic reflection coverage of the upper insular slope coupled with extensive bottom sampling the above scenario has to be considered to be in the realm of speculation. Possibly the origin of the scarp on the upper slope may have a simpler explanation. It may simply represent differences in the thickness of the avalanche deposits or a plane of separation between the head and toe of the avalanche. If it is a plane of separation it would imply that as the avalanche traveled down slope it broke apart into two or more segments with the upper one being perched on the flank of the subaerial segment of the volcanic edifice.

Another factor that controlled the physical aspects of inshore/offshore composite avalanches is the location of the volcano whose collapse contributed to the avalanche. For example the 3500 m high Pliocene Roque Nublo cone was built in the center of Gran Canaria. The Roque Nublo breccia, formed by one or more subaerial volcanic debris avalanches created by the collapse of the cone, had to travel 10 to 20 km via barrancos to reach the coast. In contrast the centers of the Central and Southern Volcanic complexes in Fuerteventura were located on the present coast and the avalanches could be deposited directly offshore. Similarly the distances from the volcanic edifices in La Palma and El Hierro to the offshore depocenter were also short as reflected by the pronounced coastal embayments in these islands. Distances from source to depocenter were somewhat greater in Tenerife and Lanzarote. In the islands where the source was near the depocenter the avalanche must have been quite coarse and contained an over overabundance of exotic blocks. As the flow was probably quite dense its travel distance offshore was short even though its momentum was much greater than the flows described below. As the distance increased, the avalanche became finer, and the exotic block content diminished. However, the finer avalanche would be able to travel a longer distance than those produced by a source located near the depocenter. As the transport distance increased the avalanche may evolve into a debris flow or a turbidity current.

An additional factor that controls the dimensions of a debris avalanche is the texture of its deposits with runout distances increasing as the amount of fine-grained sediment content increases (Iberson, 1997). This increase in distance with decreasing grain size is because fines tend to sustain high pore pressure that reduces frictional resistance. This suggests that the flows offshore La Gomera are short because they are coarse, the flows offshore La Palma and El Hierro have a lobate form because they are somewhat finer-grained, and the avalanche offshore Tenerife are linear because they are even finer grained (Masson et al., 2002). Such assumptions, however, are in need of testing. The surface features of these offshore flows are well known, but their internal structure and texture is unknown. To date the only flows mapped in detail are those on land, such as the Pliocene Roque Nublo Debris Avalanche in Gran Canaria (Mehl and Schmincke, 1999). Other flows on land that have been mapped to some degree are the avalanches in Tenerife (Bravo, 1952; Coello, 1973).

Conclusion

Volcanism in the Canary Islands appears to be rift related with mass wasting being represented by two landslides, one above sea level and the other on the submarine insular slope. The surface separating these two landslides is either the detachment surface along which the landslide broke in two or represents the transition between the subaerial volcanic rock and the slope hyaloclastic rocks. Drilling in Hawaii indicates that alternating massive basalts, subaerial flows that continued beyond the shoreline, marks this transition with the flank hyaloclastic rocks. Catastrophic failures are readily recognizable in Tenerife, La Palma and El Hierro, islands in the shield-phase of development (Figure 1). The failures are generally interpreted as avalanches; the only debris flow identified is the Canary Debris Flow off El Hierro. However, it is possible that the landslides are not solely the creation of avalanches, but are constructed by both debris avalanche and debris flow processes. Such an interpretation is not unrealistic, as these sediments were deposited during repetitive failures. Slumps have been recognized along the flanks of Lanzarote, El Hierro and Gran Canaria. It is possible that some of the features identified as avalanches on land may represent slumps that turned into avalanches on their distal ends as they broke apart as they moved across the volcano's uneven surface. The debris avalanches of the young islands in the shield-phase, Tenerife, La Palma and El Hierro, have runout distances that range from 12 to 95 km, cover 30 to 3500 km^2 of sea floor, and have volumes <50 to 800 km^3. Differences between flows off the islands are related to their texture, with the finer ones with a greater water content traveling farther. At their proximal ends, in depths shallower than 3000 m, the avalanches are in the form of flat-bottomed steep sided chutes, where erosion and down-slope transport predominate. At the heads some of the chutes are low relief ridges aligned parallel to its axis. These highs may be due either to scour or to compression created by velocity differences between adjacent flows. The distal ends of some of the chutes are characterized by a series of convex down-slope steps that may be the creation of a series of overlapping flows, or represent local exposures of massive basalt levels. At their distal ends the flows are in the form of overlapping lobes. The surfaces of the lobes are disrupted by hummocks and lows that may represent grabens/horsts resulting from their lateral displacement. Scattered over the surfaces of the lobes are exotic blocks that glided to their present sites from inland and/or the insular slope by being supported by the excess pore water pressure of the avalanche, or supported by a matrix of fine grained sediment in the avalanche. Most of the blocks retain their integrity, indicating that they were resistant to the internal shear of the flow. Some of these blocks had enough momentum to outrun the main part of the flow.

Defining the processes that sculptured the margins of the islands in the erosional phase (Lanzarote, Fuerteventura, Gran Canaria and La Gomera) from their morphology is not as clear-cut as the islands in the shield-volcano phase. If it was not for the presence of highs that resemble exotic blocks and box-shaped canyons, the margins could be interpreted as the construction of turbidity currents and hemipelagic deposition. The margin of La Gomera, for example, has the appearance of being mainly the creation of turbidity current deposition and erosion, with volcanic edifices aligned along a rift offshore the northwest corner of the island adding to the margins configuration. The western margin of Lanzarote consists of a westerly dipping platform cut by low relief channels whose divides are made of numerous small highs and bound on the landward and seaward sides by scarps. No other island displays this type of morphology. This morphology could be the consequence of a massive slump with the highs representing small volcanic cones emplaced after slumping, or a volcanic apron. We interpret the morphology of the western margin of Fuerteventura, the side slopes of Banquete and Amanay Banks, as well as the divide between Gran Canaria, as the creation of mass wasting. The terrain has been modified by post-landslide turbidity current deposition and erosion and hemipelagic deposition. That features created by mass wasting in the Miocene (Miocene) and <4 Ma (Pliocene) are still recognizable off Gran Canaria is a reflection of the intensity and magnitude of these processes. It is not surprising that such events can also be recognized in Fuerteventura. This island experienced deep erosion at the end of the shield building phase by catastrophic failures <12 Ma that reduced its relief by nearly 3000 m in 2 Ma.

In addition to being modified by turbidity currents and hemipelagic sediments, the landslides have been modified by bottom current activity. Southerly flowing NADW has created a sediment drift out of the sediments off Lanzarote and sediments waves and furrows out of the deposits in the divide between Gran Canaria and Tenerife. North flowing Antartic Deep Water (AADW) has created a sediment drift out of the landslides offshore El Hierro and La Palma. Whereas the features molded by NADW are recent, those formed by ADDW may be relict and were formed in the Pleistocene when the water mass was characterized by stronger currents.

Acknowledgements

We thank the officers and crew of the R/V Hespérides and Vizconde de Eza for their cooperation during the various cruises to the Canary Island, and. S. A. Swift and D.K. Smith of the Geology and Geophysics Department at WHOI for their suggestions during the initial preparation of this report. We also express our gratitude to P. Clift, E. de Reus, S. Day and two anonymous reviwers. Their comments were most helpful in revising a previous version of the manuscript. This is a contribution to the Spanish ZEE Program.

References

Abdel-Monem, A., Watkins, N.D. and Gast, P.W., 1971, Potassium-Argon ages, volcanic stratigraphy, and geomagnetic polarity history of the Canary Islands: Lanzarote, Fuerteventura, Gran Canaria, and La Gomera, *Am. J. Sci.*, **271**, 490–521.

Ancochea, E., Brändle, J.L., Cubas, C.R., Hernán, F. and Huertas, M.J., 1996, Volcanic complexes in the eastern ridge of the Canary Islands: the Miocene activity of the island of Fuerteventura, *J. Vol. Geoth. Res.*, **70**, 183–204.

Ancochea, E., Fuster, J.M., Ibarrola, E., Cendrero, A., Coello, J., Hernan, F., Cantagrel, J.M. and Jamond, C., 1990, Volcanic evolution of the island of Tenerife (Canary Islands) in the light of new K-Ar data, *J. Vol. Geoth. Res.*, **44**, 231–249.

Ancochea, E., Hernán, F., Cendrero, A., Cantagrel, J.M., Fúster, J. M., Ibarrola, E. and Coello, J., 1994, Constructive and destructive episodes in the building of a young oceanic island, La Palma, Canary Islands, and genesis of the Caldera de Taburiente, *J. Vol. Geoth. Res.*, **60**, 243–262.

Arthur, M.A., Cornford, C., McCoy, F.W. and Sarnthein, M., 1979, Evolution and sedimentary history of the Cape Bojador Continental margin, northwest Africa, in V. Von Rad, W.B.F. Ryan (eds.); *Init. Rep. Deep Sea Drill. Proj.*, **47**, pt. 1, 773–816.

Bosshard, E. and MacFarlane, D.J., 1970, Crustal structure of the western Canary Islands from seismic refraction and gravity data, *J. Geophys. Res.*, **75**, 4901–4918.

Bravo, T., 1952, Aportación al estudio geomorfológico y geológico de la costa de la fosa tectónica de valle de la Orotava, *Bol. Real Soc. Esp. Hist. Nat. (Sec. Geol.)*, **50**, 5–32.

Bravo, T., 1964a, Estudio geológico y petrográfico de la isla de La Gomera, I, *Inst. "Lucas Mallada" Estud. Geol.*, **20**, 1–21.

Bravo, T., 1964b, Estudio geológico y petrográfico de la isla de la Gomera, II, *Inst. "Lucas Mallada", Estud. Geol.*, **20**, 23–56.

Bravo, T., 1982, Formaciones geológicas en la isla de El Hierro, *Inst. Estud. Canarios Aulas de Cultura, Cabido de Tenerife*, 85–99.

Bugge, T., Belderson, R.H. and Kenyon, N.H., 1988, The Storegga Slide, *Phil. Trans. R. Soc. London, Ser.A.*, **325**, 357–388.

Cantagrel, J.M., Arnaud, N.C., Ancochea, E., Fúster, J.M. and Huertas, M.J., 1999, Repeated debris avalanches on Tenerife and génesis of Las Cañadas caldera wall (Canary Island), *Geology*, **27**, 739–742.

Cantagrel, J.M., Cendrero, A., Fuster, J.M., Ibarrola, E. and Jamond, C., 1984, K-Ar chronology of the volcanic eruptions in the Canarian archipelago: island of La Gomera, *Bull. Vol.*, **47**, 597–609.

Carracedo, J.C., 1994, The Canary Islands: an example of structural control on the growth of large oceanic-island volcanoes, *J. Vol. Geoth. Res.*, **60**, 225–241.

Carracedo, J.C., 1996, A simple model for the génesis of large gravitational landslide hazards in the Canary Islands, in McGuire, W.J., Jones A.P., and Neuberg, J. (eds.), Volcano Instability on the Earth and Other Planets, *Geol. Soc. Lond., spec. publ.*, **110**, 125–135.

Carracedo, J.C., 1999, Growth, structure, instability and collapse of Canarian volcanoes and comparisons with Hawaiian volcanoes, *J. Vol. Geoth. Res.*, **94**, 1–19.

Carracedo, J.C., Day, S., Guillou, H. and Torrado, F.J.P., 1997, El Hierro geological, excursion handbook. International Workshop on immature oceanic Islands, La Palma, Sept. 1997, *Est. Vol. Canarias and Uni. Las Palmas, Tenerife/Gran Canaria*, 43 pp.

Carracedo, J.C., Day, S.J., Guillou, H. and Gravestock, P., 1999a, Later stages of volcanic evolution of La Palma, Canary Islands: rift evolution, giant landslides, and the genesis of the Caldera de Taburiente, *Geol. Soc. Am. Bull.*, **111**, 755–768.

Carracedo, J.C., Day, S.J., Gillou, H. and Pérez-Torrado, F.J., 1999b, Giant Quaternary landslides of La Palma and El Hierro, Canary Islands, *J. Vol. Geoth. Res.*, **94**, 169–190.

Carracedo, J.C. and Rodríguez Badiola, E., 1993, Evolución geológica y magmática de la isla de Lanzarote (Islas Canarias), *Rev. Acad. Canar. Cienc.*, **V**, 25–58.

Carracedo, J.C., Rodríguez Badiola, E., Guillou, H., de la Nuez, J. and Pérez Torrado, F.J., 2001, Geology and volcanology of La Palma and El Hierro, western Canaries. *Estud. Geol.*, **57**, 175–273.

Cendrero, A., 1967, Nota previa sobre la geología del complejo basal de la isla de La Gomera, *Inst. 'Lucas Mallada', Estud. Geol.*, **23**, 71–79.

Coello, J., 1973, Las series volcánicas en subsuelos de Tenerife, *Estud. Geol.*, **29**, 491–512.

Coello, J., Cantagrel, J.M., Hernán, F., Fúster, J. M., Ibarrola, E., Ancochea, E., Casquet, C., Jamond, C., Diaz de Terán, J. R. and Cendredo, A., 1992, Evolution of the eastern volcanic ridge of the Canary Islands based on new K-Ar data, *J. Vol. Geoth. Res.*, **53**, 251–274.

Dañobeitia, J.J. and Collette, B.J., 1989, Estudio sísmica de reflexión de un grupo de estructuras submarinas situadas al Norte y Sur del archipiélago Canario, *Acta Geol. Hisp.*, **24**, 147–163.

Day, S.J., Carracedo, J.C. and Guillou, H., 1997, Age and geometry of an aborted rift flank collapse, the San Andres fault system, El Hierro, Canary Islands, *Geol. Mag.*, **134**, 523–537.

DePaolo, D.J., Stolper, E. and Thomas, D.M., 2001, Deep drilling into a Hawaii volcano, *EOS*, **82**, 149, 154–155.

Emery, K.O. and Uchupi, E., 1984, The Geology of the Atlantic Ocean, *Springer-Verlag, New York*, 1050 pp.

Elsworth, D. and Day, S.J., 1999, Flank collapse triggered by intrusion: the Canarian and Cape Verde archipelago, *J. Vol. Geoth. Res.*, **94**, 323–340.

Funck, T. and Lykke-Andersen, H., 1998, Seismic structure of the volcanic apron north of Gran Canaria, in P.P.E., Weaver, H.-U., Schmincke, J.V., Firth, and W., Duffield, (eds.), *Proc. Ocean Drill. Prog., Sci. Res.*, **157**, 11–28.

Funck, T. and Schmincke, H.-U., 1998, Growth and destruction of Gran Canaria deduced from seismic reflection and bathymetric data, *J. Geophys. Res.*, **103**, 15393–15407.

Fuster, J.M., Henán, F., Cendrero, A., Coello, J., Cantagrel, J.M., Ancochea, E. and Ibarrola, E., 1993, Geocronología de la Isla de El Hierro (Islas Canarias), *Bol. R. Soc. Hist. Nat. (Sec. Geol.)*, **88**, 85–97.

Gee, M.J.R., Masson, D.G., Watts, A.B. and Allen, P.A., 1999, The Saharan debris flow: an insight into the mechanics of long runout of submarine debris flows, *Sedimentology*, **46**, 317–335.

Gee, M.J.R., Masson, D.G., Watts, A.B. and Mitchell, N.C., 2001a, Offshore continuation of volcanic rifts zones, El Hierro, Canary Islands, *J. Vol. Geoth. Res.*, **105**, 107–119.

Gee, M.J.R., Watts, A.B., Masson, D.G. and Mitchell, N.C., 2001b, Landslides and the evolution of El Hierro in the Canary Islands, *Mar. Geol.*, **177**, 271–293.

Grigg, R.W. and Jones, A.T., 1997, Uplift caused by lithospheric flexure in the Hawaiian archipelago as evidence for a late Pleistocene tsunami on Molakai, Hawaii. *Mar. Geol.*, **141**, 11–25.

Guillou, H., Carracedo, J.C. and Day, S.J., 1998, Dating of the upper Pleistocene-Holocene volcanic activity of La Palma using unspiked K-Ar technique, *J. Vol. Geoth. Res.*, **86**, 137–149.

Guillou, H., Carracedo, J.C., Pérez Torrado, F. and Rodriguez-Badiola, E., 1996, K-Ar ages and magnetic stratigraphy of hotspot-induced, fast grown oceanic island: El Hierro, Canary Islands, *J. Vol. Geoth. Res.*, **73**, 141–155.

Hausen, H., 1959, On the geology of Lanzarote, Graciosa and the Isletas (Canarian archipelago), *Com. Phy. Math.*, **23**, 1–116.

Hausen, H., 1962, New contributions to the geology of Gran Canary (Gran Canaria, Canary Islands), *Com. Phy. Math.*, **27**, 1–418.

Hausen, H., 1965, Some comments on the structural geology of La Gomera (Canary Islands), *Act. Geogr.*, **18**, 1–15.

Hausen, H., 1969, Some contributions to the geology of La Palma, *Com. Phy. Math.*, **35**, 1–140.

Hausen, H., 1971, Outlines of the geology of La Gomera (Canary Islands) in relation to its surface forms, *Com. Phy. Math.*, **41**, 1–53.

Hausen, H., 1973, Outlines of the geology of El Hierro (Canary Islands), *Com. Phy. Math.*, **43**, 65–148.

Hernández Pacheco., A., 1982, Sobre una posible erupción en 1793 in la isla de El Hierro (Canarias), *Estud. Geol.*, **38**, 15–25.

Holcomb, R.T. and Searle, R.G., 1991, Large landslides from oceanic volcanoes, *Mar. Geothm.*, **10**, 19–32.

Hovland, M., Croker, P.F. and Martín, M., 1994, Fault-associated seabed mounds (carbonate knolls?) off western Ireland and north-west Australia, *Mar. Pet. Geol.*, **11**, 232–246.

Ibarrola, E., Cantagrel, J.M., Fúster, J.M., Coello, J. and Jamond, C., 1988, Geocronología de las series volcánicas Neógenas de Lanzarote, Islas Canarias, *II Congr. Geol. España*, 345–348.

Iberson, R.M., 1997, The physics of debris flows, *Rev. Geophys.*, **35**, 245–296.

Jacobi, R.D. and Hayes, D.E., 1992, Northwest African continental rise: effects of near-bottom processes inferred from high-resolution seismic data, in C.W. Poag and P.C. Graciansky (eds.), *Geologic Evolution of Atlantic Continental Rises*, Van Nostrand Reinhold, New York, 293–326.

Jacobs, C.L., 1995, Mass wasting along the Hawaiian Ridge: giant debris avalanches, in K.T., Pickering, R.N., Hiscott, N.H., Kenyon, R., Ricci Lucchi, and R.D.A., Smith, (eds.), *Atlas of Deep water Environments. Architectural Style in Turbidite systems.* Chapman & Hall, New York, 26–28.

Krastel, S., Schmincke, H.-U. and Jacobs, C.L., 2001, Formation of submarine canyons on the flanks of the Canary Islands, *Geo-Mar. Lett.*, **20**, 160–167.

Labazuy, P., 1996, Recurrent landsliding events on the submarine flanks of Piton de la Fournaise volcano (Reunion Island), in W.J., McGuire, A.P., Jones, and J., Neuberg, (eds.), *Volcano instability on the Earth and Other Planets*, *Geol. Soc. Lond. spec. publ.*, **110**, 295–306.

Le Bas, M.J., Rex, D.C. and Stillman, C.J., 1986, The early magmatic chronology of Fuerteventura, Canary Islands, *Geol. Mag.*, **123**, 287–298.

Lee, H.J., Schwab, W.C. and Twichell, D.C., 1993, Submarine slides: an introduction, In W.C. Schwab, H.J. Lee, and D.C. Twichell, (eds.), *Submarine Landslides; Selected Studies in the U.S. Exclusive Economic Zone, U.S. Geol. Survey Bull.*, **2002**, 1–13.

Lénat, J.F., Bachèlery, P., Bonneville, A., Galdeano, A., Labazuy, P., Rousset, D. and Vicent, P., 1990, Structure and morphology of the submarine flank of an active basaltic volcano: Piton de la

Fournaise (Réunion Island, Indian Ocean), *Oceanol. Acta, Vol. Spéc.*, **10**, 211–223.

Martí, J. and Gudmundsson, A., 2000, The Las Cañadas Caldera (Tenerife, Canary Islands): example of an overlapping collapse caldera generated by magma-chamber migration, *J. Vol. Geth. Res.*, **103**, 161–173.

Martí, J., Hurlimann, M., Ablay, G.J. and Gudmundsson, A., 1997, Vertical and lateral collapses on Tenerife (Canary Islands) and other volcanic ocean islands, *Geology*, **25**, 879–882.

Martí, J., Mitjavila, J. and Araña, V., 1994, Stratigraphy, structure and geochronology of the Las Cañadas caldera (Tenerife, Canary Islands), *Geol. Mag.*, **131**, 715–727.

Masson, D.G., 1996, Catastrophic collapse of the volcanic island of Hierro 15 Ka ago and the history of landslides in the Canary Islands, *Geol.*, **24**, 231–234.

Masson, D.G., Kidd, R.B., Gardner, J.V., Hugget, Q.J. and Weaver, P.P.E., 1992, Saharan continental rise: facies distribution and sediment slides, in C.W. Poag and P.C. Gracinsky (Eds.), *Geologic Evolution of Atlantic Continental Rises*, Van Nostrand Reinhold, New York, 322–343.

Masson, D.G., Van Neil, B. and Weaver, P.P.E., 1997, Flow processes and sediment deformation in the Canary Debris Flow on the NW African rise, *Sed. Geol.*, **45**, 411–432.

Masson, D.G. and Watts, A.B., 1995, Slope failures and debris avalanches on the flanks of volcanic oceanic islands; the Canary Islands, off NW Arica, *Landslide News*, **9**, 21–24.

Masson, D.G., Watts, A.B., Gee, M.J.R., Urgeles, R., Mitchell, N.C., Le Bas, T.P. and Canals, M., 2002, Slope failures on the flanks of the Canary Islands, *Earth-Sci. Rev.*, **57**, 1–35.

McGuire, W.J., Hearth, R.J., Firth, C.R., Solow, A.R., Pullen, A.D., Saunders, S.J., Stewart, I. C. and Vita-Finzi, C., 1997, Correlation between rate of sea-level change and freque ncy of explosive volcanism in the Mediterranean, *Nature*, **389**, 473–476.

Mehl, K.W. and Schmincke, H.-U., 1999, Structure and emplacement of the Pliocene Roque Nublo debris avalanche deposit, Gran Canaria, Spain, *J. Vol. Geoth. Res.*, **94**, 105–134.

Moore, A. L., 2000, Landward fining in onshore gravel as evidence for a late Pleistocene tsunami on Molokai, Hawaii. *Geology*, **28**, 247–250.

Moore, J.G., Bryan, W.B. and Ludwig, K.R., 1994, Chaotic deposition by a giant wave, Molokai, Hawaii. *Geol. Soc. Am. Bull.*, **106**, 962–967.

Moore, J.G. and Clague, D.A., 2002, Mapping the Nuuanu and Wailau lanslides in Hawaii, Hawaiian Volcanoes: Deep Underwater Perspectives, E. Takahashi, P.W. Lipman, M. O. García, J. Naka and Aramaki (eds), *Am. Geophys. Union Monogr.*, **128**, 223–244.

Moore, J.G., Clague, D.A., Holcomb., R.T., Lipman, P.W., Normark, W.R. and Terresan, M.E., 1989, Prodigious submarine landslides on the Hawaiian Ridge, *J. Geophys. Res.*, **94**, 17,465–17,484.

Moore, J.G. and Moore, G.W., 1984, Deposit from giant wave on the island of Lanai, Hawaii. *Science*, **226**, 1312–1315.

Moore, J.G., Normark, W.R. and Holcomb, R.T., 1994, Giant Hawaiian landslides, *Ann. Rev. Earth Planet. Sci.*, **22**, 119–144.

Moore, G.W. and Moore, J.G., 1988, Large-scale bedforms in boulder gravel produced by giant waves in Hawaii. *Geol. Soc. Am. Spec. Pap.*, **229**, 101–110.

Muñoz, A., Palomo, C., Acosta, J. and Pardo de Domlebum, M., 1998, Hydrographic and Oceanographic Programme for the Spanish EEZ, *EEZ Tech.*, **2**, 71–75.

Navarro, J.M. and Coello , J., 1989, Depressions originated by landslide processes in Tenerife, *ESF Mtng. Canarian Volc. Lanzarote*, 150–152.

Normark, W.R., Hess, G.R., Stow, D.A.V. and Bowen, A.J., 1980, Sediment waves on the Monterey Fan levee: a preliminary physical interpretation, *Mar. Geol.*, **37**, 1–18.

Normark, W.R., Moore, J.G. and Torresan, M.E., 1993, Giant-volcano related landslides and development of the Hawaiian Islands, in W.C. Schwab, H.J. Lee and D.C. Twichell, (Eds.), Landslides, Selected Studies in the U.S. Exclusive Economic Zone, *U.S. Geol. Survey Bull.*, **2002**, 184–196.

Prior, D.B., Bornhold, B.D., Coleman, J.M. and Bryan, W.R., 1982, Morphology of a submarine slide. Kitimat Arm, British Columbia, *Geol.*, **10**, 588–592.

Ridley, W.I., 1971, The origin of some collapse structures in the Canary Islands, *Geol. Mag.*, **108**, 477–484.

Robertson, A. H. and Stillman, C. J., 1979, Submarine volcanic and associated sedimentary rocks of the Fuerteventura Basal Complex, Canary Islands, *Geol. Mag.*, **116**, 203–214.

Rothwell, R.G., Alibés, B. and Weaver, P.P.E., 1998, Seismic facies of the Madeira abyssal Plain: a correlation between seismic reflection profile and borehole data, in Weaver, P.P.E., Schmincke, H.-U., Firth, J.V. and Duffield, W. (Eds.), *Proc. Ocean Drill. Prog. Sci. Res.*, **157**, College Station, TX, 473–498.

Schmincke, H-U, 1968, Faulting versus erosion and the reconstruction of the Mid-Miocene shield volcano of Gran Canaria, *Geol. Mitt.*, **8**: 23–50.

Schmincke, H.-U. and Segschneider, B., 1998, Shallow submarine to emergent basalt shield volcanism of Gran Canaria: evidence from drilling into volcanic apron, in P.P.E. Weaver, H.-U. Schmincke, J.V. Firth and W. Duffield, (Eds.), *Proc. Ocean Drill. Prog., Sci. Res.*, **157**, College Station, TX, 141–182.

Schmincke, H.-U. and Sumita, M., 1998, Volcanic evolution of Gran Canaria reconstruction from apron sediments: synthesis of VICAP project drilling, in P.P.E. Weaver, H.-U. Schmincke, J.V. Firth and W. Duffield, (Eds.), *Proc. Ocean Drill. Prog. Sci. Res.*, **157**, 443–469.

Schmincke, H.-U. and Von Rad, U., 1979, Neogene evolution of Canary Island volcanism inferred from ash layers and volcaniclastic sandstones of DSDP site 397 (leg 47A), In Von Rad, U., Ryan, W.B.F., et al., *Init. Rep. Deep Sea Drill. Proj.*, **XLVII, Part 1**, U.S. Govt. Printing Office, Washington, 703–725.

Shipboard Scientific Party, 1995a, Site 953, in H.-U. Schmincke, P.P.E. Weaver, J.V. Firth, et al., *Proc. Ocean Drill. Prog., Init. Rep.*, **157**, 317–394.

Shipboard Scientific Party, 1995b, Site 954, in H.-U. Schmincke, P.P.E. Weaver, J.V. Firth, et al., *Proc. Ocean Drill. Prog., Init. Rep.*, **157**, College Station, TX, 395–432.

Shipboard Scientific Party, 1995c, Site 955, in H.-U. Schmincke, P.P.E. Weaver, J.V. Firth, et al., *Proc. Ocean Drill. Prog. Init. Rep.*, **157**, College Station, TX. 433–496

Shipboard Scientific Party, 1995d, Site 956, in H.-U. Schmincke, P.P.E. Weaver, J.V. Firth, et al., *Proc. Ocean Drill. Prog., Init. Rep.*, **157**, 497–561.

Soriano, C., Zafrilla, S., Martí, J., Bryan, S., Cas, R. and Ablay, G., 2002, Welding and rheomorphism of phonolitic fallout deposits from Las Cañadas caldera, Tenerife, Canary Islands, *Geol. Soc. Am. Bull.*, **114**, 883–895.

Stillman, C.J., 1999, Giant Miocene landslides and evolution of Fuerteventura, Canary Islands, *J. Vol. Geoth. Res.*, **94**, 89–104.

Szérémeta, N., Guillou, H., Kissel, K., Mazaud, A. and Carracedo, J.C., 1999, Geomagnetic paleosecular variation in the Brunhes period, from the island of El Hierro. *Earth and Planet. Sci. Lett.*, **165**, 241–253.

Teide Group, 1997, Morphometric interpretation of the northwest and southeast slope of Tenerife, Canary Islands, *J. Geophys. Res.*, **102**, 20,325–20,342.

Uchupi, E., Emery, K.O., Bowin, C.O. and Phillips, J.D., 1976, Continental margin off western Africa: Senegal to Portugal, *Am. Assoc Pet. Geol. Bull.*, **60**, 809–878.

Urgeles, R., Canals, M., Baraza, J. and Alonso, B., 1998, Seismostratigraphy of the western flanks of El Hierro and La Palma (Canary Islands): a record of Canary Island volcanism, *Mar. Geol.*, **146**, 225–241.

Urgeles, R., Canals, M., Baraza, J., Alonso, B. and Masson, D.G., 1997, The most recent megaslides on the Canary Islads: the El Golfo debris avalanche and the Canary debris flow west of El Hierro Island, *J. Geophys. Res.*, **102**, 20, 305–323.

Urgeles, R., Masson, D.G., Canals, M., Watts, A.B. and Le Bas, T., 1999, Recurrent large-scale landsliding on the west flank of La Palma, Canary Islands, *J. Geophys. Res.*, **104**, 24,331–25,348.

Van den Bogaard, P. and Schmincke, H.-U., 1998, Chronostratigraphy of Gran Canaria, in H.-U. Schmincke, J.V. Firth ans W. Duffield (Eds.), *Proc. Ocean Drill. Prog., Sci. Res.*, **157**, 122–140.

Varnes, D.J., 1958, Landslide types and processes, *Pub. U.S. Natl. Acad. Sic.*, **544**, 20–47.

Voight, T., Blacken, H., Jana, R.J. and Douglass, P.M., 1981, The 1980 eruptions of Mount St. Helens, Washington, *U.S. Geol. Survey Prof. Paper*, **1250**, 347–377.

Watts, A.B. and Masson, D.G., 1995, A giant landslide on the north flank of Tenerife, Canary Islands, *J. Geophys. Res.*, **100**, 24,487–24,498.

Weaver, P.P.E., Jarvis, I., Lebreiro, S.M., Alibés, B., Baraza, J., Howe, R. and Rothwell, 1998, Neogene turbidite sequence on the Madeira Abyssal Plain: basin filling and diagenesis in the deep ocean, in P.P.E. Weaver, H.-U. Schmincke and W. Duffield, *Proceedings of the Ocean Drilling Program, Scientific Results*, **157**, 619–634.

Weaver, P.P.E. and Rothwell, R.G., 1987, Sedimentation on the Madeira Abyssal Plain over the last 300,000 years, *Geol. Soc. Lon. Sp. Publications*, **31**, 71–86.

Weaver, P.P.E., Rothwell, R.G., Ebbing, J., Gun, D. and Hunter, P.M., 1992, Correlation, frequency of emplacement and source directions of megaturbidites on the Madeira abyssal Plain, *Mar. Geol.*, **109**, 1–20.

Wynn, R.B., Masson, D.G., Stow D.A.V. and Weaver, P.P.E., 2000, Turbidity current sediment waves on the submarine slopes of the western Canary Islands, *Mar. Geol.*, **163**, 185–198.

Zazo, C., Goy, J.L., Hillaire-Marcel, C., Gillot, P.Y., Soler, V., Gozález, J.A., Dabrio, C.J. and Ghaleb, G., 2002, Raised marine sequences of Lanzarote and Fuerteventura revisited a reappraisal of relative sea-level changes and vertical movements in the eastern Canary Islands during the Quaternary, *Quat. Sc. Rev.*, **21**, 2019–2046.

Marine Geophysical Researches (2003) 24: 41–57
DOI 10.1007/s11001-004-1161-7

Salt Diapirs, Salt Brine Seeps, Pockmarks and Surficial Sediment Creep and Slides in the Canary Channel off NW Africa

J. Acosta[1,*], E. Uchupi[2], A. Muñoz[1], P. Herranz[1], C. Palomo[1], M. Ballesteros[1] & ZEE Working Group[3]
[1]*Instituto Español de Oceanografía. Grupo de Cartografía Multihaz. Corazón de María, 8, 28002 Madrid, Spain*
[2]*Woods Hole Oceanographic Institution, Woods Hole, MA 02543, USA*
[3]*A. Carbó, A. Muñoz-Martín, Univ. Complutense, Madrid, Spain; J. Martín-Dávila, M. Catalán and J.A. Marín, Real Observatorio de la Armada. S. Fernando, Cádiz, Spain; F. Pérez-Carrillo, C. Maté, Instituto Hidrográfico de la Marina. Cádiz, Spain*
Corresponding Author (E-mail: juan.acosta@md.ieo.es)

Key words: Salt diapirism, mounds, Canary Channel, multibeam mapping, fluid seepage

Abstract

Circular to elliptical mounds in the Canary Channel with reliefs of 75 to 375 m and diameters of 4 to 8 km partially surrounded by moats with reliefs of 25 to 75 m, were formed by piercement of the seafloor by Mesozoic evaporites. Several long gullies, < 1 km wide, with abrupt terminations and pockmarks associated with these mounds were probably eroded by dense brine and hydrocarbon seeps. The salt brines that eroded the gullies were formed where salt diapirs intersect the seafloor, or in the subsurface by circulating ground water heated by igneous activity along the Canary Ridge. If the brines originated in the subsurface they reached the seafloor along faults. Displacement of the surficial sediments by sliding and creep is probably the result of the expulsion of hydrocarbons and/or vertical motion of the Mesozoic evaporites. Microtopographic features along or near the east flank of the Canary Ridge are the creation of uplift of the ridge, hydrothermal activity, mass wasting processes and turbidity currents.

Introduction

Cold seeps in the marine environment have been reported in historical times as far back as 10 A.D. when the people of Aradus, an island city 4 km offshore Syria, obtained fresh water from a submarine spring located between the islet and the mainland during sieges (Emery and Uchupi, 1984). It was not until the mapping of pockmarks on the Scotian Shelf by King and MacLean (1970), however, did we came to appreciate the importance of fluid discharge in the marine environment. There is no doubt that the expulsion of fresh water, salt brines, or gas and oil plays a significant role in recycling processes and in sculpturing the morphology of continental margins.

In active margins the diffused discharge of pore fluids from a tectonically dewatered and degassed accretionary wedge has led to the construction of mud volcanoes and diatremes (formed by rapid flow of pore fluids through a sediment mass which becomes fluidized and entrained into the discharged fluids) (Brown, 1990). In addition, such dewatering has caused precipitation of authigenic carbonates in the form of thin crusts, slabs, irregular edifices and chimneys with reliefs of 1 to 2 m, the grown of mats of chemoautotrophic oxidizing bacteria, and colonies of benthic invertebrates that obtain their nutrients from the oxidizing bacteria (Cadet et al., 1987; Le Pichon et al., 1987; Ohta and Labuier, 1987; Pautot et al., 1987; Kennicut et al., 1989; Henry et al., 1990; Kulm and Suess 1990; Lewis and Cochrane, 1990; Wang et al., 1995; von Rad et al., 1996). Along active transform plate margins, such as the Eurasian/African boundary in the Gulf of Cádiz, mud and fluid eruptions have led to the construction of mud volcanoes and carbonate pipes, chimneys and mounds (Maldonado et al., 1999; Somoza et al., 2002).

Seeps in passive margins have been reported in environments ranging from lakes, intertidal/shallow tidal areas, bays, fjords, marginal seas, gulfs, continental shelves and continental slopes, submarine canyons and deep-sea fans (Hovland and Judd, 1988; Uchupi et al., 1996; Stakes et al., 1999). They are particularly abundant in offshore evaporite, or hydrocarbon regions where salt brines, crude oil, fluid mud and gases are common seep products (Roberts et al., 1990). Examples of such regions include the Gulf of Mexico, the Gulf of Cadiz, the Kattegat Strait between Denmark and Sweden, the southern Skagerrak between Norway and Denmark, the North Sea, the northern Norwegian shelf, and the Persian Gulf (Behrens, 1988; Kennicutt and Brooks, 1990; MacDonald et al., 1990; Sassen et al., 1991; Hovland, 1992; Aharon, 1994; Roberts and Aharon, 1994; Baraza and Ercilla, 1996; Uchupi et al., 1996; Aharon et al., 1997). In most of these regions the deep-seated hydrocarbons are linked to salt-induced faults that provide the conduits for the migration and seepage onto the seafloor.

The most common seep related-features in continental margins are pockmarks, although mud volcanoes and mud diapirs formed by the upwelling of gaseous muds are also quite common; e.g. Nigeria; (Graue, 2000), in the Gulf of Mexico (Neurauter and Bryant, 1990) and Gulf of Cadiz (Maldonado et al., 1999; Somoza et al., 2002). Associated with some of the pockmarks are black-sulphide rich sediments, mats of sulphur oxidizing bacteria and invertebrates, which obtain their nutrition from species with symbiotic sulphur-oxidizing bacteria and symbiotic methane-oxidizing bacteria (Dando and Hovland, 1992; Dando et al., 1991). Degradation of hydrocarbons seeps has also resulted in authigenic carbonate build-ups (Roberts et al., 1989). This combination of authigenic carbonate formation and biological accumulation can lead to the formation of mounds (von Bitter et al., 1990; Dando et al., 1991; Hovland et al., 1994; Beauchamp and Savard, 1992) or pseudobioherms, i.e. bioherm-like features not rising above the seafloor and lacking reef builders (Gaillard et al., 1992).

According to von Rad et al. (1996) pockmarks in the Makran accretionary wedge, offshore Pakistan, were formed by focused fault-controlled expulsion of fluids. At or near some plate boundaries of transform/compression type, such as those in the Mediterranean and New Zealand, some of the pockmarks are related to volcanism and may have been formed by the expulsion of hydrothermal gases and water (Pickrill, 1993; Acosta et al. 2001). Other pockmarks are due to biogenic gas formed by the decay of organic matter. Such organic-rich sediments are abundant within the Pliocene-Quaternary sediments of the Mediterranean (Cramp and O'Sullivan, 1999; Thomson et al., 1999; Coleman and Ballard, 2001).

Where salt domes pierce the seafloor brine pools may form, such as the pool seen within a fractured carbonate cap in the Gulf of Mexico (Bright et al., 1980). In another part of the Gulf, flowing brine has been trapped on the axis of a circular basin formed by the coalescing of several salt domes. The brine in the basin (Orca Basin) merges from flows into small gullies cut on the flanks of the salt domes (Shokes et al., 1977). Brines may also fill some pockmarks by lateral seepage through the pockmark walls (MacDonald et al., 1990; MacDonald, 1992). Brine seepages along the base of the Florida Escarpment in the Gulf of Mexico have also led to the establishment of methane-oxidizing bacteria that serve as a food source for a densely packed population of invertebrates via symbiosis (Paull et al., 1984). Paull and Neumann (1987) further inferred that the acid generated by the oxidation of the sulfide in the brines might undercut the scarp causing its collapse and retreat.

In this study we describe gullies associated with salt diapirs from the Canary Channel, which are inferred to be formed by brine seepages. This is the first report of such occurrences off the African margin. They were discovered in the course of an investigation of the Canary Island archipelago.

Material and methods

Three data sets were used in this study. One set consists of single channel seismic reflection profiles recorded during the International Decade of Ocean Exploration investigation of the west African margin. The profiles included in this report were collected aboard the R/V Atlantis II in 1973. These profiles were recorded using a 300 in^3 air gun fired every 10 or 12 seconds and two 30 m hydrophone arrays, whose signals were summed, amplified and recorded in a dry-paper printer; ship's position was obtained via satellite at two hour intervals. Uchupi et al. (1976) published line interpretations of these profiles in a manuscript regarding the morpho-tectonic setting of the northwest African and south Iberian continental margins.

The second data set corresponds to the bathymetry of the Canary Channel based on swath data recorded by the Instituto Español de Oceanografía and the Instituto Hidrográfico de la Marina between 1998 and 2002. The data were collected during an investigation of the Canary archipelago aboard the R/V Hesperides and R/V Vizconde de Eza. The multi-beam sounding systems used separately or in combination included Simrad EM 1000, EM 1002 and EM 300, in conjunction with GPSD and inertial navigation systems, were used to survey shallow waters and an EM12S in deep waters. A DTM and contour software (Cfloor from Roxar) and a Geographic Information System (IberGIS from ICI) was used to create a bathymetric map of the Canary Channel and the shaded relief and 3D block diagrams of the surveyed area.

The third data set was obtained during a cruise aboard R/V Vizconde de Eza in October 2002. During this cruise three of the mounds in the Canary Channel were sampled and photographed. Sampling was by means of a biological dredge and photography with a Benthos model 372 submarine camera capable of operating to depths of 11000 m. The photos were acquired with the vessel drifting over the mounds at a speed of 0.5–1.0 knots using a Dynamic Positioning System. The photographs were taken every 10 seconds along five paths with the camera 'flying' 6 m above the sea floor. More than 3000 photographs were recorded during the cruise with each covering an area of about 21 m^2. The morphology and structure of the mounds during the R/V Vizconde de Eza cruise was defined with the aid of a SIMRAD EM-300 multibeam echosounder, a Topas 018 high resolution seismic profiler and a marine Geomag SMM with 0.01 nT precision magnetometer. The ship's position during the survey was determined with a D-GPS integrated with a inertial navigation system Seapath 200.

Morpho-tectonic setting of the Canary Channel

Stratigraphic setting

Making up the flanks of the Canary Channel, offshore northwest Africa, are the African passive margin on the east and the Canary Ridge on the west (Figure 1). Construction of the passive margin is due to syn-rift deposition prior to 190 Ma (Late Triassic-Early Jurassic) and post-rift sedimentation since sea-floor spreading began about 190–185 Ma latest Early Jurassic (Toarcian-Present; Hinz et al., 1982; Emery and Uchupi, 1984; Steiner et al., 1998). The syn-rift sequence roughly consists of continental clastic rocks, volcanic rocks, evaporites and carbonates. The post-rift or drift strata are made up of Mesozoic platform and distal carbonates, regressive clastics, transgressive carbonates and euxinic and clastic sediments, Paleogene manganese oxide and phosphatic and siliceous deposits and Neogene clastic sequences. Drift deposition was influenced by events such as the massive gravitational sliding off the Suess Trough in the Late Cretaceous, possibly due to uplift of the western High Atlas and motion along the South Atlas Fault (Price, 1980). Volcanic construction of the Canary Ridge in the Early Cretaceous-Cenozoic, establishment of polar circulation and the trade winds in the Paleogene, Cenozoic Pyrenean-Alpine orogenies and glacially induced transgressions and regressions during the Cenozoic and Quaternary also contributed to the formation of the west African margin.

Construction of the Canary Ridge along the west side of the Canary Channel apparently took place along the continental/oceanic crustal boundary (Coello et al., 1992; Ancochea et al., 1996; Steiner et al., 1998; Martínez del Olmo and Buitrago Borrás, 2002). Price (1980) also proposed that volcanism along the ridge may have take place along a sinistral shear fault, the curved seaward continuation of the South Atlas Fault. This fault merges southwestward with the east-west trending fracture zone along the trend of Gran Canaria, Tenerife, La Gomera and El Hierro islands. Others have ascribed a hot-spot origin for the volcanism in the Canary Islands (Holik et al.,1991; Hoernle and Schmincke, 1993; Carracedo et al., 1998; Dañobeitia and Canales, 2000). With construction of the north trending Canary Ridge, the African continental rise was split in two and sediment input to the distal part of the African continental rise became drastically reduced.

Results

Neogene acoustic stratigraphy

The acoustic stratigraphy displayed by the seismic reflection profiles acquired in the Canary Channel aboard the R/V Atlantis II is probably limited to the Neogene. The presence of diapirs along some of the profiles do indicate, however, that the Upper Triassic/Lower Jurassic evaporites in the coastal basins extend offshore. The wide line spacing, the lack of tie

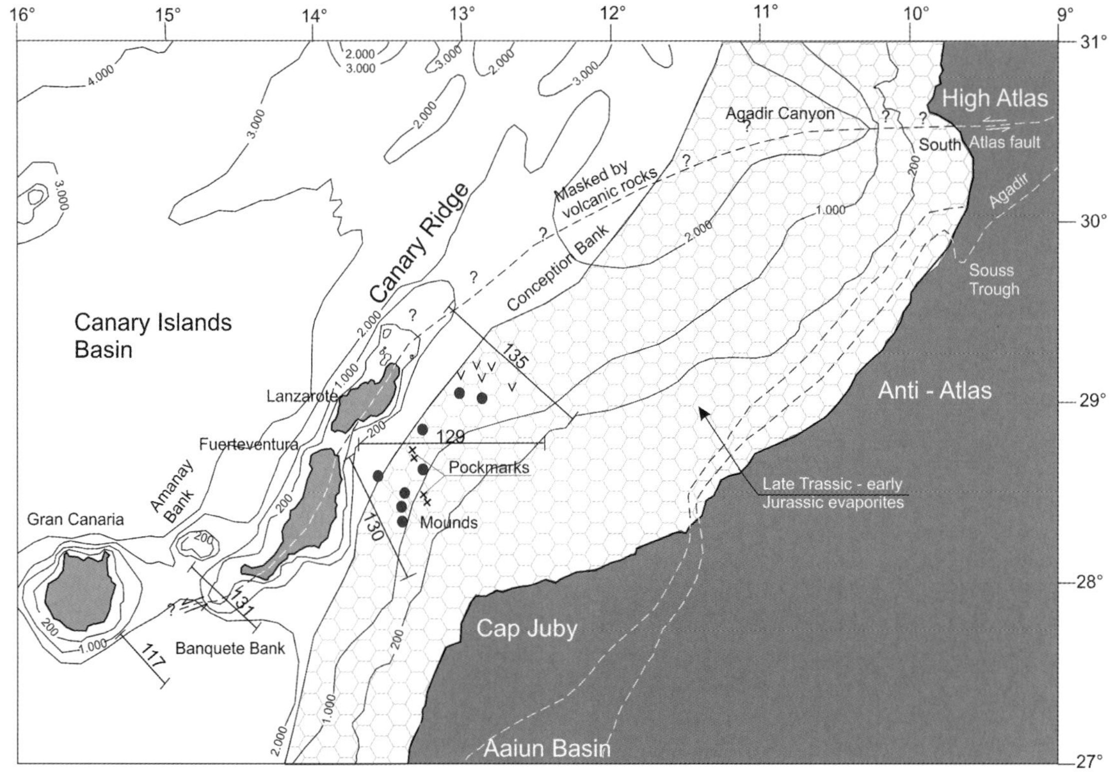

Figure 1. Morpho-tectonic map of the Canary Channel showing location of the seismic reflection profiles shown in Figures 2 and 3. Compiled from Uchupi et al. (1976) and Hinz et al. (1982). Hexagonal pattern = Extent of Mesozoic evaporites. Water depths in meters.

lines and the presence of a massive diapiric field makes it difficult to define the Neogene acoustic stratigraphy of the Canary Channel as a unit. In the absence of stratigraphic data is impossible to assign ages to the units imaged by the seismic profiles. Each of these lines displays an acoustic facies suggesting that sediment input from the Canary Ridge was important during the Neogene and that input from the African margin and the Canary Ridge varies along the strike of the channel. This variation coupled with salt diapirism, bottom current and vertical oscillations of the Canary Ridge has led to a complex geometric facies pattern in the Canary Channel.

Line 117 (Figure 2) extends southeastward from the flank of Gran Canaria to a water depth in excess of 2000 m. It is only along this line that stratigraphic data from Ocean Drilling Program (ODP) Site 955 makes it possible to determine the lithology and ages of the seismic units imaged by the seismic reflection profiles described in this report. ODP Shipboard Scientific Party (1995) identified five units, A–E, at ODP Site 955 at a depth of 2865 m (3.8 seconds) on the volcanic apron surrounding Gran Canaria. The upper sequence made up of units A–C consists of Pleistocene to late Pliocene nanno ooze interbedded with clayey silt and sand and early Pliocene to late Miocene nannofossil ooze. The lower sequence made up of units D to E consists of middle to upper Miocene nannofossil ooze interbedded with quartz silt and sand at the top and a mixture of nannofossil ooze, volcaniclastics and siliciclastics at the base. The volcaniclastic detritus is correlative with the Miocene Fataga and Mogan Group volcanism in Gran Canaria. The upper sequence thickens basinward and displays lateral changes in seismic facies and the lower sequence thins basinward. Acoustic basement at the base of unit E is the top of the lower volcanic sands correlative with the Mogan Formation of Gran Canaria. These two sequences could be correlative with units 1–3 imaged by the profiles in the Canary Channel. The absence of a tie line and location of ODP Site 955, on the Gran Canaria volcanic apron 109 km southwest of Fuerteventura and 125 km west of the African margin (Shipboard Scientific Party, 1995), make such

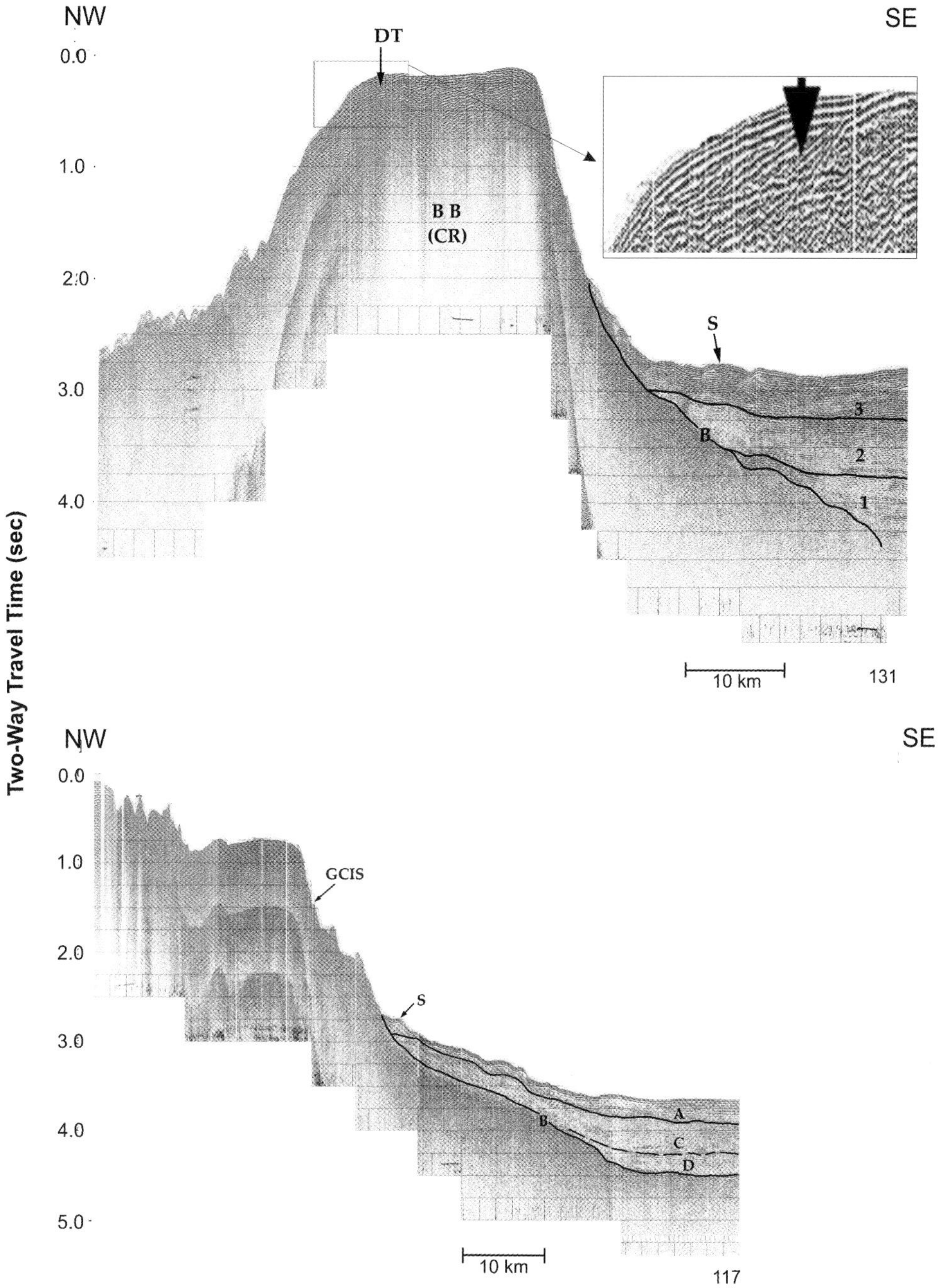

Figure 2. Seismic reflection profiles 131 and 117. See Figure 1 for location of profiles. B = Basement; BB = Banquete Bank; CR = Canary Ridge; DT = Delta; GCIS = Gran Canaria Insular Slope; S = Slump or Sediment waves. Units A, C and D along profile 117 are from Shipboard Scientific Party (1995).

correlation tenuous at best. Thus we use a distinct stratigraphic nomenclature for the lines in the Canary Channel and Line 117.

Its possible that the Neogene stratigraphic sequence in the lines in the Canary Channel is more compatible with that at Deep Sea Drilling Project (DSDP) Site 397 drilled on the continental rise at a depth of 2914 m (Arthur et al., 1979). This site is located over 100 km southwest of the Canary Channel. If so, then Neogene deposition in the channel was influenced by massive erosion by a geostrophic bottom currents and slumping during the late Eocene to late Oligocene and by early-late Miocene hemipelagic deposition in an upwelling environment. Subaerial volcanism during the middle Miocene in Fuerteventura and Lanzarote was another event that influenced deposition in the channel. This event is represented at DSDP Site 397 by volcaniclastic debris flows and rhyolitic ash layers. Late Neogene deposition at the site is characterized by high carbonate productivity under upwelling conditions and minor input of terrigenous influx. At other sections of the margin, however, the terrigenous input was high. A seismic reflection profile linking DSDP Sites 369 on the slope and 397 on the continental rise has been described by Hinz (1979). This profile clearly images the massive erosion that the slope off northwest Africa underwent from late Eocene to late Oligocene.

The Neogene sequence along this profile on the upper rise (Hinz's acoustic sequence CB1b) is well stratified. Its lower boundary, the erosional surface carved by geostrophic bottom current erosion in late Eocene to late Oligocene, is not marked by a strong reflector. The Cenozoic sequence (Hinz's acoustic sequence CB1a) on the lower slope consists of a prograding unit downlaping on a lower unit and is not continuous with CB1b on the rise. Apparently part of this unit was eroded before the deposition of sequence CB1b. No such geometry is displayed by the Neogene acoustic succession in the Canary Channel and none of the profiles in the channel display the massive erosion displayed by this profile. Thus, the correlation of the Canary Channel acoustic succession with that of DSDP Sites 397 and 369 also is not feasible. These sites are too far from the channel, too deep and the influence of the Canary Ridge on the depositional history of DSDP Site 395 was of secondary importance. This ridge, not only contributed detritus to the channel, but also acted as a dam to continental detritus preventing its dispersion to the deep-sea.

The presence of oceanic crust south of Gran Canaria (Emery and Uchupi, 1984) suggests that line 117, southwest of the Canary Channel, is located over oceanic crust and explains the lack of piercement structures along the line (Figure 2). Of the seismic units described by ODP Shipboard Scientific Party (1995) in the vicinity of ODP Site 955, line 117 on the Gran Canaria insular slope only displays Units A, C and D. ODP Shipboard Scientific Party (1995) interpreted an undulating reflector on top of Unit A as possible mud waves. Unit C is irregularly stratified upslope and is characterized by at least one lens-like sediment build-up on the distal end of the profile. Unit D also displays undulating reflectors. The acoustic basement, the top of Gran Canaria's clastic flank facies, can be traced the length of the profile. Data from ODP Site 955 suggests that Units A and C along 117 correlate with Quaternary and lower Pliocene-upper Miocene nannofossil ooze mixed with foraminifers. Unit D is constituted of upper to middle Miocene nannofossil clay toward the top and Miocene volcaniclastic sediments toward the bottom.

Line 131 at the southwest end of the Canary Channel is either on oceanic crust or continental crust. If the line is over continental crust the lack of diapirs along this line could be due to masking by volcanic rocks. Line 131 cuts obliquely across El Banquete Bank located at the southwest end of the Canary Ridge. The bank has a flat-top beneath which there is evidence of stratification with the superficial sediments along the northwest edge of the bank displaying a geometry suggestive of shelf progradation (DT, Figure 2). The three sediment units (1–3) along this line rest on the east flank of the Canary Ridge with the upper one (Unit 3) displaying an acoustic character typical of mass movement or bottom current activity (S, Figure 2). Southeastward the bedding within this unit is suggestive of the merging of two sources, one from the ridge and another one from the African margin.

Line 130 (Figure 3), cutting obliquely across the embayment separating Fuerteventura and Lanzarote, ends just east of the axis of the Canary Channel. Whereas along Lines 135 and 129 the flank of the Canary Ridge (probably the top of the islands clastic flank facies) plunges rapidly eastward, the ridge's east flank along line 130 descends more gradually and can be traced to, at least, the axis of the Canary Channel. The absence of diapirs along this line most probably results from acoustic energy not being able to penetrate the volcanic apron off Fuerteventura. Line 130

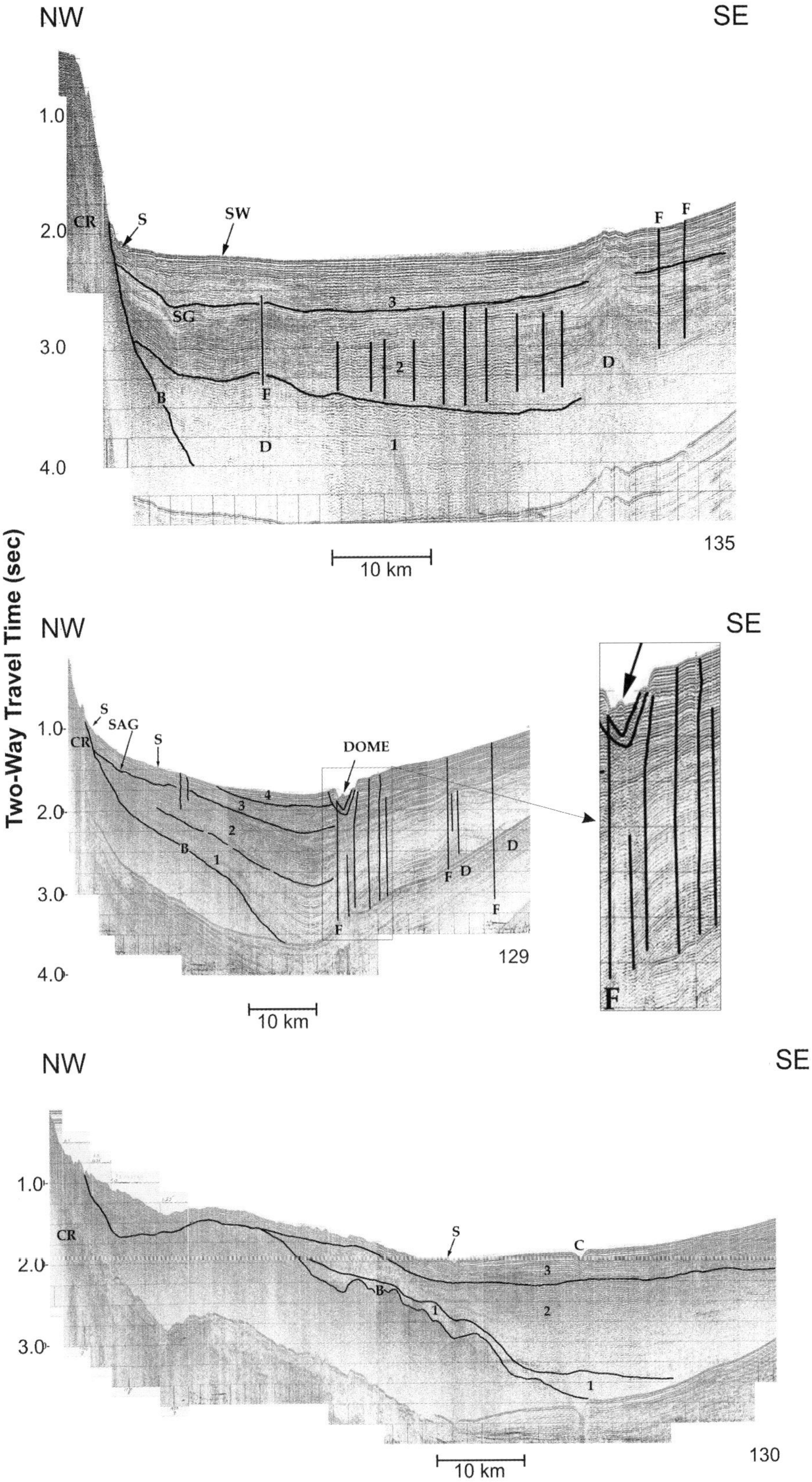

Figure 3. Seismic reflection profiles 135, 129 and 130 across the Canary Channel. See Figure 1 for locations of profiles. B = Acoustic Basement; C = Channel; CR = Canary Ridge; D = Diapir; F = Fault; S = Slump; SG = Sag; SW = Sediment Waves. Sag may be due to uplift of Canary Ridge. Diapirs along profiles 135 and 129 were formed by the plastic flow of Mesozoic evaporites.

also displays three units with a reversed in the dip of strata in Unit 2 suggests that either part of the sequence was derived from the Canary Ridge or is due to the uplift of the ridge after its deposition. Unit 3 is cut by a low relief channel (C, Figure 3) near the axis of the Canary Channel at a water depth of 750 m. West of this channel the surface of the unit is rough, with irregularities that may reflect mass movement or bottom current activity. This unit also reverses its dip west of the axis of the Canary Channel suggesting that part of the unit may have been derived from the Canary Ridge.

Two diapirs and a 3.5 km wide and 225 m deep low disrupt the stratigraphic succession along line 129 (Figure 3). The fault network disrupting the strata has led us to infer that the low is a graben that structurally controls the erosional channel. A 300 m wide and 400 m high mound in the center of the channel rests on a horizon displaying sag beneath the mound and the acoustic character below this horizon is chaotic. Strata beneath this chaotic sequence and immediately east of the low, in the vicinity of two diapirs (D, Figure 2), are diffused, a texture possibly resulting from the formation of the piercement structures and fault network seen in the profile. West of the low the dip of Units 1–3 changes abruptly from west to east. We infer that this change in inclination is due to subsidence of the Canary Channel. It is this subsidence that also may be responsible for the faulting along Line 129. No such subsidence is displayed by the Canary Channel to the north (Line 135) nor to the south (Line 130) suggesting that this subsidence is not a regional phenomena. Possibly the subsidence and associated faulting may be due to salt withdrawal that led to the formation of the salt massif on the eastern end of Line 129.

On Line 135, extending from the flat-topped northern end of the Canary Ridge to the African margin (Figures 1 and 3), the acoustic basement of the Canary Ridge is covered by a stratified unit (1) whose eastward dip indicates that it was either derived from the ridge or is due to uplift of the ridge. Units 2 and 3 dip westward indicating that they were derived from the African margin. Unit 2 near the ridge displays a sag (SG, Figure 3) west of which the strata's dip changes to the east. Although the reversal in dip could indicate a change in sediment source from east to west (Canary Ridge), we infer that this change in dip results from the tectonic uplift of the Canary Ridge. Such deformation is not surprising as Martínez del Olmo and Buitrago Borrás (2002) have inferred from the stratigraphy of a well in Lanzarote that the Canary Ridge has experienced an uplift of 3500 to 2000 m in the Cenozoic. The surficial sediments of Unit 3 at the base of the Canary Ridge display features that appear to be the result of slumping (S, Figure 2). Further down slope the seafloor is undulating (SW, Figure 3), a morphology that we infer to be due to bottom currents. Immediately below this undulating seafloor, the subsurface strata display a slight up-doming that also is reflected in the seafloor. This slight upwarp imaged throughout the width of the profile is due to vertical flow of the Mesozoic evaporites at the base of the sediment apron (D, Figure 2). The eastern end of the profile shows a prominent piercement structure with seafloor expression.

Swath Bathymetry

The swath bathymetric data provides additional constraints on the geologic setting of the Canary Channel (Figures 4–6). The topographic map shows a broad swell (S, Figure 4) with a minimum water depth of 475 m, centered near 28° 30′ N and 13° 0′ W. This high plunges to the northwest in the direction of the axis of the Canary Channel. Cutting the northwest flank of the swell is a northwest-trending narrow trough whose relief ranges from 25 m at its northwest end to 250 m at its southeast end. The low is asymmetrical in cross section with its steeper side on its southwest side. This is the graben imaged by Line 129 (Figure 3). Two channels dominate the axis of the Canary Channel, one drains northeast and the other southwest. The channels are separated by a high with a relief of > 275 m extending from 28° 18′ N to 28° 38′ N. The northern Canary Channel is slightly convex westward as it flows around the base of the continental slope. This valley can be traced for a distance of 87 km to a water depth of 1550 m at the northern limit of the swath survey. The southern Canary Channel is slightly convex eastward as it flows around a southeast trending buttress attached to Fuerteventura (Figure 4). This channel has a topographic expression for a distance of 40 km where it debouches into the southwest trending trough south of El Banquete Bank at a water depth of 1650 m. The buttress attached to Fuerteventura is made up of two topographic features, a shallow platform less than 200 m deep at its northwest end and a series of circular highs suggestive of volcanic construction at its southeast end (Figure 4). Ancochea et al. (1996) interpreted the buttress as a submarine prolongation of the sheeted

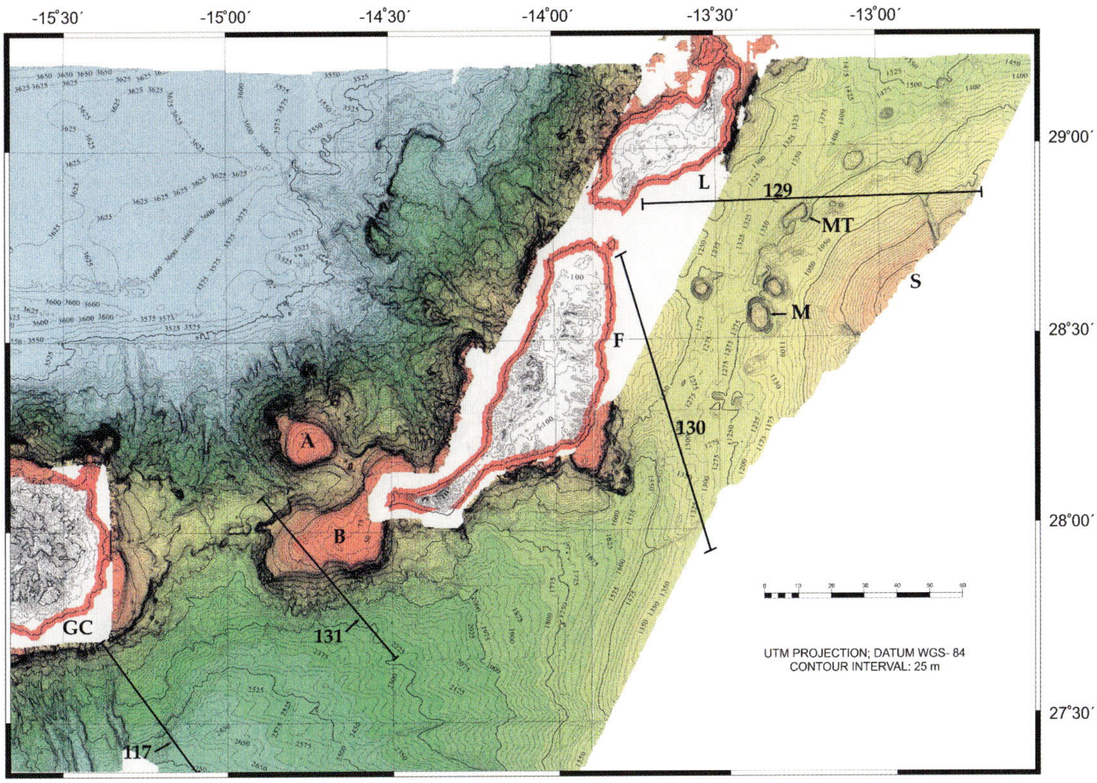

Figure 4. Swath bathymetric map of the Canary Channel showing locations of seismic reflection profile 129, 130, 131 and 117. A = Amanay Bank; B = Banquete Bank; F = Fuerteventura; GC = Gran Canaria; L = Lanzarote; M = Mound; MT = Moat; S = Swell.

dike swam intruded into the Central Volcanic Complex in Fuerteventura.

Along the base of the continental slope, centered near 28° 40′ N and 13° 19′ W, is a northeast trending chain of circular to elliptical mounds. Those at the southern end of the chain have reliefs of about 100 m and diameters between 2 and 10 km. These are features that we have named mounds M2 and M3, and were surveyed aboard R/V Vizconde de Eza (Figures 7 and 8). Those in the center of the chain display reliefs of 100 to 375 m and diameters between 6 and 10 km and those at the northern end of the chain have heights between 75 and 100 m and diameters of 4 to 8 km. The mounds in the center of the chain are surrounded by moats depressed between 25 and 75 m with respect to the surrounding seafloor. West of the center of the chain, near the flank of the Canary Ridge, is another mound (Mound M1, Figure 7). This high lacks a moat, appears to have a crater on its crest, has a relief of 275 m and a diameter of 6 km (Figures 4, 6 and 7). This is the third high that was surveyed aboard R/V Vizconde de Eza.

The mounds and surrounding moats are clearly imaged on the relief map (Figure 5A). A 3D diagram (Figure 6) constructed from the swath data not only images the morphology of the mounds and their surrounding moats, but also a series of small pockmarks, the crater on the apex on Mound 1 located near the base of the Canary Ridge and a system of gullies. The largest of these gullies originates at the base of continental slope from where it meanders down slope dying out northward among two of the smaller mounds (Figure 5A). Other gullies appear to form along the side of one of the mounds with their heads truncated by the moat surrounding the dome.

The shaded relief image (Figure 5A) shows that the Fuerteventura slope where the trend of the insular slope changes from northeast-southwest to east-west, is scalloped, a feature indicative of mass wasting. At the base of this slope segment are apron-shaped sediment accumulations resembling a landslide. At the southwest tip of Fuerteventura is a ramp-like feature that may represent a slump (Figure 5B). Along the west side of this slump is a gully whose head is

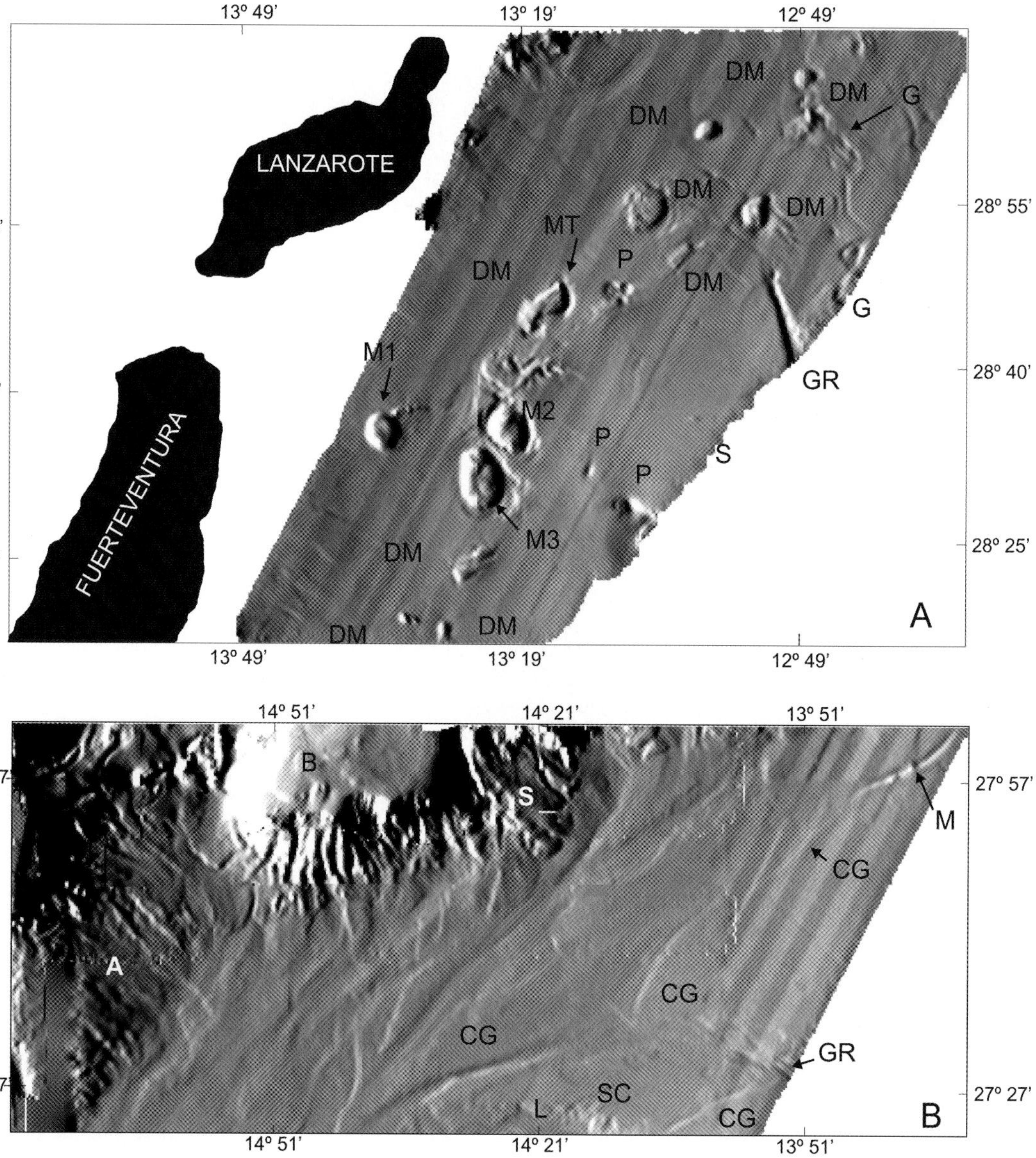

Figure 5. A. Shaded relief diagram of the of the Canary Channel imaging the salt diapir mounds (DM), Volcanic Mound? (M1), Moat (MT), Gullies (G), Graben (GR), Pockmarks (P) and Swell (S) on the seafloor of the channel. B. Shaded relief diagram of the southern approach to the Canary Channel showing the turbidity current generated channel system and mass wasting features. GR = Grooves; L = Landslide; SC = Scar. Note curved groove (CG) on right side with two small mounds (M).

embayed and that terminates abruptly seaward; the gully may represent the pathway of a debris flow with the flow located at its mouth. The southern side of El Banquete Bank is indented by two small embayments that may represent the detachment planes of landslides. West of El Banquete Bank (B, Figure 5B) is an apron (A, Figure 5B) whose surface is cut by gullies that can be traced to the distal end of the apron. Beyond this apron is a braided system of gullies that originate on the southern slope of El Banquete Bank and the southwest tip of Fuerteventura.

The most unusual features in the Canary Channel are the channels located east of the braided system described above. One of these channels or grooves is convex southeastward at its northern end and northwestward at its southern end. Two small highs occur along its axis. Farther south there is another channel that is convex northward. To the east of this channel

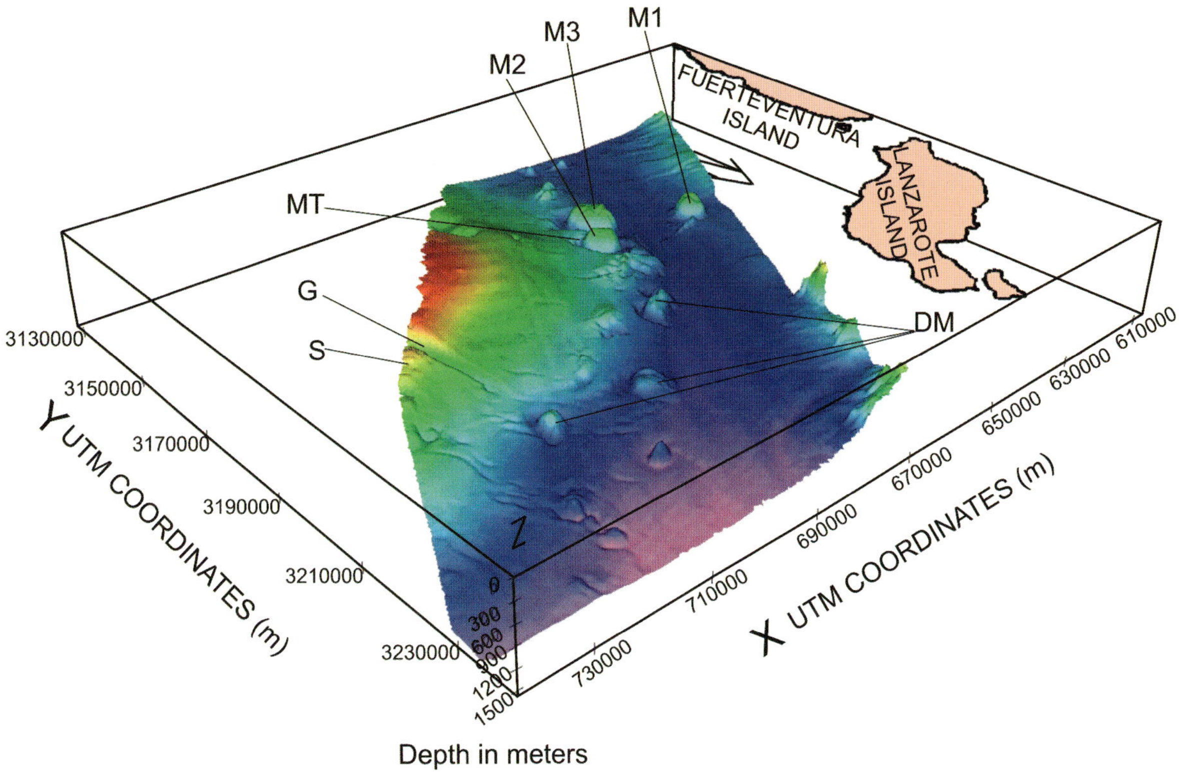

Figure 6. 3D diagram of the mounds in the Canary Channel. DM = Diapiric Mounds (cold seep); MT = Moat; G = Graben; S = Swell; M1 = Volcanic Mound (hot seep). Note that diapiric mounds are located along the base of continental slope.

or groove are two subparallel grooves that divert at their western end. To their south is another curved groove convex northward that terminates on an irregular south-facing low relief scar. None of these channels or grooves displays morphology typical of a turbidity current regime.

Discussion

The seismic reflection Profiles 135 and 129 and those recorded by Hinz et al. (1982) demonstrate that the Canary Channel is underlained by Mesozoic evaporites whose plastic flow has led to the formation of diapirs and piercement structures. These evaporites were deposited in a basin along the northwest African continental margin from Senegal at 10° N to the southern Iberian peninsula at 37° N (Uchupi et al., 1976). Within the Canary Channel the evaporite province extends westward from the northwest African margin to the vicinity of the Canary Ridge terminating southwestward at about the same latitude than the most southern tip of Fuerteventura. However, since evaporites may be masked by Miocene volcanic rocks it is possible that the evaporite province extends farther south and west. Of the lines described in this report, only on Line 135, at the northern end of the Canary Channel, do the diapirs have seafloor expression. Another seafloor feature that may be the result of diapirism is the broad swell on the east side of the Canary Channel centered about 28° 40′ N. Only its western side lies within the area surveyed by swath bathymetry. As defined by the 1000 m contour the feature is about 90 km long in a northeast direction and 20 km wide in a northwest direction. Seismic reflection Profile 129 across the northeast edge of the swell suggests that there are at least two diapirs beneath the swell. The rest of Line 129, extending eastward beyond the shelf's edge (not included in this report), reveals several diapirs beneath the upper continental slope and continental shelf (Uchupi et al., 1976); thus the swell may be a salt massif.

One of the features imaged by the swath bathymetric map, as well as the relief and 3D diagrams in the Canary Channel are pockmarks formed by the expulsion of fluids. Their occurrence within a salt diapiric field suggests that the fluids are of thermogenic origin.

Another feature imaged by the swath data are low relief circular to elliptical mounds (Figures 4–8). One of the mounds is within a graben and is associated with faults that propagate to the seafloor. Two other small mounds are located within a convex shaped groove. The others are located along the base of the salt massif. Except for the mounds in the convex groove that have reliefs of less than 10 m and diameters of tens of meters, mounds have heights of 75 to 375 m, diameters of 4 to 10 km and are partially surrounded by moats with reliefs of 25 to 75 m. As there is evidence of bottom current activity in the region (Jacobi and Hayes, 1992; this study) it is possible that the moats surrounding the mounds are due to current scour (Figure 8). However, their reliefs suggest that they may be structural in origin.

None of the seismic reflection profiles described in this study crosses these features, thus, it is not possible to determine their relationship to the salt diapirs imaged in the seismic reflection profiles. The high immediately east of the Canary Ridge (Mound M1, Figures 5A and 7) does not have a moat and displays a crater on its crest, a feature not observed on the other highs (Figures 6 and 8). It is also located on top of toe thrusts and overhangs along the east flank of the Canary Ridge described by Martínez del Olmo and Buitrago Borrás (2002). We infer from its proximity to the Canary Ridge and its location over the overhangs that this mound is probably the consequence of the expulsion of volcanogenic gas, hot venting (Figure 7). Such a conclusion is also supported by geomagnetic studies (Catalán et al., this issue) that show that Mound 1 has a 300 nT absolute magnetic maxima (peak to peak) showing a characteristic induced dipolar aspect. In contrast Mounds 2 and 3 display a small amplitude short wavelength maxima (smaller than 20 nT).

Analytical signal studies (magnetic source boundaries without remnant magnetism) further support our contention that Mound 1 is of volcanic origin and Mounds 2 and 3 are not. The other highs, located within the diapiric field along the base of the salt massif centered at 28° 45′ N and 13° 19′ W may be the result of salt diapirs. Possibly hydrocarbon and oil seeps (cold venting) along the crest of the salt diapirs may be responsible for much of the relief of the mounds. In the Gulf of Mexico such seeps along faults on the crests and flanks of salt diapirs has led to the creation of small depressions, mud vents or carbonate mounds (Roberts et al., 1989, 1990). In addition to carbonate build-ups the topographic highs in the Gulf of Mexico formed by diapiric activity are also capped by carbonate-rich, lithified sediments and yellowish to orange nodules of thermogenic gas hydrate associated with lithified carbonate rubble and crude oil. Dredge samples recovered from Mound 3 are of similar composition consisting of carbonate pavement and manganese oxide coated yellowish nodules.

The gullies or rills (Figures 5 and 6) associated with the mounds and pockmarks north of 27° 30′ N tend to begin and end abruptly. The largest of these gullies originates in the swell centered near 28° 30′ N and meanders down its side terminating in the vicinity of two other mounds. The other gullies occur in the vicinity of the mound and appear to originate from the sides of the mounds. The morphology of these gullies is suggestive of erosion by a dense flow, possibly salt brine. As noted by MacDonald (1992) such brines are produced where the salt comes in contact with seawater along fault planes. According to MacDonald (1992) these brines typically have salinities that exceed 200 ppt and density that is at least 125% that of salt water. This density difference tends to preserve the salt brines as a distinct fluid on the sea floor, a fluid that can flow down-slope eroding gullies on the surficial sediments. Another possible mechanism for brine formation in the Canary Channel region may have been igneous activity on the Canary Ridge. That the ridge could have served as a thermal source is not unrealistic as Lanzarote has been active between 1730 and 1736 when extrusives covered much of the southwest side of the island. (Carracedo and Rodríguez Badiola, 1993) and hot springs are present in the Montaña del Fuego, Lanzarote (Calami and Ceron, 1970). If interstitial waters warmed by the heat produced by this volcanicity propagated eastward and deep into the sediments in the Canary Channel it may have led to the dissolution of the Mesozoic salt and formation of brines. Once formed the brines would have migrated upward along faults created by salt diapirism.

Other channels in the region tend to differ from those that we ascribe to erosion by brine seeps.The channels south of 27° 57′ N and west of 14° 51′ W display characteristics typical of turbidity currents. Those channels south of 27° 57′ W and east of 14° 51′ W display an unusual morphology. They tend to be sharply curved with two small mounds located along the axis of one of the lows. Their morphology demonstrates that they are not the creation of brine seeps or turbidity currents. The grooves are associated with a scalloped scarp south of 27° 27′ S that may represent the failure plane or head of scar of a landslide involving the uppermost sediments (Figure 5B). Possibly the

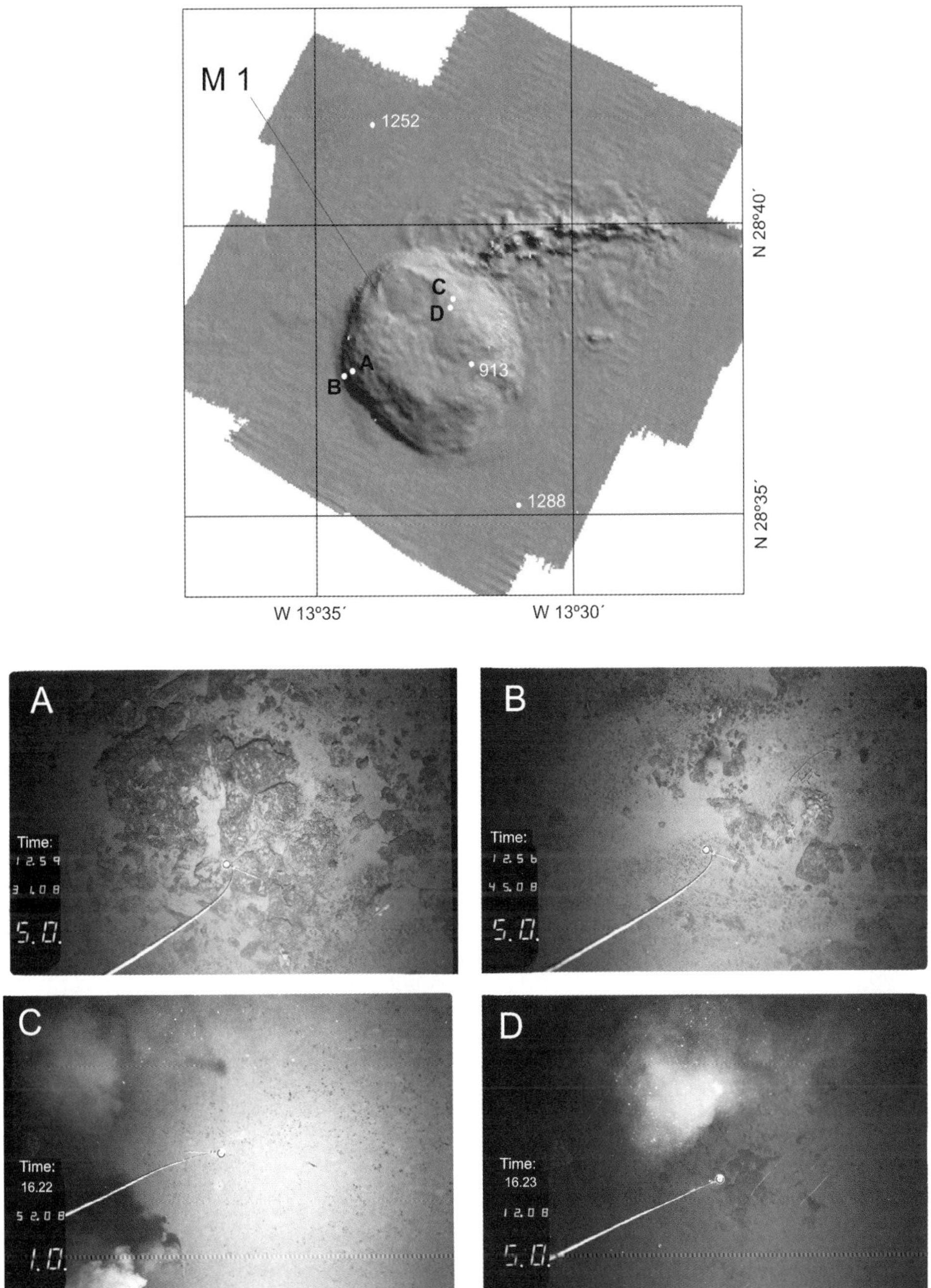

Figure 7. Upper Panel: Shaded relief diagram of Mound 1. A–D = Positions of bottom photographs; White numbers = water depth in meters. Lower Panel: A–B: Flank of Mound 1 showing outcrops surrounded by sediments. The outcrops may represent carbonate pavement or lava. The twigs in B may be deep-water coral. C–D = White smoker on crest of Mound 1. Area covered by each photographs is approximately 5 × 4 meters.

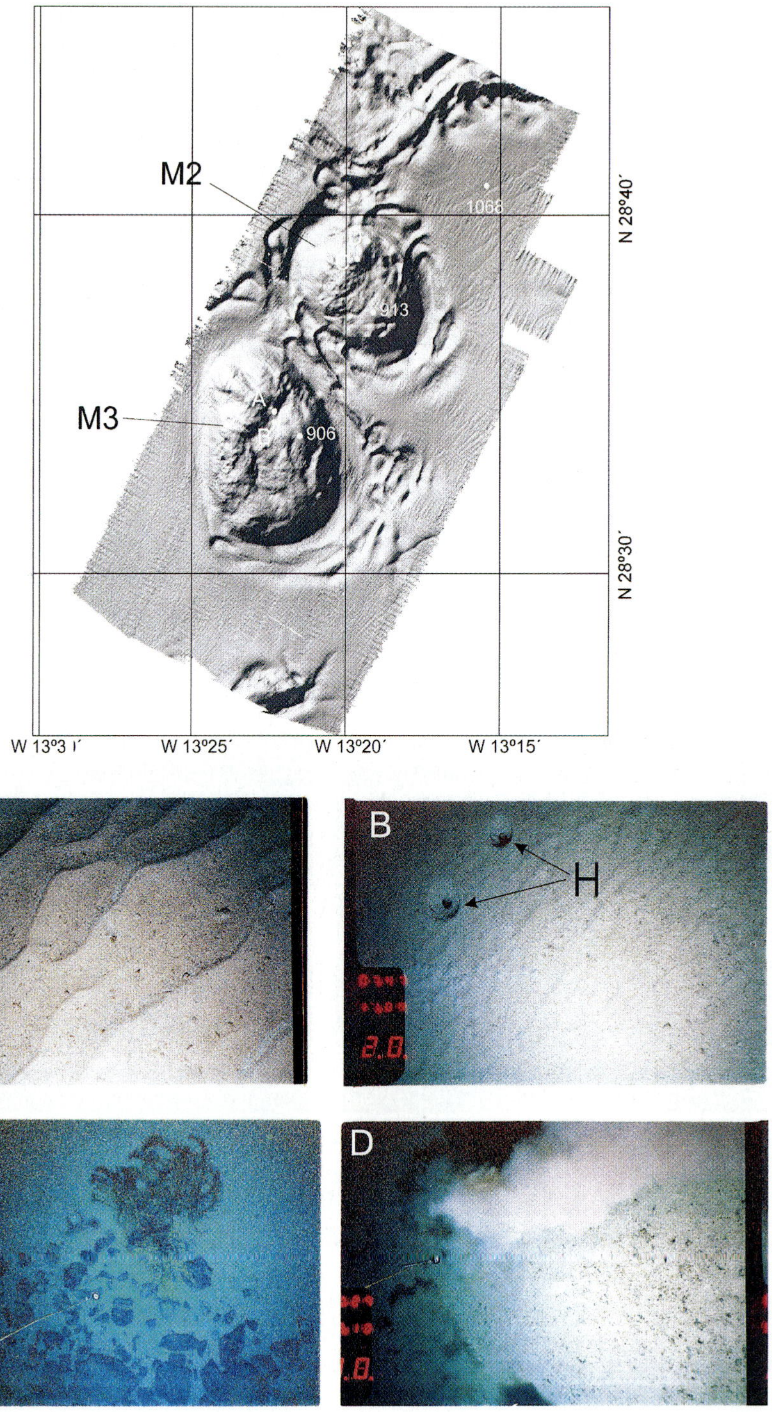

Figure 8. Upper Panel: Shaded relief diagram of Mounds 2 and 3. A–D = Positions of bottom photographs. White numbers = water depth in meters. Lower Panel: A–B: Photographs on crest of Mound 3, showing sediment waves. H: Holes created by bioturbation or fluid seepage. C: Photograph on crest of Mound 2 showing crinoids (dark), deep-water corals (branches below the crinoid) and rock out crops that may represent carbonate pavement. D: Photograph in crest of Mound 2. 'Sediment cloud' in photo was generated by the impact of compass on seafloor. Area covered by each photograph is approximately 5 × 4 m.

curved grooves north of the landslide scar may represent cracks on the surficial sediment section that is being displaced downslope by creep. A possible scenario for their evolution may involve sediment creep down slope creating the grooves until part of the sediment becomes detached to form the slide scar south of 27° 27′ S. The presence of two small mounds within one curved groove (Figure 5B) suggests that fluid expulsion has played a role in sediment instability. In the ultimate instance, these gravitational micro-structures also may have been triggered by fluid expulsion and vertical motion of the Mesozoic evaporites.

Conclusion

Features found during a swath bathymetric survey of the Canary Channel include a mound near the east flank of the Canary Ridge, landslide scars and landslide deposits along the base of the Canary Ridge, a line of mounds aligned in a northeast direction along the axis of the channel, gullies associated with mounds that have abrupt terminations, pockmarks and curved grooves. The features along or near the east flank of the Canary Ridge are probably the creation of tectonism and turbidity current activity. The mound, a smoker, was formed by hydrothermal activity. The features in the Canary Channel have a different origin. The mounds in the channel have diameters of 4 to 10 km, reliefs of 75 to 375 m and are partially surrounded by moats. They occur within a Mesozoic evaporitic basin. One of the mounds is a smoker. We infer that the mounds may represent the crests of salt diapirs capped by carbonate pavement, mud vents, and yellowish to orange nodules of thermogenic gas hydrate associated with lithified carbonate rubble and crude oil. These seeps may have provided nutrients to bacteria that in turn served as a food chain for carbonate-secreting organisms. We further speculate that the gullies found in the vicinity of the mounds were eroded by dense salt brines created where salt in the diapirs are in contact with sea water at the seafloor or were formed in the subsurface by water heated by igneous activity along the Canary Ridge. From there the brines migrated upward along the fault system that served as conduits for the hydrocarbons. The pockmarks associated with the diapirs are probably the result of expulsion of thermogenic fluids and the curved grooves may reflect sediment displacement by creep of the surficial sediments and gravitational sliding triggered by salt and hydrocarbon seeps.

Acknowledgements

The authors wish to thank the officers and crew of the R/V Hesperides and R/V Vizconde de Eza for their cooperation during the cruises to the Canary Archipelago. Thanks also are due to Peter Clift, and two anonymous referees. Their constructive comments were of considerable help in revising the manuscript. This paper is a contribution of the Spanish ZEE Program.

References

Aharon, P., 1994, Geology and biology of modern and ancient submarine hydrocarbon seeps and vents: an introduction, *Geo-Mar. Lett.* **14**, 69–73.

Aharon, P., Schwarz, H. P. and Roberts, H. H., 1997, Radioelectric dating of submarine hydrocarbon seeps in the Gulf of Mexico, *Geol. Soc. Am. Bull.* **109**, 568–579.

Acosta, J., Muñoz, A., Herranz, P., Palomo, C., Ballesteros, M., Vaquero, M. and Uchupi, E., 2001, Pockmarks in Ibiza Channel and western end of the Balearic Promontory (western Mediterranean) revealed by multibeam mapping, *Geo-Mar. Lett.* **21**, 123–130.

Ancochea, E., Brändle, J. L., Cubas, C. R., Hernán, F. and Huertas, M.J., 1996, Volcanic complexes in the eastern ridge of the Canary Islands: the Miocene activity of the island of Fuerteventura, *J. Vol. Geoth. Res.* **70**, 183–204.

Arthur, M. A., Cornford, C., McCoy, F. W. and Sarnthein, M., 1979, Evolution and sedimentary history of the Cape Bojador continental margin, northwestern Africa, in von Rad, U. Ryan, W. B. F. et al. (eds.), *Init. Repts, Deep Sea Drill. Prog. College Station, TX* **47**, 773–816.

Baraza, J. and Ercilla, G., 1996, Gas-charged sediments and large pockmarck-like features on the Gulf of Cadiz slope (SW Spain), *Mar. Pet. Geol.* **13**, 253–261.

Beauchamp, B. and Savard, M., 1992, Cretaceous chemosynthetic carbonate mounds in the Canadian Arctic. *Palaios* **7**, 4, 434–450.

Behrens, E. W., 1988, Geology of a continental oil seep. Northern Gulf of Mexico, *Am. Assoc. Pet. Geol. Bull.* **72**, 105–114.

Bright, T. J., La Rock, P. A., Lauer, R. D. and Brooks, J. M., 1980, A brine seep at the east Flower Garden Bank, northwestern Gulf of Mexico, *Int. Rev. Ges. Hydrobiol.* **65**, 4, 535–549.

Brown, K. M., 1990, The nature and hydrogeologic significance of mud diapirs and diatremes for accretionary wedges, *J. Geophys. Res.* **95**, 8969–8982.

Cadet, J. P., Kobayashi, K., Lallemand, S., Jolivet, L., Aubouin, J., Boulègue, J., Dubois, J., Hotta, H., Ishii, T., Knishi, K., Niitsuma, N. and Shiamura, H., 1987, Deep scientific dives in the Japan and Kuril trenches, *Earth Planet. Sci. Lett.* **83**, 313–328.

Calami, A. and Ceron, P., 1970, Air convection within "Montaña del Fuego" (Lanzarote Island, Canary Archipelago), *Geothermics* Special Issue **2**, 611–614.

Carracedo, J. C. and Rodríguez Badiola, E., 1993, Evolución geológica y magmática de la isla de Lanzarote (Islas Canarias), *Rev. Acad. Canar. Cienc.* **V**, 25–58.

Carracedo, J. C., Day, S., Guillou, H., Rodríguez, E., Canas, J. A. and Pérez, F. J., 1998, Hotspot volcanism closet o a passive continental margin, *Geol. Mag.* **135**, 591–604.

Catalán, M., Martín-Davila, J. and ZEE Working Group, A magnetic anomaly study offshore Canary archipelago, this issue.

Coello, J., Cantagrel, J. M., Hernán, F., Fúster, J. M., Ibarrola, E., Ancochea, E., Casquet, C., Jamond, C., Diaz de Téran, J. R. and Cendredo, A., 1992, Evolution of the eastern volcanic ridge of the Canary Islands based on new K-Ar data, *J. Vol. Geoth. Res.* **53**, 251–274.

Coleman, D. F. and Ballard, R. D., 2001, A highly concentrated region of cold hydrocarbon seeps in the southeastern Mediterranean Sea, *Geo-Mar. Lett.* **21**, 162–167.

Cramp, A. and O'Sullivan, G., 1999, Neogene sapropels in the Mediterranean: a review, *Mar. Geol.* **153**, 11–28.

Dando, P. R. and Hovland. M., 1992, Environmental effects of submarine seeping natural gas, *Contl. Shelf. Res.* **12**, 1197–1207.

Dando, P. R., Austen, M. C., Burke, Jr., Kendall, M. A., Kennicutt III, M. C., Judd, A. G., Moore, D. C., O'Hara, S. C. M., Schmaljohann, R. and Southward, A. J.,1991, Ecology of North Sea pockmark with an active methane seep, *Mar. Ecol. Prog. Ser.***70**, 49–63.

Dañobeitia, J. J. and Canales, J. P., 2000, Magmatic underplating in the Canary Archipelago. *J. Volc. and Geotherm. Res.* **103**, 27–41.

Emery, K. O. and Uchupi, E., 1984, The Geology of the Atlantic Ocean, *Springer-Verlag, New York*, 1050 pp.

Gaillard, C., Rio, M. and Rolin, Y., 1992, Fossil chemosynthetic communities related to vents or seeps in sedimentary basins: the pseudobioherms of southeastern France compared to other world examples, *Palaios* **7**, 451–465.

Graue, K., 2000, Mud volcanoes in deepwater Nigeria, *Mar. Pet. Geol.* **17**, 959–974.

Henry, P., Le Pichon, X., Lallement, S., Foucher, J. P., Westbrook, G. and Hobart, M., 1990, Mud volcano field seaward of the Barbados accretionary complex: a deep-towed side scan sonar survey, *J. Geophys. Res.* **95**, 8917–8929.

Hinz, K., 1979, Seismic sequences of Cape Bojador, northwest Africa, in Von Rad, U., Ryan, W. B. F. et al. (eds.), *Init. Repts. Deep Sea Drill. Proj.* **47**, 485–490.

Hinz, K., Dostman, H. and Frisch, M., 1982, The continental margin of Morocco; seismic sequences, structural elements and geological development, in Von Red, U., Heinz, K., Earthen, M., and Seabed, E. (eds.), *Geology of the Northwest African Continental Margin* Springer-Verlag, New York, 34–60.

Hoernle, K. and Schmincke, H. U., 1993, The role of partial melting in the 15-Ma Agrochemical evolution of Gran Canaries: a blob model for the Canary hotspot, *J. Petrol.* **34**, 599–626.

Holik, J. S., Rabinowitz, P. D. and Austin, J. A., 1991, Effects of Canary hotspot volcanism on structure of oceanic crust off Morocco. *J. Geophys. Res.* **96**, 12039–12067.

Hovland, M., 1992, Hydrocarbon seeps in the northern marine waters, their occurrence and effects, *Palaios* **7**, 376–382.

Hovland, M. and Judd, A. G., 1988, Seabed seepages. Impact on Geology, Biology and Marine Environment, *Graham and Trotman, London* **293**, 203 pp.

Hovland, M., Croker, P. F. and Martin, M., 1994, Fault-associated seabed mounds (carbonate knolls ?) off western Ireland and north-west Australia, *Mar. Pet. Geol.* **11**, 232–246.

Jacobi, R. D. and Hayes, D. E., 1992, Northwest African continental rise: effects of near-bottom processes inferred from high-resolution seismic data, in Poag, C. W., and de Graciansky (eds.), *Geologic Evolution of Atlantic Continental Rises*, Van Nostrand Reinhold, New York, 137–156.

Kennicutt, M. C. and Brooks, J. M., 1990, Recognition of areas affected by petroleum seepages: northern Gulf of Mexico continental slope, *Geo-Mar. Lett.* **10**, 221–224.

Kennicutt, M. C. II, Brooks, J. M., Bidigare, R. R., MaDonald, T. J., Adkinson, D. L. and Macko, S. A., 1989, An upper slope "cold" seep community, northern California, *Lim. Oceanog.* **34**, 561–566.

King, L. H. and MacLean, B., 1970, Pockmarks on the Scotian Shelf, *Geol. Soc. Am. Bull.* **81**, 3141–3148.

Kulm, L. and Suess, E., 1990, Relationship between carbonate deposits and fluid venting: Oregon accretionary wedge, *J. Geophys. Res.* **96**, 8899–8915.

Le Pichon, X., Iiyama, T., Lallement, S., Okada, H., Rangin, K., Taira, A., Urabe, T. and Uyeda, S., 1987, Nankai Trough and Zenisu Ridge: a deep-sea submersible survey, *Earth Planet. Sci. Lett.* **83**, 285–299.

Lewis, B. T. R. and Cochrane, G. C., 1990, Relationship between location of chemosynthetic benthic communities and geologic structure on the Cascadia subduction zone, *J. Geophys. Res.* **95**, 8783–8793.

MacDonald, I. R., 1992, Sea-floor brine pools affect behavior, mortality, and preservation of fishes in the Gulf of Mexico: Lagerstätten in the making, *Palaios* **7**, 383–387.

MacDonald, I. R., Reilly II, J. F., Guinasso, Jr., N. L., Brooks, J. M., Carney, R. S., Bryant, W. A. and Bright, T. J., 1990, Chemosynthetic mussels at a brine-filled pockmark in the northern Gulf of Mexico, *Science* **248**, 1096–1099.

Maldonado, A., Somoza, L. and Pallarés, L., 1999, The Betic orogen of the Iberian-African boundary in the Gulf of Cádiz; geological evolution (central North Atlantic), *Mar. Geol.* **155**, 9–43.

Martínez del Olmo, W. and Buitrago Borrás, J., 2002, Sedimentación y volcanismo al este de las isles de Fuerteventura y Lanzarote (Surco de Fúster Casas), *Geogaceta* **32**, 51–54.

Neurauter, T. W. and Bryant, W. R., 1990, Seismic expression of sedimentary volcanism on the continental slope, northern Gulf of Mexico, *Geo-Mar. Lett.* **10**, 225–231.

Ohta, S. and Laubier, L., 1987, Deep biological communities in the subduction zone of Japan from bottom photographs taken during the "nautili" dives in the Kaiko Project, *Earth Planet Sci. Lett.* **83**, 329–342.

Paull, C. K. and Neumann, A. C., 1987, Continental margin brine seeps: their geological consequences, *Geology* **15**, 545–548.

Paull, C. K., Hecker, B., Commeau. R., Freeman-Lynde, R. P., Neumann, A. C., Corson, W. P., Golubic, S., Hook, J., Sikes, E. and Currray, J., 1984, Biological communities at Florida Escarpment resemble hydrothermal vent communities, *Science* **226**, 965–967.

Pautot, G., Nakamura, K., Huchon, P., Angelier, J., Bourgeois, J., Fujioka, K., Kanazawa, T., Nakamura. Y., Ogawa, Y., Séguret, M. and Takeuchi, A., 1987, Deep-sea survey in the Sugura and Japan trenches: preliminary results of the 1985 Kaiko Cruise, Leg 2, *Earth Planet Sci. Lett.* **83**, 300–312.

Pickrill, R. A., 1993, Shallow seismic stratigraphy and pockmarks of a hydrothermally influenced lake, Lake Rotoiti, New Zeland, *Sedimentology* **40**, 813–828.

Price, I., 1980, Gravity tectonics on a passive margin.: Deep-Sea Drilling Project site 415 in the light of regional seismic data, in Y. Lancelot, E. L. Winterer et al., *Init. Repts. Deep Sea Drill. Proj.* **50**, 759–771.

Roberts, H. H., Sassen, R., Carney, R. and Aharon, P., 1989, Carbonate buildups on the continental slope off Louisiana, *Proc. Off. Tech. Conf.* **OTC 5953**, 655–662.

Roberts. H. H., Aharon, P., Carney, R., Larkin, J. and Sassen, R., 1990, Sea floor responses to hydrocarbon seeps, Louisiana continental slope, *Geo-Mar. Lett.* **10**, 232–243.

Roberts, H. H. and Aharon, P., 1994, Hydrocarbon-derived carbonate buildups of northern Gulf of Mexico continental slope: a review of submersible investigations, *Geo-Mar. Lett.* **14**, 135–148.

Sassen, R., Grayson, P., Cole, G., Roberts, H. H. and Aharon, P., 1991, Hydrocarbon seepage and salt-dome related carbonate reservoir rocks of the U.S. Gulf Coast, *Trans. Gulf Coast Assoc. Geol. Soc.*, 570–578.

Shipboard Scientific Party, 1995, Site 955, in Schmincke, H.-U., Weaver, P. P. E., Firth, J. V. et al., *Proc. Ocean Drill.Prog. Init. Rep.* **157**, 433–496.

Shokes, R. F., Trabant, P. K., Preseley, B. J. and Reid, D. F., 1977, Anoxic, hypersaline basin in the northern Gulf of Mexico, *Sci.* **196**, 1443–1446.

Somoza, L., Gardner, J. M., Diaz del Río, V., Vázquez, J. T., Pinheiro, L. M. and Hernández-Molina, F. J., 2002, Numerous methane gas-related sea floor structures identified in the Gulf of Cádiz, *EOS* **83**, 541–549.

Stakes, D. S., Orange, D., Paduan, J. B., Salamy, K. A. and Maher, N., 1999, Cold-seeps and authigenic carbonate formation in Monterrey Bay, California, *Mar. Geol.* **159**, 93–109.

Steiner, C., Hobson, A., Favre, P., Stampfli, G. M. and Hernandez, J., 1998, Mesozoic sequence of Fuerteventura (Canary Islands): witness of early Jurassic sea-floor spreading in the central Atlantic, *Geol. Soc. Am. Bull.* **110**, 1304–1317.

Thomson, J., Mercone, D., De Lange, G. J. and Van Sanvoort, P. J. M., 1999, Review of recent advances in the interpretation of eastern Mediterranean sapropel S1 from geochemical evidence, *Mar. Geol.* **153**, 77–89.

Uchupi, E., Emery, K. O., Bowin, C. O. and Phillips, J. D., 1976, Continental margin off western Africa: Senegal to Portugal, *Am. Assoc Pet. Geol. Bull.* **60**, 809–878.

Uchupi, E., Swift, S. A. and Ross, D. A., 1996, Gas venting and late Quaternary sedimentation in the Persian (Arabian) Gulf, *Mar. Geol.* **129**, 237–269.

Von Bitter, P. H., Scott, S. D. and Schenk, P. E., 1990, Early Carboniferous low temperature hydrothermal vent communities from Newfoundland, *Nature* **344**, 145–148.

Von Rad, U., Rösh, H., Berner, U., Geyh, M., Marchig, V. and Schultz, H., 1996, Authigenic carbonates derived from oxidized methane vented from the Makran accretionary prism off Pakistan, *Mar. Geol.* **136**, 55–77.

Wang, C. Y., Shi, Y, Hwang, W. T. and Chen, H., 1995, Hydrogeologic processes in the Oregon-Washington accretionary wedge, *J. Geophys. Res.* **95**, 9009–9023.

Marine Geophysical Researches (2003) 24: 59–90
DOI 10.1007/s11001-004-1162-6

Comparison of volcanic rifts on La Palma and El Hierro, Canary Islands and the Island of Hawaii

J. Acosta[1], E. Uchupi[2], D. Smith[2], A. Muñoz[1], P. Herranz[1], C. Palomo[1], P. Llanes[3], M. Ballesteros[1] & ZEE Working Group[3]
[1]*Instituto Español de Oceanografía. Grupo de Cartografía Multihaz. Corazón de María, 8, 28002 Madrid*
[2]*Woods Hole Oceanographic Institution, Woods Hole, MA 02543, USA*
[3]*A. Carbó, A. Muñoz-Martín, Univ. Complutense, Madrid; J. Martín-Dávila, M. Catalán and J.A. Marín, Real Observatorio de la Armada. S. Fernando, Cádiz; F. Pérez-Carrillo, C. Maté, Instituto Hidrográfico de la Marina, Cádiz*

Key words: Canary Islands, Hawaii, La Palma and El Hierro, multibeam bathymetry, volcanic rifts

Abstract

The meso-scale (km) morphology of the well-studied volcanic rift zones on the Island of Hawaii is compared to the morphology of the lesser known rift zones of La Palma and El Hierro, Canary Islands. We find that there are both differences and similarities in their morphologic characteristics. In general, the rift zones on La Palma and El Hierro are shorter (a few tens of km in length) than those on Hawaii (ranging up to >100 km in length), perhaps reflecting both magma supply and composition. Many of the rift zones on Hawaii have well defined axial zones, both on-and offshore. In contrast, the rift zones on La Palma and El Hierro display various geometries ranging from linear ridges having smooth to irregular crests to structures with a broad fan-like morphology in plan view. The pronounced fanning may be a reflection of: 1) the stress field within the rift being insufficient to trap dikes within a narrow region, 2) dike injection and volcanism shifting laterally through time, 3) volcanoes building nearly one atop of another in the Canary Islands, superimposing the stress field of one structure on the other and thus yielding a more complex distribution of gravitational stresses, and 4) low rate of magma supply producing low magma pressures and thus randomly oriented dike injections. Irregularities and curvature along the axes of the rifts on La Palma and El Hierro may be a reflection of differences in the rate of magma production. Unlike the volcanoes on the Island of Hawaii there may be insufficient volumes of lavas erupted on La Palma and El Hierro to smooth out irregularities. The superposition of rifts from different volcanoes may also add to topographic irregularities in the Canary Islands, especially if eruption rates are low.

Introduction

Of all the oceanic island chains inferred to have been generated by hotspot activity, the best known are the Canary and the Hawaiian Islands (Figures 1 and 2). Although both are formed by magmatism associated with a hotspot, the volcanoes within the two island groups display both similar and, of primary interest to us, distinctly different geologic and geophysical characteristics. We are not sure why their characteristics vary and thus the aim of this work is to illuminate possible controls on the formation, interaction, and evolution of the volcanoes and rift zones that make up these two island groups.

Previous work comparing the morphology of the Canary and Hawaiian Islands includes that by Carracedo (1999), who examined the structure and evolution of individual volcanoes, and Mitchell et al. (2002), who investigated the effect of landslides on submarine flank morphology and the role of volatiles in controlling the meso-scale morphology. In this paper we focus on the meso-scale (km) morphology of the subaerial and submarine sections of the volcanic rift zones of the two youngest Canary Islands, La

Palma and El Hierro, and compare their characteristics with those of the well-studied rift zones on the youngest of the Hawaiian Islands, Hawaii.

Our work shows that the rift zones on La Palma and El Hierro are typically shorter (10 s of km) than those on Hawaii (ranging up to >100 km). In addition, the rift zones on La Palma and El Hierro display various geometries ranging from linear ridges with smooth to irregular crests to ridges with a fan-like morphology in plan view. In contrast, the rift zones on Hawaii tend to consist mostly of well-defined linear ridges. Although the toes of some of the Hawaiian rifts do partially fan out, they do not to the extent displayed by the Canary Island rifts. Irregularities and curvature along the axes of the rift zones is also more common on La Palma and El Hierro than on Hawaii. We examine the possible causes for these observations.

Data

Comparison of the subaerial sections of the volcanic rifts in La Palma, El Hierro, and the island of Hawaii is based on published sources. Investigation of the offshore sections of the rifts in the Canary Islands is based on swath bathymetric data collected by the Instituto Español de Oceanografia and the Instituto Hidrográfico de la Marina in 1998 to 2002. Description of the submarine rifts in Hawaii is based on available swath bathymetry data from Monterey Bay Aquarium Research Institute (MBARI, 2000), side scan sonar data from a deep towed vehicle and bottom photographs (Smith et al., 2002a), and a multitude of published sources as cited in that study.

Background

Regional setting

The Canary archipelago is located on the continental rise off Cape Juby, northwest Africa. Fuerteventura and Lanzarote, at the eastern end of the chain, are 100 km from the African coast, and El Hierro and La Palma at its western end are 500 km from the coast (Figure 1). The inner islands, Lanzarote and Fuerteventura, are built on seafloor at a water depth of about 2000 m, just west of the transition from attenuated continental to oceanic crust (Emery and Uchupi, 1984). Gran Canaria, Tenerife, La Gomera, La Palma and El Hierro, built on seafloor with water depths of 2000 to 4000 m, are located on oceanic crust of Jurassic age (Uchupi et al., 1976). The boundaries of this crust are defined by magnetic anomaly S1 (180 Ma; along the transition from continental to oceanic crust) east of Lanzarote-Fuerteventura (Canary Ridge) and anomaly M-25 (156 Ma) west of La Palma and El Hierro (Emery and Uchupi, 1984).

Volcanism in the Canary Islands ranges from Late Cretaceous in the eastern islands to the present (Le Bas et al., 1986; Carracedo, 1994). Carracedo (1994) has divided the islands into three groups, those that have had eruptions in historic times (<500 yr; Tenerife, La Palma; Lanzarote and probably El Hierro), those with a history of Quaternary volcanism (Fuerteventura and Gran Canaria) and those lacking evidence of Quaternary volcanism (La Gomera). Lanzarote, Fuerteventura and Gran Canaria are in the post-erosional phase, La Gomera is in the repose stage and Tenerife, La Palma and El Hierro are in the shield stage of development (Carracedo, 1999). In Gran Canaria the post-erosional phase is also characterized by catastrophic mass wasting (Mehl and Schmincke, 1999). The seamount phase of construction is represented in the Canary Islands by the Basal Complex that is formed of basic plutonic rocks, marine sediments and volcanic materials, cut by a very dense dike network. It crops out in Fuerteventura (Stillman, 1999), in La Gomera (Bravo, 1982; Cendrero, 1970), and in the Caldera de Taburiente and the Barranco de las Angustias in La Palma (Staudigel and Schmincke, 1984).

At the southeast end of the Hawaiian Ridge are the islands of the Hawaiian chain (Figure 2), built on crust older than magnetic anomaly 34 and younger than anomaly M0, between 90 and 80 Ma (Figure 2) (Clague and Dalrymple, 1989). Some of these islands are formed by a single volcano and others consist of two or more coalescing volcanic edifices (Langenheim and Clague, 1987). The ages of the volcanoes associated with the islands get older to the northwest with historical volcanic activity being limited to Maui and Hawaii (Figure 2). Hawaiian volcanoes, like those in the Canary Islands, go through four stages of evolution: a seamount phase, the best example being Loihi Seamount sitting at the leading edge of the island chain, a shield phase during which a caldera may form, a long period of volcanic quiescence and erosion, and a post erosional volcanic phase during which a caldera also may form (Clague and Dalrymple, 1989). During the shield phase and post-shield phase eruptions occur at the summit of the volcano and along rift zones extending down the flanks of the volcanoes.

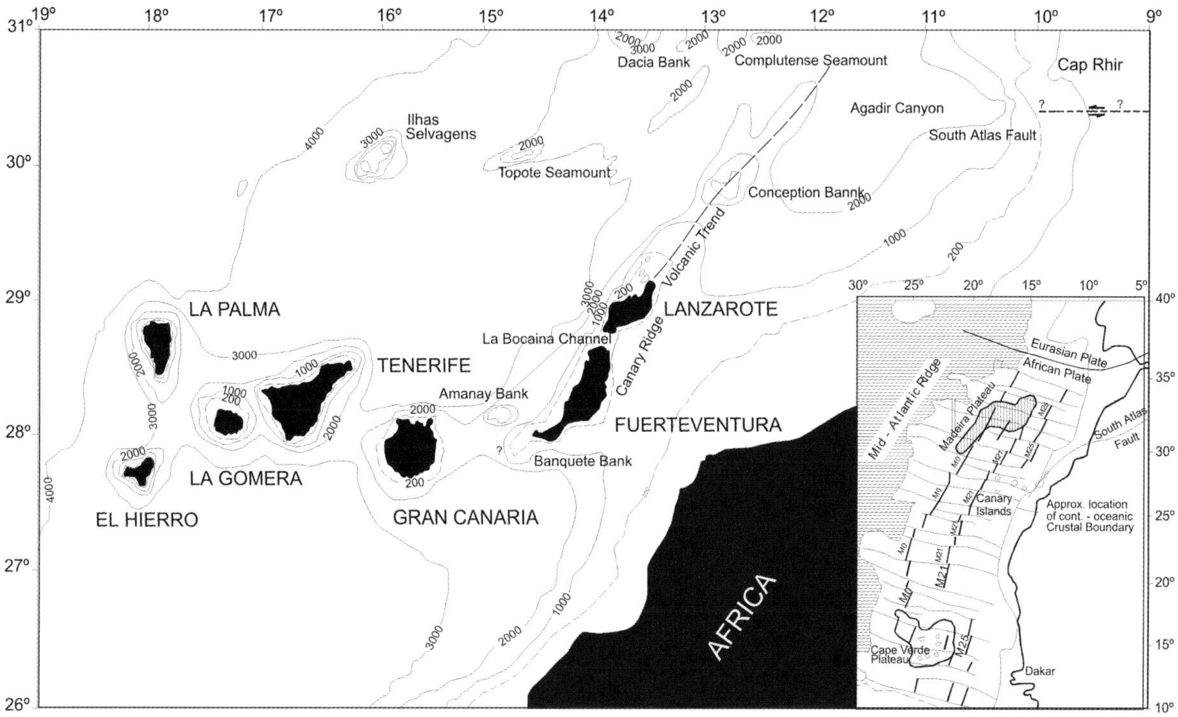

Figure 1. Geographic setting of the Canary Islands. Compiled from Emery and Uchupi (1984) and Uchupi et al. (1976).

Age progression

The volcanoes associated with the Emperor Seamounts-Hawaiian Ridge are an obvious hotspot trace superimposed on the fast moving Pacific plate ~70 mm/yr (Morgan, 1972). In contrast, the Canary Islands, on the slow-moving African plate (~25 mm/yr), form a less convincing hotspot trace (Carracedo, 1999). The Canary Islands do display a general age progression, from oldest in the east to the youngest in the west, but this trend displays several anomalies in the age-distance plot. Carracedo (1999) proposed that these anomalies may be a consequence of the islands having been created by a combination of a low-activity hotspot and a slow-moving plate. He noted that a similar age anomaly occurs in the Cape Verde Islands, an island group in the Eastern Atlantic also formed by hotspot activity. In the Cape Verde Islands, Quaternary volcanism shifted back to the east from Brava to Fogo, a situation comparable to the late Miocene shift in volcanism from La Gomera east to Tenerife. Carracedo (1999) also pointed out that Lanzarote is not really a separate island, but a northeast extension of Fuerteventura. The islands are separated by only the narrow 50-m-deep La Bocaina Channel. Differences in the age of volcanic activity in these two islands, therefore, may simply reflect the duration of their composite construction.

The lack of a simple east-west age progression in the older radiogenic age data led Anguita and Hernán (1975) to question a hotspot origin for the Canary Islands. They proposed, instead, that the islands were formed by magmatic intrusion along propagating fracture zones associated with seafloor spreading. The age of the islands (Late Cretaceous-Quaternary) is much younger than the age of the crust on which the islands are superimposed (Jurassic; Uchupi et al.,1976), however, and is not consistent with this theory. Another possibility is that magmatic activity along existing fracture zones may have been triggered by Alpine tectonism in western Africa (Uchupi et al., 1976). Price (1980), for example, proposed that volcanism along the Canary Ridge, capped by Lanzarote and Fuerteventura, was due to motion along a southwest extension of the sinistral South Atlas Fault (Figure 1). Another idea is that Lanzarote and Fuerteventura were emplaced on a structural weakness along the continental-oceanic crustal boundary. To date there are no data to verify any of these models.

New radiometric dating, however, appears to support a plume model for the Canary Islands with Fuerteventura-Lanzarote forming 20–21 Ma, Gran

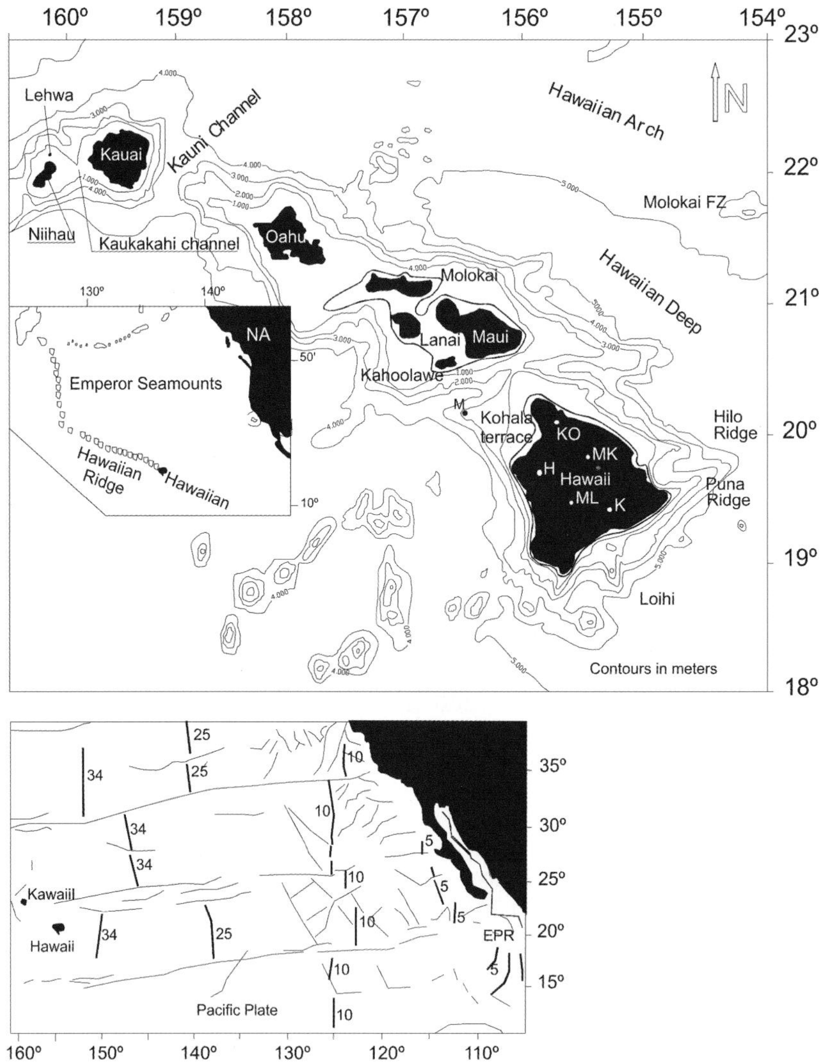

Figure 2. Top panel: geographic setting of the Hawaiian Islands. Modified from Fornari and Campbell (1986). Volcanoes are labeled as: KO-Kohala; H-Hualalai; MK-Mauna Kea; ML-Mauna Loa; KI-Kilauea. Lower panel: tectonic map of the eastern Pacific Ocean. Compiled from Atwater and Severinghaus (1989) and Decker (1989). Insert Map show the relationship of the Hawaiian Ridge and the Emperor seamounts. Modified from Clague and Dalrymple (1989).

Canaria 14–15 Ma, Tenerife 11–12 Ma and La Gomera 9–10 Ma. After the formation of La Gomera the east to west hotspot activity split into a north-south component, with dual hotspot activity forming La Palma and El Hierro < 2 Ma. Such dual volcanism is generally associated with changes in plate motion, but to date no such change in plate motion has been noted in the Canaries (Carracedo et al., 2001). On the basis of the new radiometric data Carracedo et al. (2001) concluded that the extension of the Atlas Fault System to the Canary Islands is a geographic coincidence and did not control the formation of the islands. They further proposed that the hotspot produced the islands, and that the hotspot is either spreading westward beneath the lithosphere or is fixed beneath the plate, which is slowly moving eastward.

Magma supply

The Hawaii and Canary Islands groups differ in their rates of magma supply. For example, eruption rates in Kilauea Volcano, Hawaii, from 1956 to 1983 suggest that the magma supply rate to this volcano is on the order of 86 km^3/1000 yr (e.g., Tilling and Dvorak,

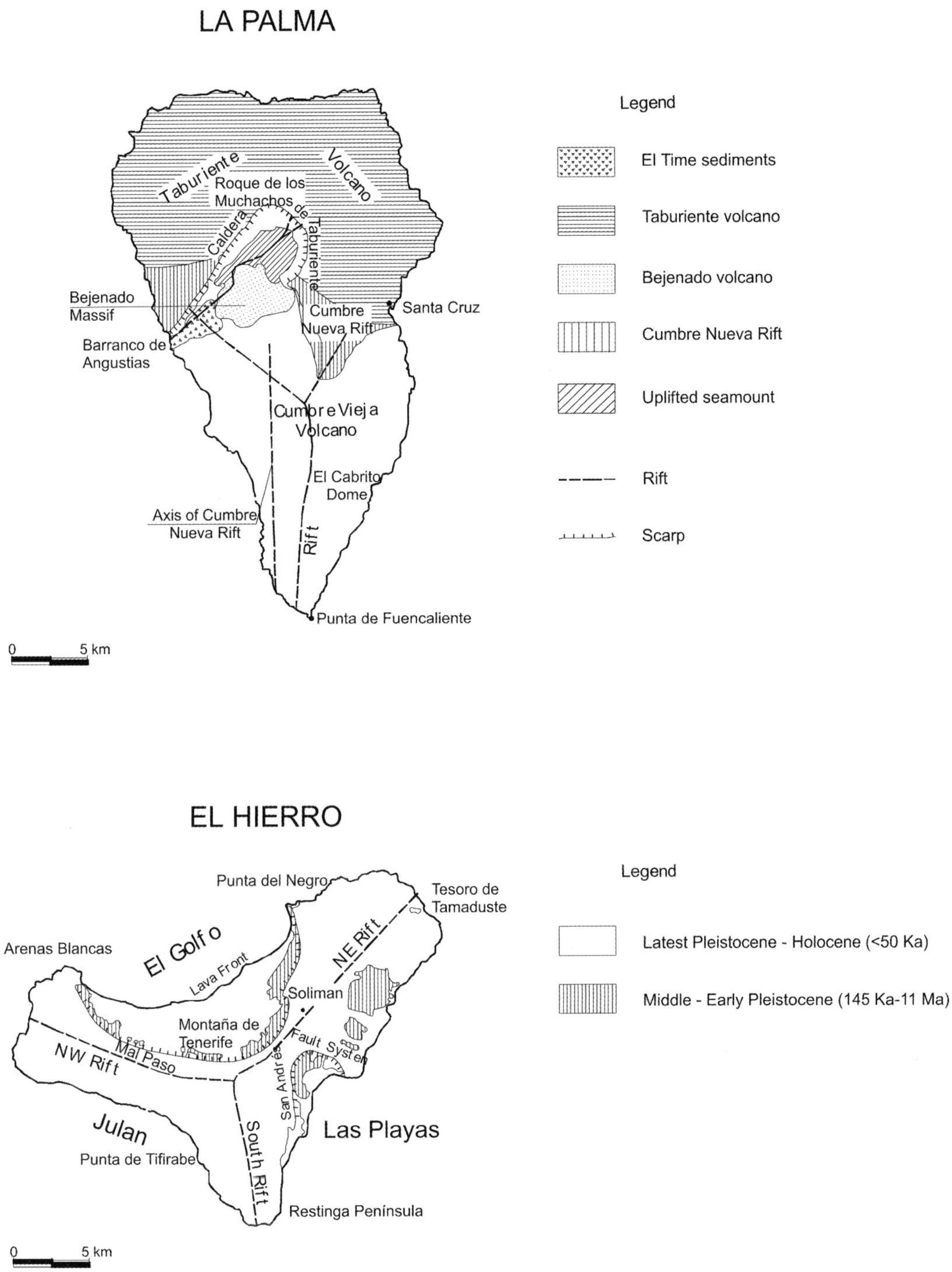

Figure 3. Geologic maps of La Palma and El Hierro, Canary Islands. Compiled from Hausen (1969), Fuster et al. (1993), Guillou et al. (1998) and Carracedo et al. (1999a).

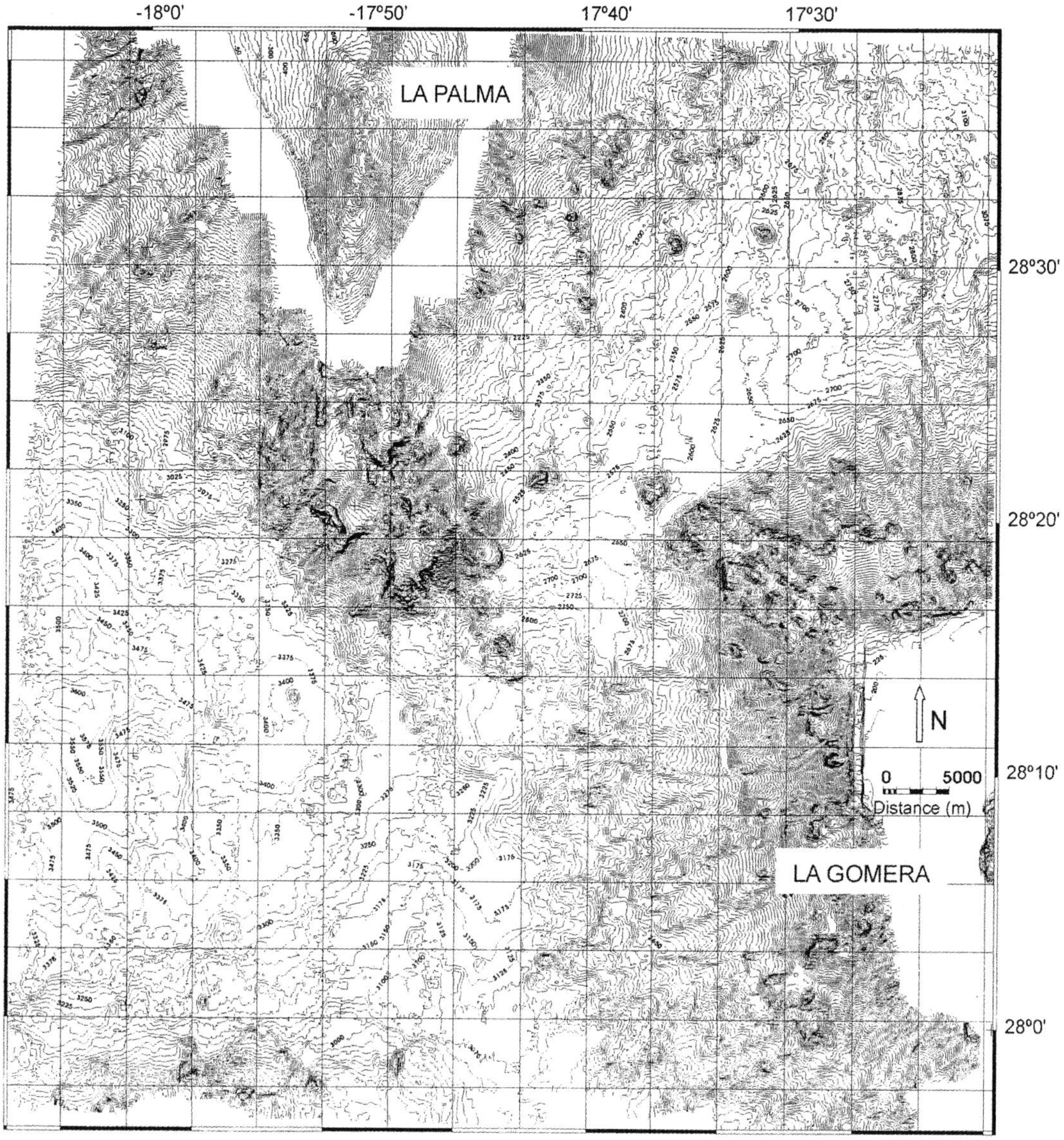

Figure 4. Topography of southern La Palma and swath bathymetry of the rift south of La Palma.

1993 and references therein). Magma supply rate can be estimated for El Hierro, which is probably the youngest and the best constrained geochronologically of any the Canary Islands (Guillou et al., 1996). The present emerged volume of the island, of about 140–150 km^3, has been produced in the last 1.2–1.5 Ma, giving an apparent average magma production rate of 0.12–0.13 km^3/1000 yr (Carracedo, 1999). Carracedo took into account the volume lost during the three giant lateral collapses that have affected the island, and concluded that the volume of the island would only be increased by ~100 km^3 for each collapse, yielding only a slightly greater magma supply rate of about 0.40 km^3/1000 yr. The fact that magma supply is much greater in Hawaii is reflected in the differences in the areas of the islands. The Island of Hawaii has an area of 10,458 km^{2}whereas La Palma and El Hierro have areas of 687 and 273 km^2 respectively and even the whole Canary Archipelago only has a total area of 7,273 km^2 representing 70% of the area of the Island of Hawaii.

Cooling and subsidence

The Canary and Hawaii Islands vary in their rate of subsidence due to lithospheric cooling as they move

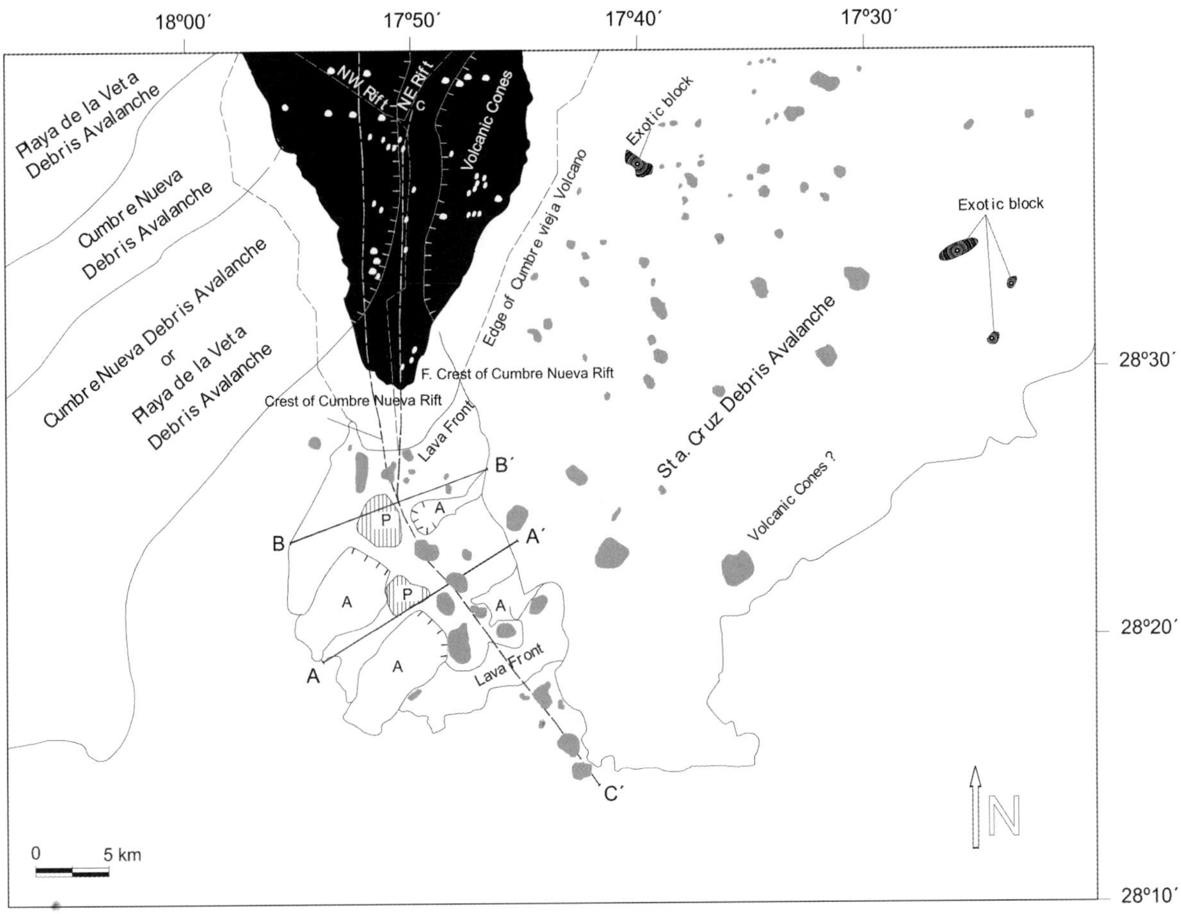

Figure 5. Morphologic interpretation of the swath bathymetry in Figure 4. A = Avalanche; P = Platform; A-A′, B-B′ and C-C′ = Profiles.

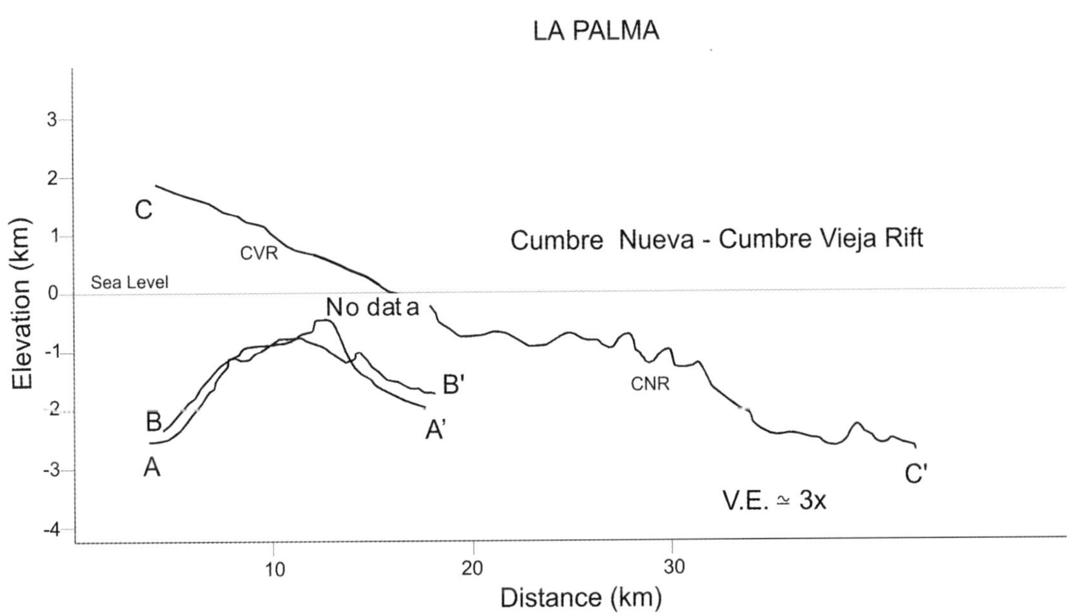

Figure 6. Profiles of the rift (Cumbre Vieja) south of La Palma. Compiled from Figure 4. See Figure 5 for locations of profiles.

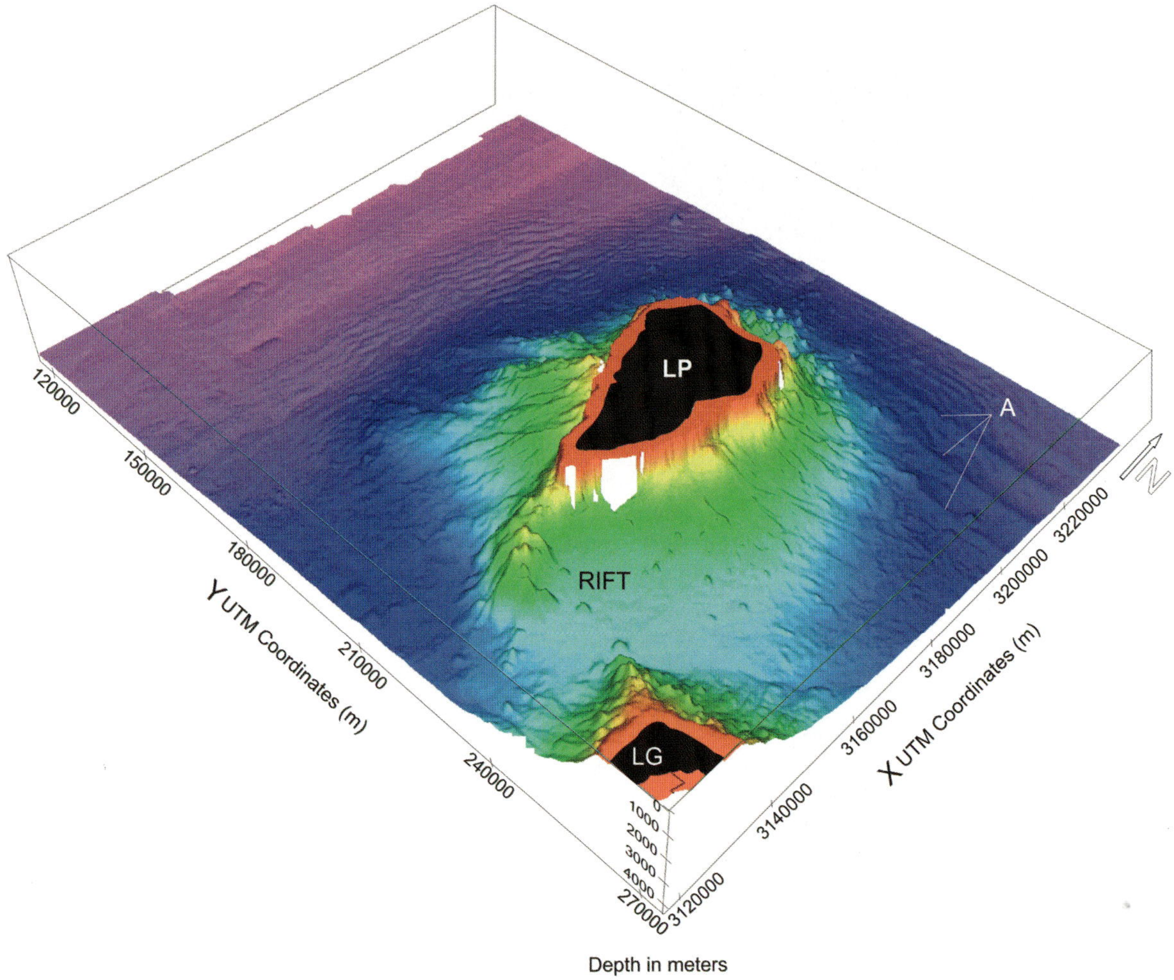

Figure 7. Relief diagram of the Cumbre Vieja Rift south of La Palma. Compiled from swath bathymetry. LG = La Gomera; LP = La Palma; A = Artifacts.

away from the hotspot (Morgan, 1970; Clague and Dalrymple, 1989 and references there in). In the Hawaiian Islands, there is a gradual progression from active volcanoes, Mauna Loa and Kilauea, to erosional remants (Niihau, Nihoa, Necker), growing atolls, (French Frigates Shoal and Midway Islands) to submerged guyots, (Ojin and Suiko) (Clague et al., 1988). Moore and Campbell (1987) estimated that most of the Hawaiian volcanoes have subsided 2–4 km and Moore and Campbell (1987) proposed that the bulk of this subsidence is rapid and concludes within a million years of the end of the shield phase. A much reduced subsidence rate due to regional processes continues beyond this phase. Dating of reefs on the gentle slope northwest of the Island of Hawaii indicates that it has been subsiding at a rate of 2.6 ± 0.4 mm/yr during the last 463 k.y. (Moore and Fornari, 1984; Moore and Clague, 1992). This rate of subsidence is such that one of the oldest islands in the chain, Kauai, is less than 6 Ma (McDougall, 1979).

No such subsidence trend is seen in the Canary Islands, with over 20 m.y. of volcanic rocks being exposed on some of the islands (Carracedo, 1999). The islands did, however, experience subsidence during the shield phase. Bathymetric data indicate a change from gently sloping ($\sim$87 m/km) subaerial lavas to more steeply sloping ($>$176 m/km) submarine lavas off shore Gran Canaria. The depth of the now submerged subaerial lavas suggests that during the shield building phase this island subsided 600–800 m (Funck and Schmincke, 1998). In the post-shield phase of island development, Gran Canaria and Tenerife have remained relatively stable with respect to sea level since about 14 Ma (Funck and Schmincke, 1998; Carracedo, 1999). Funck and

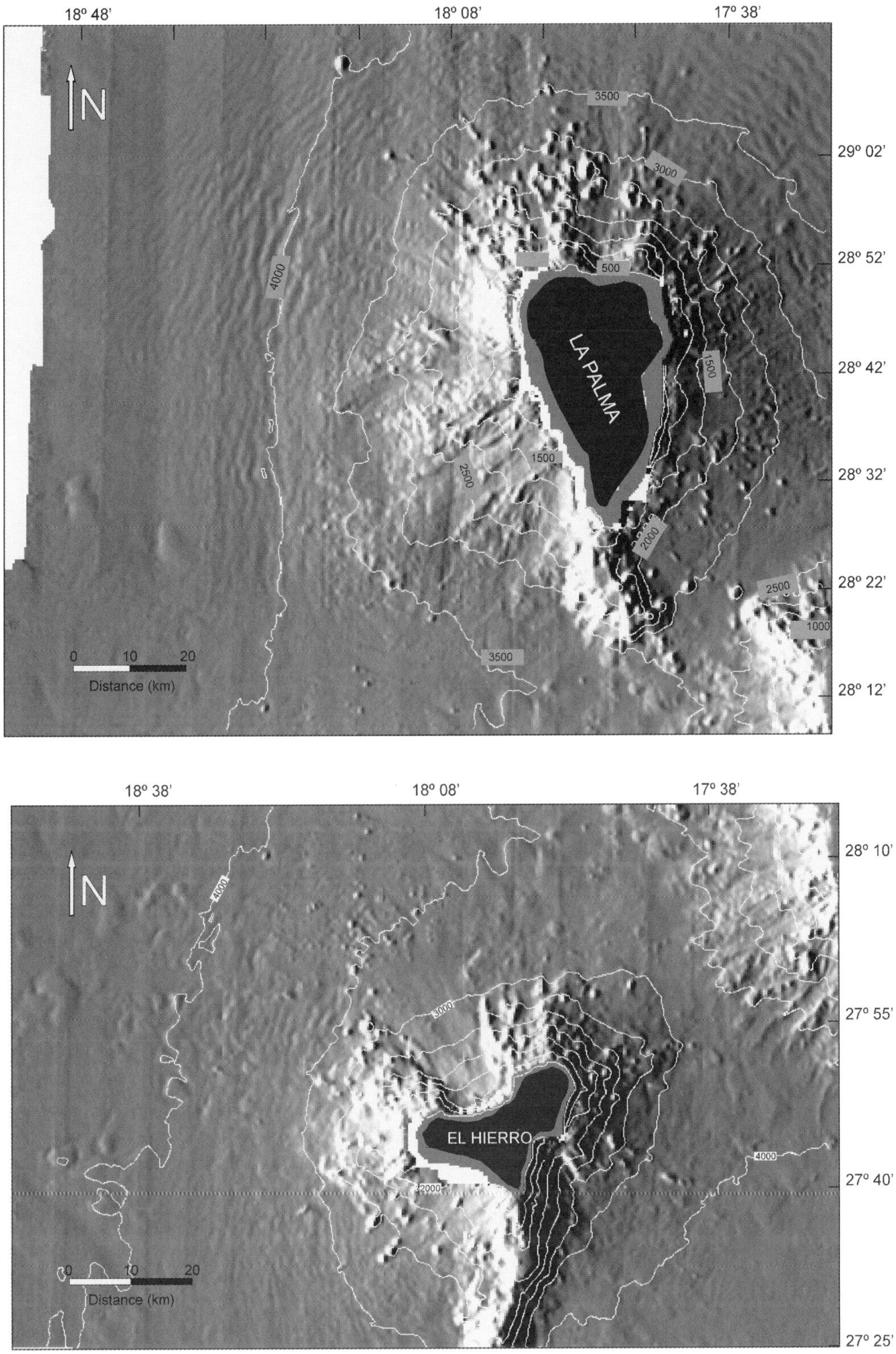

Figure 8. Relief diagrams of La Palma and El Hierro compiled from swath bathymetry. From Acosta et al. (this volume).

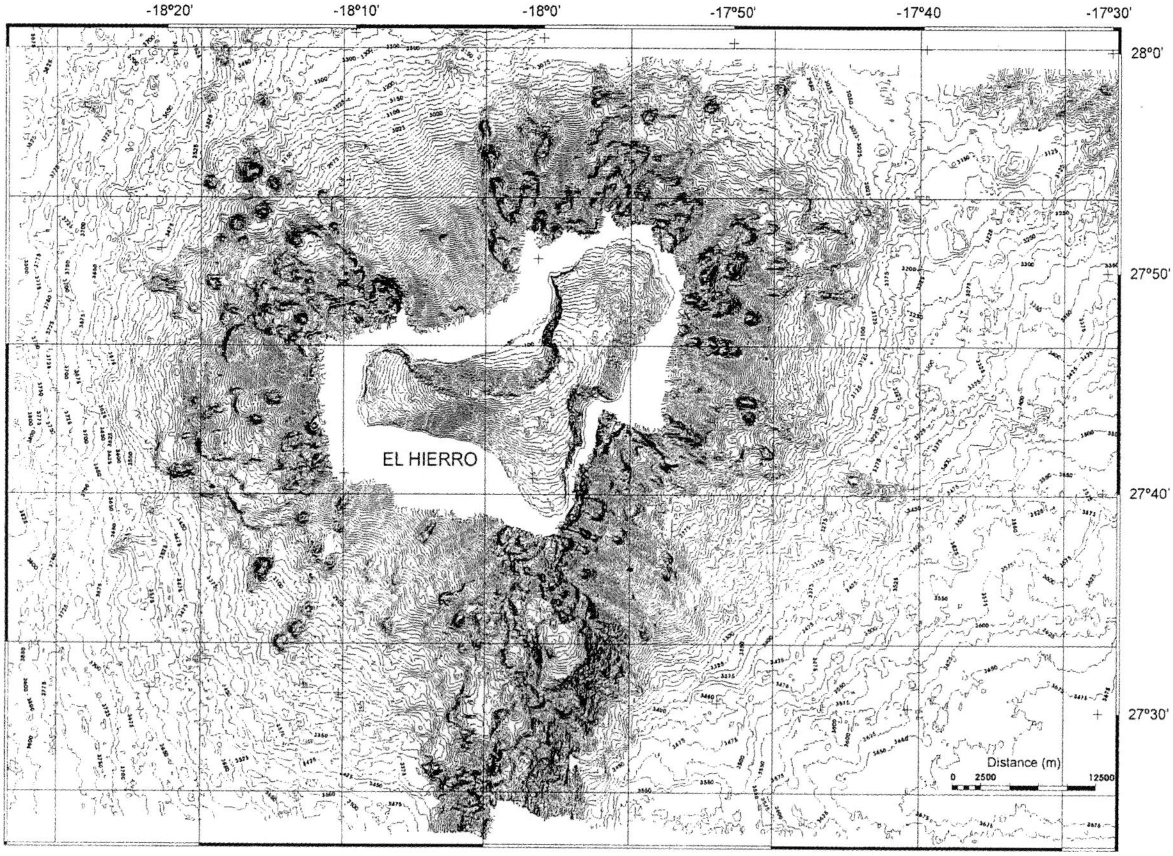

Figure 9. Bathymetry of the El Hierro offshore area based on swath bathymetry.

Schmincke (1998) proposed that this stability is due to a combination of minor volcanic loading during the post-shield phase and the strength of the Jurassic lithosphere, beneath the islands.

There is also evidence that at least two of the older islands, Lanzarote and Fuerteventura, have undergone recent uplift, as documented by raised marine terraces, beaches and Tertiary marine and lacustrine limestone at elevations of 55–60 m, 20 m and 10 m above sea level (Hausen, 1959; Coello et al., 1992; Stillman, 1999; Zazo et al., 2002). The raised Pleistocene terraces indicate an average uplift of 1.7 cm/1000 years since 1 Ma, with the present elevation of the last interglacial deposits suggesting that during the last 300,000 years Fuerteventura has been stable and that Lanzarote has subsided at about 0.7 cm/1000 years.

Swell and deep

The regional bathymetry of the Hawaiian Islands displays a moat, the Hawaiian Deep, with a maximum depth of over 5500 m. Beyond the moat is a broad swell, the Hawaiian Arch, a zone of tension whose crest is less than 4500 m deep, 500–1000 m shallower than the adjacent deep. These features on Cretaceous oceanic crust are the manifestations of lithospheric flexure caused by volcanic loading. Straddling the arch off the island of Oahu is the 24,000 km^2 active North Arch Volcanic Field (Clague et al., 2002). According to Clague et al. this field surrounds Cretaceous volcanic ridges, flat-topped volcanoes, and low relief sediment covered seafloor. They inferred that the volcanic field is younger than the 0.5–1.5 Ma age previously assigned to the field. South of the Island of Hawaii the arch is covered by flat sheet flows whose thin sediment cover and thin palagonite rinds on the lava surface suggest that the flows were erupted 1–10 Ka (Lipman et al., 1989).

No such topographic moat and swell or volcanic features are exposed on the seafloor surrounding the Canary Islands. Canales and Dañobeitia (1998), who studied the coherence between gravity and topography, proposed that a swell may be present, but likely masked by a thick sedimentary and volcanic cover.

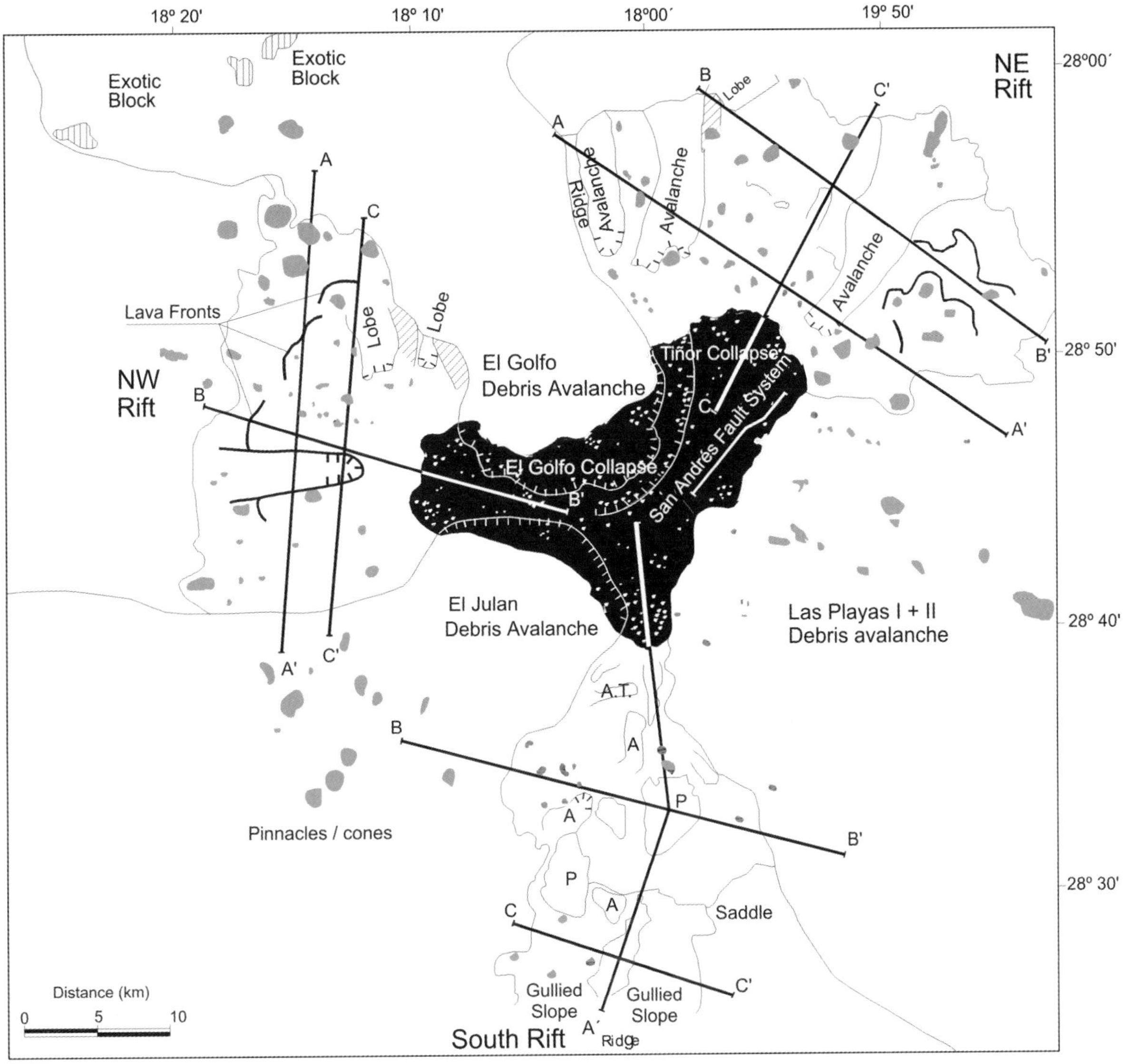

Figure 10. Morphologic interpretation of swath bathymetry in Figure 9. A = Avalanche; P = Platform; A–A′, B–B′ and C–C′ = Profiles.

This was verified by seismic reflection profiles (Watts et al., 1997) which showed a buried moat off Tenerife with a relief of 2–3 km and filled with sediments in part derived from the islands (Watts et al., 1997).

Volcanic and tectonic features of volcanoes

The geologic similarities between the Canary Archipelago and the Hawaiian Islands are as striking as their differences. As mentioned, both island groups underwent four volcanic phases: a seamount phase, a shield-building submarine and subaerial phase, characterized by rapid growth and massive slope failures, a period of quiescence and deep erosion (erosional gap), and a post-erosional stage of volcanic activity.

Volcanic rift zones in Hawaii tend to form during the seamount phase and grow above sea level as the edifice matures (e.g., Moore and Chadwick, 1995). They tend to form along zones of extension propagating from a magmatic center with the three principal rift arms arranged with angles of about 120° between them, possibly controlled by tangential stresses produced by a rising plume (Wyss, 1980). This origin also has been proposed for the rift zones in the Canary Islands (Carracedo, 1994). However, not all the volcanoes in the Hawaiian chain and the Canary Islands

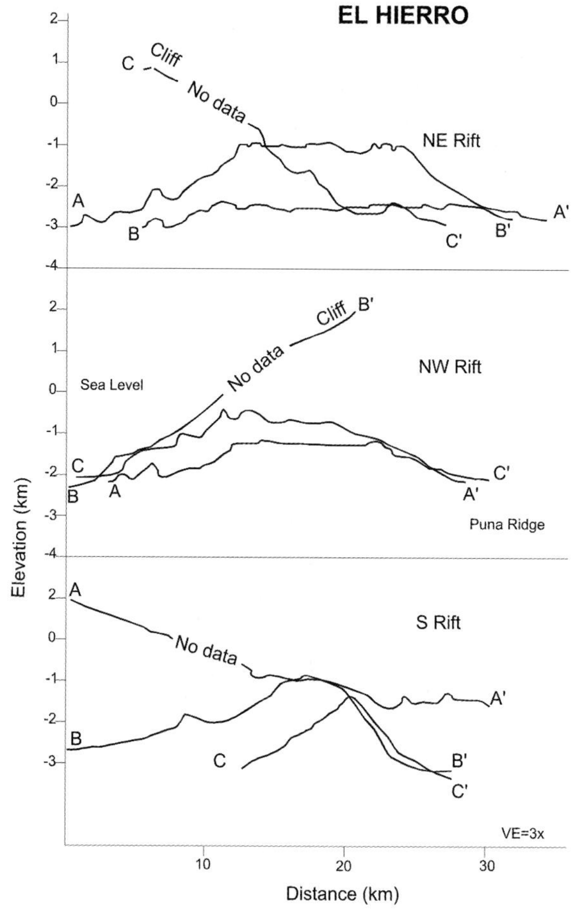

Figure 11. Profiles of Northeast, Northwest and South rifts of El Hierro. Compiled from Figure 9. See Figure 10 for location of profiles.

have three rift arms. Fiske and Jackson (1972) suggested that the locations of rift zones in Hawaii, and supposedly those in the Canary Islands, also are controlled by gravitational stresses within the volcanic edifice, which are influenced by the shapes and positions of preexisting volcanoes whose flanks serve as foundations for the younger structures.

On a small scale, rift zones in both the Hawaii and the Canary Islands are marked by eruptive vents and fissures aligned along the rift, open cracks, pit craters and grabens that confirm that the rift zones are the location of dike intrusions and fissure eruptions. It is thought that the development of the rift systems and the subsequent concentration of dikes injected into individual rifts promotes destabilization of the volcano flanks. Magma overpressure and mechanical and thermal overpressure of pore fluids leads to gravitationally unstable flanks (Carracedo, 1994, 1996; Elsworth and Day, 1999). In Hawaii, the rifts form the boundaries for zones of slumping that may ultimately lead to catastrophic collapse (Moore et al., 1989). Typically, however, the headwalls of the landslides are seaward of the rift zone axis (e.g., Moore and Clague, 1992). The wide arcuate landslide depressions in the Canary Islands often tend to be located between the two most active rifts, some with the third rift acting as a buttress for the landslide.

At a larger scale caldera formation at the summit of the volcanoes is common in both island groups. In the Canary Islands the calderas have been ascribed to several processes including: 1) collapse due to mass wasting (Ancochea et al., 1994; Masson et al., 2002), 2) fluvial erosion (Carracedo et al., 1999b), 3) magma chamber collapse (Martí et al., 1994, 1997), 4) a combination of collapse and post-collapse incision and retrogressive erosional collapse (Carracedo et al., 2001), and 5) a complex interplay of volcano inflation and deflation cycles (Troll et al., 2002). In Hawaii it has long been recognized that the formation of summit calderas is linked primarily to magma movement in

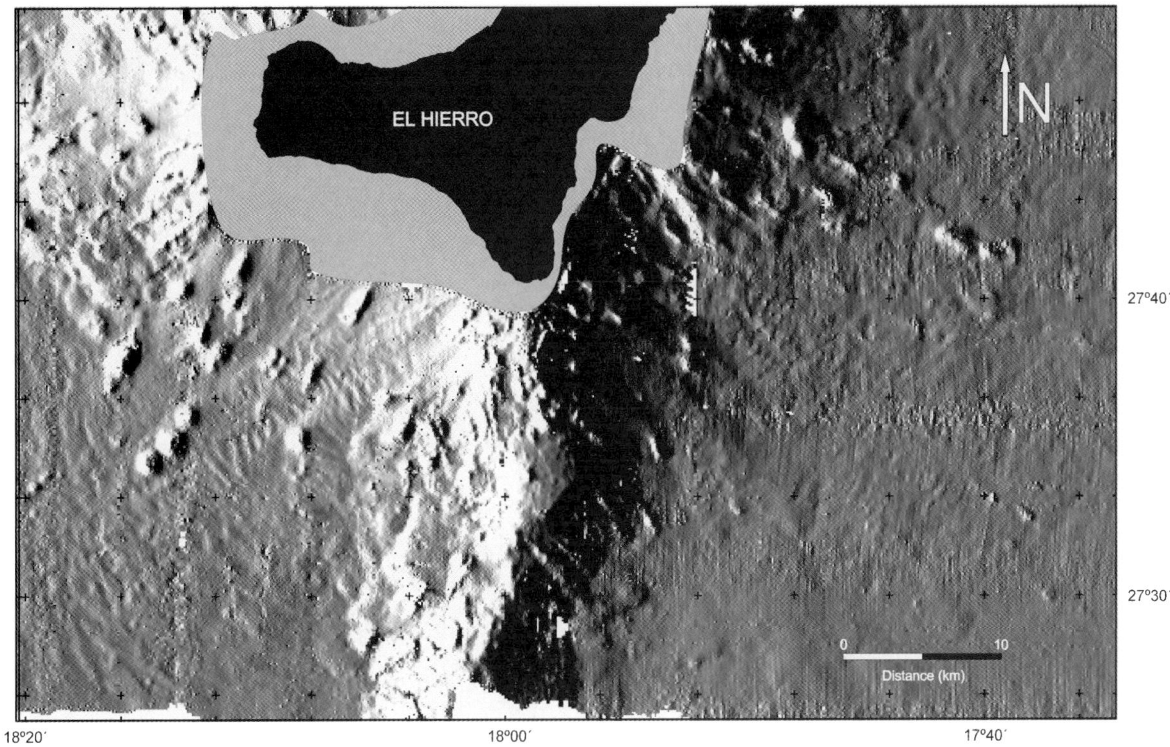

Figure 12. Relief diagram of South Rift off El Hierro. Compiled from swath bathymetry.

the summit reservoir. The roof of the reservoir is left unsupported when the magma is withdrawn leading to caldera collapse (e.g. Peterson and Moore, 1987; Tilling and Dvorak, 1993). Collapse may be incremental and recurring (Peterson and Moore, 1987). Walker (1988) pointed out, however, that the relationship between caldera subsidence and magma movement in Hawaiian volcanoes may not be as straightforward as we think. He suggested that subsidence may be caused locally by the excess load of intrusives. Either way it is clear that caldera formation in Hawaii is associated with magmatic processes compared to caldera formation in the Canary Islands, which can be ascribed to several different processes.

Canary Islands

In this section we examine the characteristics of La Palma and El Hierro, Canary Islands, and the Island of Hawaii. We then focus on the volcanic rift zones and compare and contrast their characteristics. The results will be used to help constrain the magmatic process that form these islands.

La Palma

General geology

The pendant-shaped, north-south trending island of La Palma is the second youngest island of the Canary Island archipelago (Figure 3). It has a maximum width of 28 km in an east-west direction, a maximum length of 47 km in a north-south direction, a maximum elevation of 2426 m above sea level at Pico de Los Muchachos, along the rim of Caldera de Taburiente, and an area above sea level of 687 km^2. The surface of the island is cut by many barrancos, the largest of which is the westward-draining Barranco de las Angustias (Figure 3). Barranco de las Angustias originates in the Caldera de Taburiente, whose vertical walls are up to 800 m high (Hausen, 1969) and whose floor has a maximum elevation of 1500 m above sea level (Hausen, 1969; Ancochea et al., 1994). The outer flanks of the caldera are incised by a radial system of barrancos, draining way from the center of the dome whose collapse, incision and retrogressive erosion (Carracedo et al., 2001) formed the Caldera de Taburiente. Extending southward from the rim of the caldera to Punta de Fuencaliente at the southern tip of La Palma is a narrow ridge, Cumbre Vieja, whose

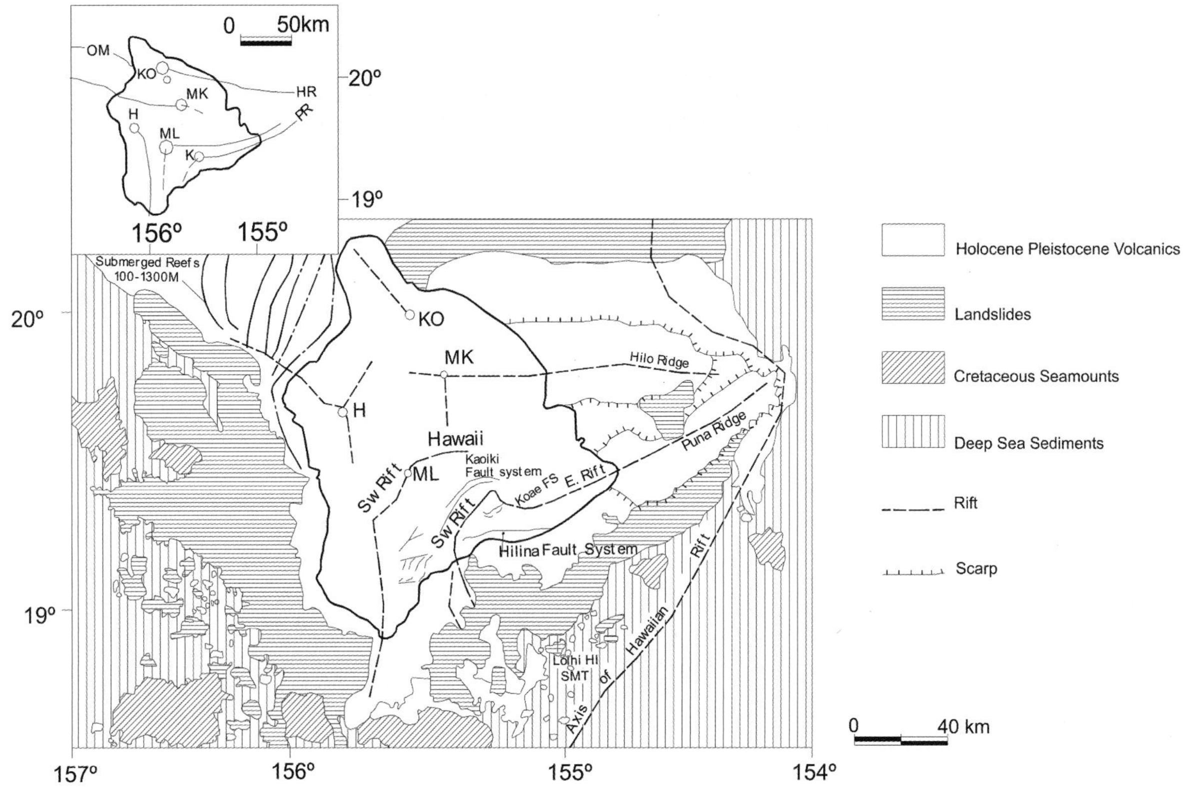

Figure 13. Geologic map of Hawaii compiled from Moore and Campbell (1987), Moore and Chadwick (1995) and D.J. Fornari (written communication, 2003). Insert shows distribution of rift zones postulated by Holcomb et al. (2000). Volcanoes are labeled as: KO-Kohala; H-Hualalai; MK-Mauna Kea; ML-Mauna Loa; KI-Kilauea. The east rift zone (ERZ) and southwest rift zone (SWRZ) of Kilauea are labeled.

crest is at an elevation of nearly 2000 m above sea level (Figure 3). Along the crest and slopes of Cumbre Vieja are many relatively young cinder cones. Lava flows originating from the cones extend to the east and west coasts of La Palma, filling many of the barrancos in the southerly slopes. The Cumbre Vieja has had seven eruptions in historical times, with the last taking place ~500 years ago (Guillou et al., 1996). La Palma's coasts are generally steep and rocky and except for Santa Cruz (Figure 3) on the east coast, which to a great extent is artificial, the rocky coasts lack natural harbors.

Geologically, La Palma can be divided into two sectors (Figure 3). The northern two-thirds of the island consists of the basement complex and the lavas of the Cobertera Series (1.7 to 0.8 Ma according to Ancochea et al., 1994 and 1.77 to 1.20 Ma according to Carracedo et al., 2001). The southern third of the island is dominated by the Cumbre Vieja volcano whose older lavas have been dated to 123 Ka (Carracedo et al., 2001). The northern complex consists of a basal plutonic sequence made up of small intrusions of gabbro, leucogabbro and cumulative ultramafics overlained by sills, and an alkali basaltic dike swarm cutting through the complex. Also included within and resting on the basement complex also are four debris avalanches. Staudigel and Schmincke (1984) inferred that the section was formed during the building of a seamount as it rose from deep-water to above sea level. The age of the seamount remains to be resolved. Hernández Pacheco and Fenández-Santin (1975) described Miocene microfossils in the seamount's submarine series, Staudigel and Schmincke (1984) found Pliocene fossils in the complex and Feraud et al.(1985) reported an age of 9 Ma and younger for the dikes in the complex.

The Garafía Volcano, with a diameter of ~20 km and a relief of 2500 and 3000 m, was constructed in the northern part of La Palma over the basement complex during the emplacement of the Lower Series Basalts (2.0–1.3 Ma according to Ancochea et al. (1994) and 1.77 to 1.20 Ma according to Carracedo et al. (2001). About 1.2 Ma, a paleocaldera formed on Garafía Volcano, as a consequence of the gravitational sliding of the southern slope or collapse and subsequent subaerial erosion (Carracedo et al., 2001). Avalanches created by the collapse of the Garafía Volcano are represented inshore by a several hundred meter-thick breccia (Carracedo et al., 1999a). An uplift of ~1000 m of the central part of La Palma and westward tilting of large blocks of the basement complex were associated with the caldera formation. The uplift was subdued by emplacement of the 0.89 Ma Cumbre Nueva Series of Taburiente Volcano. This volcanic activity ceased around 0.4 Ma having formed three coalescing volcanoes to make up the northern shield of La Palma (Carracedo et al., 2001).

The Cumbre Nueva Series formed a volcanic edifice and a north-south trending rift, the Cumbre Nueva Ridge, along the southern flank of the volcano during its last stages of growth (Figure 3) (Day et al., 1999; Carracedo et al., 2001). Ancochea et al. (1994) inferred that the Cumbre Nueva Ridge rift only extended a short distance south of its present outcrop next to the Caldera de Taburiente. In contrast, Day et al. (1999) proposed that the rift extended further southward to nearly the southern tip of La Palma. Whether it extends offshore is not known. As it grew, the Cumbre Nueva Ridge became progressively steeper until its western flank collapsed and initiated the formation of the Caldera de Taburiente. The Cumbre Nueva Avalanche off the west coast of La Palma represents the debris avalanches created during this collapse (Urgeles et al., 1999; Masson et al., 2002). Another extensive debris flow, the Santa Cruz Debris Avalanche, off the east coast of La Palma, also could be a product of this collapse, or possibly the flow is older and is due to the collapse of the Garafía Volcano, unfortunately Carracedo et al. (2001) stated that there is no geological evidence for this debris on land.

Ancochea et al. (1994) proposed that the Caldera de Taburiente was formed by large landslide events. They suggested that the debris avalanches at the bottom of the Caldera de Taburiente, the age of the lava flows at the top and within the caldera, and the absence of materials from the caldera wall, between the Basal Complex and the Bejenado Massif, a volcano that fills the intersection of the Cumbre Nueva and Taburiente Calderas (Figure 3), supports such an origin. Bejenado Volcano has been dated by Ancochea et al. (1994) as being ~0.70–0.71 Ma. In contrast, Carracedo (1999) speculated that Bejenado Volcano began to form soon after the formation of the Caldera de Taburiente about 560 Ka and ended 400 Ka. Guillou et al. (1998) proposed that the Caldera de Taburiente was formed from fluvial erosion by a drainage system trapped between the growing Bejenado volcano and the 558 Ka Cumbre Nueva collapse scar. Carracedo et al. (2001) speculated further that the morphology of the Caldera de Taburiente was formed by series of events: (1). Gravitational collapse of the Cumbre Nueva Volcano, (2) Development of the Bejenado volcano in the collapse structure, and (3) Incision of the Barranco de Las Angustias to form the initial Caldera de Taburiente.

Cumbre Vieja Rift: Onshore section

Dominating the southern third of La Palma is the north-south trending Cumbre Vieja rift whose construction took place in the last 123 Ka (Figure 3) (Carracedo et al., 2001). Day et al. (1999) inferred that the Cumbre Vieja Ridge volcanic rocks rest on the collapse scar formed by the failure of the western flank of the Cumbre Nueva rift about 560 Ka. Some authors have inferred that the Cumbre Vieja is a single rift somehow associated with the older volcanoes in northern La Palma (Ancochea et al., 1994). Others workers, (Carracedo (1994) and Carracedo et al. (1999a, b), have proposed that the construction of the Cumbre Vieja Volcano was controlled by a triple rift system, a north-south and northeast and northwest rift system that was re-organized into a single north-south rift at 7 Ka.

During the eruptive cycles the 24 km long and 17 km wide Cumbre Vieja Volcano grew into a steep sided ridge with side slopes of 300–400 m/km and a crest nearly 2000 m above sea level. Its west slope is partially covered by basaltic streams originating upslope, some of which date to historical times and its eastern side is strewn with cinder cones extending to the coast (Hausen, 1969). At its southern extremity is the 1860 m high El Cabrito lava dome from where the ridge descends to sea level and a field of cones (Hausen, 1969). According to Carracedo et al. (1999a, b) a Pleistocene glacially induced regression led to the formation of seacliffs on the western flank of the rift that in places are as high as 700 m. During the subsequent Holocene rise in sea level, the cliffs were partially buried between 20 to 15 Ka by scree-forming lavas. We question such a scenario, as seacliffs are indicative of transgressions, not regressions and in fact they are one of the features that are used to distinguish shores of submergence from ones of emergence. Thus it seems more geologically reasonable to infer that the cliffs were formed during a Pleistocene transgressive cycle and the lavas cascading over the cliffs were emplaced during a subsequent regression.

The Cumbre Vieja Ridge appears to be unstable and in state of collapse. During the 1949 eruptions, for example, normal fault ruptures developed along its crest (Klügel, 1997). Day et al. (1999), Moss et al. (1999) and Ward and Day (2001) infer that these ruptures are not the surface expression of a dike, but may represent surfaces of separation along which Cumbre Vieja volcano may collapse westward in the near future. Ward and Day (2001) inferred that this collapse could occur in historical time.

Cumbre Vieja Rift: Offshore section

The offshore ridge on strike with the onshore Cumbre Vieja Ridge is made up of two topographic segments. From the coast to a depth of about 800–110 m the offshore extension of Cumbre Vieja Ridge takes the form of a 15 km long platform that terminates along a 300 m high scarp. The scarp has a gradient of 200 m/km and probably represents a lava flow front overlapping the main part of the offshore ridge. In the absence of chronological data we have inferred from morphologic data that the ridge segment south of the platform is an older, lower part of the Cumbre Vieja rift. If so then the Cumbre Vieja rift may consist of a series of volcanic sequences overlapping one another in a southward direction. Such a model presupposes that rifting associated with the Cumbre Vieja propagated southward, first along the Cumbre Nueva Volcano and further south on oceanic crust. However if Day et al. (1999) contention that the Cumbre Nueva extends in the subsurface as far south as the southern tip of La Palma is correct, then it is not geologically unreasonable to suggest that the section of the offshore rift south of the platform is part of Cumbre Nueva, not the Cumbre Vieja Volcano. If so then the seaward edge of the platform is the southern end of the Cumbre Vieja rift resting on the older rift. That the Cumbre Nueva rift is much longer than the Cumbre Vieja also makes sense geologically as construction of Cumbre Nueva rift took place since of 330 Ka, whereas the Cumbre Viejo rift was constructed during the last 123 Ka (Carracedo et al., 2001).

We assume that individual eruption cycles are superimposed on one another. This implies that the main offshore extension of the Cumbre Vieja Ridge, an older segment of the ridge, is about 38 km long and has a maximum width of about 27 km, with its crest plunging 800 m southeastward for a distance of about 15 km. It is convex westward in plan view (Figures 4, 6 and 7). Its morphology appears to be the creation of mass wasting events and volcanic construction along its crest and northeast, south, and southwest flanks. The east flank of Cumbre Vieja Ridge has a topographic gradient ranging from 300 m/km between 1300 to 1700 m water depth, 350 m/km from 1800 to 2400 m and 300 m/km from 2600 to 3200 m. Its west flank has a gradient that ranges from 300 m/km from 1600 to 1900 m water depth and 200 m/km at depths of 1900 to 2100 m. Its southeast flank has a relief of 1200 m and a gradient of 500 m/km.

Along the crest of Cumbre Vieja are 21 volcanic cones (Figures 5, 7 and 8) ranging in relief from 25 to 300 m and surrounded by aprons suggestive of lava flows. The volcanic structures are conical in shape, with diameters of about 1 km to 100 m. The crest of the ridge is also disrupted by a 200 m high step at a water depth of 1900–2100 m (Figure 8) that we interpret as a lava front. Southeast of this step are two volcanic edifices separated by a saddle whose relief is about 50 m. The high south of the saddle has a tail-like structure on its west side that resembles a lava flow. To the east of this high is another volcanic edifice whose construction has led to a pronounced protrusion of the Cumbre Vieja Ridge's eastern flank.

The flanks of the ridge have the appearance of being mass wasted, that is, characterized by embayments separated by narrow divides and volcanic lobes. On the ridge's east side northeast progradation of one of these volcanic lobes led to the construction of an amphitheater low constrained to the north by one of the north-south trending ridges. The relatively steep southeast side of the Cumbre Vieja Ridge resembles a lava front whose slope may have been enhanced by mass wasting. Two short ridges extending from the southwest of the front, one 75 m high trending south and the other, 300 m high trending southwest. Near the southeast edge of the 300 m tall ridge is another southeast-trending high made up seamounts with heights ranging between 100 to 300 m. East of the ridge are three widely spaced seamounts forming a chain extending to the northwest tip of the La Gomera's margin (Figures 4, 7 and 8).

El Hierro

General geology

El Hierro is the youngest of the Canary Islands and may be located over the present site of the hotspot (Holik et al., 1991; Hoernle et al., 1991). However, Carracedo et al. (2001) inferred that volcanism in La Palma and El Hierro alternated between them, a scenario that may be due to changes in the stress fields triggered by gravitational collapses at the peak of volcanic construction. Carracedo et al. (2001) also concluded that the volcanic shields in El Hierro overlap concentrically, whereas those in La Palma are aligned in a north-south direction as independent shield volcanoes. As a consequence of this large overlap El Hierro formed over a stationary magma source characterized by a well defined rift system.

El Hierro has an area of 273 km^2, a maximum relief of 1500 m above sea level at Mal Paso (Figure 3), and displays a few barrancos (Hausen, 1973; Fuster et al., 1993). In plan view the island has three lobes whose intersection forms a central tableland on which are superimposed numerous extinct cinder cones (Figure 3). The island's periphery consists of steep cliffs indented by three embayments (Figures 3 and 8): El Golfo in the island's northwestern sector, Las Playas opening to the southeast, and El Julan opening to the southwest. These cliff-bound embayments resulted from large mass wasting events (Ridley, 1971; Hausen, 1973; Bravo, 1982; Holcomb and Searle, 1991; Carracedo, 1994; 1996).

El Hierro was built during the Tiñor and El Golfo volcanic phases. The Tiñor volcano at the eastern part of the island was built between about 1.12 and 0.882 Ma and El Golfo from about 550 to 130 Ka (Guillou et al., 1996; Carracedo et al., 2001). The Intermediate and Holocene rift volcanic series are dated at 158-76 and < 50 Ka respectively (Fuster et al., 1993; Guillou et al., 1996). During the collapse of Tiñor, about 882 Ka, more than half of the northwest flank of the edifice was removed (Carracedo et al., 1999a, b). This collapse was followed by the construction of El Golfo Volcano, which filled the depression between 550-130 Ka (Guillou et al., 1996) or 545 to 126 Ka (Carracedo et al., 2001). Carracedo et al. (2001) also inferred that the construction of El Golfo coincided with the maximum development of the Cumbre Nueva rift zone on La Palma.

The collapse of El Golfo Volcano gave rise to the El Golfo Embayment (Figure 9). This embayment has 1400 m high walls partially covered by younger lavas that run down to the sea (Hausen, 1973). Offshore is the El Golfo Debris Avalanche (Figure 10). The age of the collapse of the El Golfo Volcano is yet to be resolved. Masson (1996) and Urgeles et al. (1997) have proposed that the collapse took place between 9–15 Ka or 10–17 Ka. Carracedo et al. (1999b) suggested that the failure occurred much earlier, ~130 Ka.

Holcomb and Searle (1991) interpreted the El Julan Embayment as a collapse structure associated with an event that Masson (1996) suggested took place about 500 to 300 Ka. Carracedo et al. (1999a, b) proposed that failure took place when El Golfo Volcano was well developed at 130 Ka. More recently Carracedo et al. (2001) argued the presence of lavas belonging to the rift volcanism in the water galleries in the Julan Embayment places the minimum age of its collapse about 150 Ka.

Masson et al. (2002) suggested that Las Playas Embayment is not the result of a landslide, but of massive slumping along the San Andrés Fault system (Figure 10) between 545 and 261–176 Ka. Although inactive, the lateral collapse structure is constrained to have formed between the last emissions of El Golfo volcano at about 176 Ka and the rift lavas cascading down the collapse scarp dated at 145 Ka (Guillou et al., 1996).

The ages of the offshore extensions of the rift zones of El Hierro are determined from the debris avalanches constrained by them. We assume that Carracedo (1999a, b) is correct in stating that the El Golfo and El Julan failures took place about 130 Ka and that the Playas I and II slumps and debris avalanches took place at 546–178 Ka and between 176 and 145 Ka respectively (Masson et al., 2002). These avalanches appear to onlap the offshore rift extensions and do not extend beneath them. This then implies that the submarine sections of the rift zones must have been in existence by at least 130 Ka and possibly as early as 546 Ka or even earlier. Such ages suggest that the rift zones originated during the formation of Tiñor Volcano at 1.12–0.882 Ma, and were subsequently rejuvenated during the construction of El Golfo edifice 550 to 130 Ka.

Tiñor and El Golfo Volcanoes may have been built under different stress regimes, with the older one affecting the gravitational stress field of the younger one. Processes associated with the collapse of Tiñor and post-collapse erosional events also possibly affected the morphology of the submarine rifts. Since rift zone volcanism continued after the collapse of El Golfo at 130 Ka, the main magma chamber associated with Tiñor and El Golfo Volcanoes must continue to persist and marks the location of the supposed Canary Island hotspot.

The greatest concentration of eruptive vents in the Canary Islands is observed on the central tableland formed by the intersection of the three rift zones on El Hierro. All of the cones consist of loose volcanoclastic material. There are no obvious set of fractures that controlled the sites of extrusion. The cones are found at different elevations, and most are of Pleistocene age with only one of them having been active in historical time (Hernández Pacheco, 1982). Carracedo et al. (2001) dated the building of the cone at 145 Ka to 2.5 Ka with some activity extending to historical time. They further suggested that this period of volcanism is coeval with the maximum development of the Cumbre Vieja rift zone in La Palma.

The Quaternary volcanic cones are on average 100 m high, have flanks that are furrowed by erosion, and lavas and tuffs that are covered by small bushes or have been converted into tilled fields. Scattered over the tableland are a few cones free of vegetation and a few young cones such as Montaña de Tenerife and Mal Paso (Figure 3). Most of the latter cone, located along the scarp of El Golfo Embayment is missing, having slid into deep water (Hausen, 1973).

Northeast and Northwest Rifts: Onshore sections

The central volcanic highland developed at the intersection of the three rift zones, descends westward to a plain inclined to the north. This plain represents the flank of the former El Golfo Volcano. It is covered by lavas originating from cones in the central upland to the south and is cut by the 1400 m high cliffs (Figure 3) on the west. The east slope of the highland is divided into steps that are covered with cones and descend to the sea. According to Hausen (1973) the concentration of cones on the 12 km long by 7 km wide northeast rift attests to the intensity of eruptions along this rift. Most of the cones are Quaternary and have undergone erosion so that relief of the region is a composite of volcanism, erosion and faulting. Hausen (1973) speculated that there is evidence of more recent volcanism along the northeast rift. Among these recent volcanic edifices is Tesoro de Tamaduste (Figure 3) located along the north coast. Lavas from this double-crater volcano, open to the north, cascaded down the seacliff covering a platform at the foot of the cliff. Another young volcano is Solimán (Figure 3) found along the crest of the rift. A carbonized pine chunk enclosed in the lavas from on the crest of the rift yielded an age of 2900 ± 130 years B.P. Guillou et al. (1996) obtained an age of 2500 ± 70 years B.P. for another volcano along the crest stated that there are many vents of similar and younger ages in the rifts of El Hierro.

The southern side of the 10 km long x 4 km wide northwest rift is composed of ash, lapilli, and young lavas that flowed south from vents located in the central tableland. These flows extended to the shoreline, accounting for the smooth slopes of El Julan Embayment. The absence of barrancos and weathering indicate that the lavas of El Julan are young (Hausen, 1973). According to Carracedo et al. (2001) the lava flows filling the El Julan Embayment have been dated between 41 and 31 Ka. West of the ash-covered southern slope of the northwest rift is an extensive field of volcanic cones of Quaternary and Holocene age extending from the crest of the central tableland to the rocky west coast of El Hierro. Between the cones are lavas, the youngest of which are to the west. The rift slopes down to the west coast along a series of steps of older lava sheets in the upper steps and younger lavas and cones in the lower ones. Hausen (1973) stated that a cone near the coast looks so fresh that it must be recent; possibly from the time the Spaniards came to the region in the 15th century. A southeast valley in the region of this cone is filled with recent lavas and spatter cones whose outflows have caused progadation of the west coast. At the northwest corner of the northwest rift is the western end of the cliff of the El Golfo Embayment. The cliff is composed of lavas erupted along fractures at its crest. Here the cliff is 200 m high and lavas from young cones along its crest have cascaded down the cliff to the terrace at its foot. One of the cones along the top of the cliff has been so eroded by mass wasting that only the cone's lava-filled conduit has survived. The southwestern peninsula of El Hierro descends toward the coast and on reaching it is broken into a steep cliff. Hausen (1973) speculated that this cliff was formed prior to the construction of the volcanic cones scattered throughout the region. Also present in the region are boulders displaced from the adjacent central tableland.

Northeast and Northwest Rifts: Offshore sections

The offshore extensions of the Northeast and Northwest Rifts display a different morphology from the submarine South Rift and the submarine rift off southern La Palma. The Northeast and Northwest Rifts are lobate in plan view compared to the others, which are well defined ridges (Figures 9 and 10). They both widen down slope, becoming more diffuse with distance from the shoreline.

The submarine section of the Northeast Rift is about 26 km wide and 15 km long and extends to a water depth of 3100 m. It has an average along-axis topographic gradient of about 400 m/km compared to the onshore section that has a gradient of half that, 200 m/km (Figure 11). Local steeper gradients range from 600 m/km to 300 m/km and occur in a series of steps that extend down to about 2600 m water depth. Scattered over the rift surface are 30 volcanic cones whose diameters range from 300 m to 3 km and heights range from 25 to 250 m. Pinnacles, less than 50 m wide and up to 150 m in height, also occur along the rift. According to Gee et al. (2001) these pinnacles, which are sometimes in groups of two or three elongated or aligned down the rift, represent dike injections radiating from a central volcano. Wide expanses of sea floor lacking any topographic irregularities are also observed in the submarine rift (Figure 9). In the absence of bottom photographs or side-scan sonar it is impossible to discern the nature of this terrain, whether it represents a sedimented seafloor or lava flows. However, its embayment-like morphology at their heads suggest that their form is due to gravitational mass-wasting processes.

The flank of the submarine Northeast Rift, adjacent to the El Golfo Debris Avalanche, is linear and has the appearance of a multi-peaked, narrow, steep-sided ridge with a flank gradient of 400 m/km. In contrast the contact between the Northeast Rift and Las Playas Debris Avalanche is more discontinuous, with one ridge section extending from a water depth of 600 to 2200 m, separated by a gap from another ridge section offset to the north that extends from a depth of 2200 to 3100 m. The rift lacks a well defined front and instead takes the form of a series of lobes. Most of them appear to represent lava flows similar to the lava terraces seen along the flanks of Hawaiian rift zones (Smith and Cann, 1999). Others may be debris flows associated with landslides.

The submarine Northwest Rift zone is 26 km wide and 16 km long with its distal end located at a depth of 3100 m. Beyond this front is a several-kilometer-wide apron containing what we infer to be a few volcanic cones. The along-axis slope of the offshore extension of the Northwest Rift has a gradient ranging from 300 m/km near the coast, a gradient of 200m/km in its mid-section, and a gradient of 200–300 m/km at its distal end. Its onshore segment has a gradient of only 200 m/km. The flank of the submarine northwest rift bordering the El Golfo Debris Avalanche has a gradient of 300 m/km, whereas its gradient toward El Julan Debris Avalanche (Figure 10) to the south is much gentler with a gradient of 200 m/km. Twenty five volcanic cones ranging in diameter from 0.5 to 1.8 km and ranging in height from 50 to 275 m have formed within the rift zone. As in the submarine Northeast Rift, the volcanic edifices string together to form irregular lineations downslope. Most of the edifices are concentrated on the north side of the rift, off the field of recent lavas at the northwest tip of the island. As mentioned, Hausen (1973) inferred that the lavas on land were probably erupted along fractures associated with the western end of the cliff located along the south side of El Golfo Embayment. The lobes associated with the volcanic edifices can be traced to a water depth of 3500 m (Figure 9).

The topography of the southern side of the submarine rift is smoother than its northern side and has fewer cones and those that are present are smaller in size. The north side of the rift is characterized by northwest-trending lineaments capped by volcanic cones hundreds of meters high. A west-trending, poorly defined landslide that has a topographic expression to a water depth of 3100 m separates the terrains along the north and south sides of the rift. The distal edge of this side of the rift is marked by flat-topped lobes trending southwestward ending at a scarp with a gradient of 300 m/km.

South Rift: Onshore section

The 6 km wide by 8 km long subaerial South Rift is dominated by the Restinga Peninsula, a broad highland whose elevation decreases southward. The Las Playas Scarp forms the eastern edge of the peninsula and the El Julan Scarp marks the western edge (Figure 3). The peninsula is covered by cinder cones and lava fields (Hausen, 1973). The volcanic features vary in age, with some being weathered and in part forested, whereas others have a fresh appearance with black lava and ash fields. The northern part of the Restinga Peninsula is overgrown with bushes and tilled fields, but farther south the peninsula becomes more desert-like and still farther south it is covered by barren fresh lava fields. To the west the lavas have reached the coast, and to the east eruptions occurred before the final formation of the Las Playas Scarp (Hausen, 1973). Apparently the peninsula was created by repeated lava flows from the north. Hausen (1973) speculated that the numerous volcanic extrusions were a consequence of fractures that served as passageways for the emerging magmas.

South Rift: Offshore section

The ridge-like submarine South Rift, also known as the Southern Ridge, curves southwestward from its mid-section outwards and is asymmetrical in cross-section with its steeper side facing southeast (Figure 11). The width of the rift ranges from 3 km on its proximal end to 18 km at its distal end. Gee et al. (2001) described the crest of the South Rift as a remnant dike swarm. According to Gee et al. the tip of the south rift zone at 27°20′ N is over 1 km high, its flanks have slopes of over 600 m/km and its base is at a depth of 3700 m. Morphologically it resembles the southern tip of the submarine Cumbre Vieja Rift off La Palma.

A saddle near 27°30′ N divides the South Rift in two. The ridge section north of the saddle consists of a sequence of narrow volcanic lobes trending southeast to southwest that extend to a depth of 2500 m (Figure 12). These lobes have prograded over a probable lava terrace exposed on the southeast edge of the north segment. This terrace probably represents an older rift construction comparable in age to the southern segment, whereas the lobes are recent lava flows, probably offshore extensions of the young lava flows at the southern tip of the Restinga Peninsula. Both of these flow units superimposed one upon another are probably connected with the same source in El Hierro. Southeast of the lava terrace is a southeast trending apron whose base is at a water depth of 3200 m. Extending southwestward from the southern edge of the lava terrace is a several kilometer wide ridge, with a minimum depth of less than 1500 m.

The east-west trending, 400 m wide saddle at 27°30′ S is flanked on the east side by a smooth slope with a gradient of 500 m/km and whose base is at a depth of 3200–3400 m. On its western side the saddle is blocked by a 6 × 3 km wide terrace whose western side is gullied. South of the saddle is a 1.0–1.5 km wide ridge whose flank gradients are on the order of 400–500 m/km. The western side of the ridge is irregular with a <1 km wide east-trending ridge that can be traced from the west flank of the South Rift at a depth of 1700 m to a depth of 2600 m, over a distance of 3 km. The eastern side of the submarine South Rift is cut by gullies about 1.5 km long and 75 m wide.

Hawaiian Islands

General geology of the Island of Hawaii

The dimensions of Hawaii are roughly 127 km by 148 km; it has a surface area of about 10,000 km^2, and according to Moore and Clague (1992) the island has grown at a rate of 0.02 km^2/yr in the past 600 k.y. The Island of Hawaii consists of five coalescing volcanoes that in chronological order are: Kohala at the northwest tip of the island that completed its shield phase 245 Ka, Hualalai along the west coast which completed its shield phase 130 Ka, Mauna Kea southeast of Kohala which completed its shield phase 130 Ka, Mauna Loa slightly south of the center of the island which rose above sea level about 300 Ka and Kilauea located east of Mauna Loa which rose above sea level about 200 Ka (Moore and Clague, 1992). Hualalai, Mauna Loa and Kilauea volcanoes are historically active.

The rift zones on the Island of Hawaii (both on- and offshore components) have been well studied compared to these of the Canary Islands. Therefore, more types of data and information exist for the Hawaii rifts and more rigorous conclusions can be made about the nature of the morphologic features we observe. The locations of the rift zones (Figures 2 and 13) associated with the five volcanoes that make up Hawaii are controlled by gravitational stresses within each individual volcano, which in part are controlled by the morphologies and positions of pre-existing volcanoes (Fiske and Jackson, 1972). The characteristics of individual rift zones are variable since they are influenced by the distribution of neighboring volcanoes (Fornari, 1986). For example, the Southwest Rift Zone of Kilauea is sandwiched between Loihi Seamount and Mauna Loa's Southwest Rift Zone. It is only 32 km long on land and poorly developed offshore (Holcomb, 1987). As another example, the 55 km long subaerial portion of the East Rift Zone of Kilauea is buttressed to the north by Mauna Loa's flank, but free to spread to the south. Thus intrusion of magma into Kilauea's currently active East Rift Zone (ERZ) is leading to slumping and displacement of the south flank of the island towards the sea (Swanson et al., 1976; Morgan et al., 2000; Hills et al., 2002 and references there in).

As in the Canary Islands, the subaerial rift zones in Hawaii have seaward extensions forming salients extending from the shoreline to the base of the volcano at abyssal depths. The offshore sections of the rifts can vary in length: 55% of the 130 km long ERZ is below sea level; 61% of the 107 km long south rift of Kohala is submerged; ~28% of the 108 km long south rift of Mauna Loa also is below sea level (Moore and Chadwick, 1995). As in the Canary Islands the longitudinal and lateral slopes of the submarine extensions of the rift zones are steeper than their subaerial counterparts (Figures 6, 11 and 14).

Along the crest of the onshore portions of the rifts, volcanic cones are commonly built and then eroded by subsequent eruptions. Lavas transported down the flanks of the subaerial rift zones form smooth,

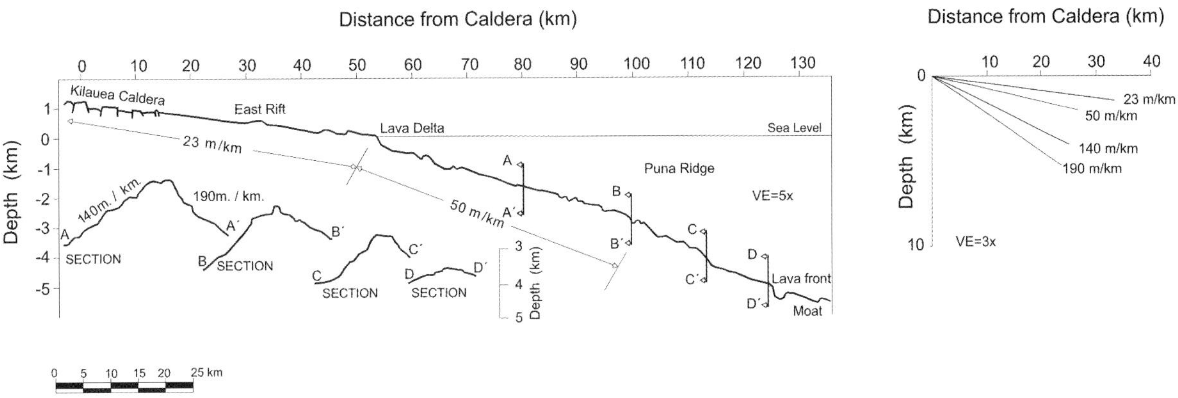

Figure 14. Profiles of Kilauea's subaerial East Rift and its seaward extension, the Puna Ridge. Vertical exaggeration 5X. Modified from Lonsdale (1989). Also included is a diagram showing the gradients displayed by the rift at a vertical exaggeration of 3X, an exaggeration similar to that of the profiles f the rifts of La Palma and El Hierro in Figures 6 and 11.

low-angle slopes, while lavas that cross the shoreline are believed to feed submarine debris flows (e.g., Moore et al., 1973). By contrast, the crest and flanks of the submarine extensions are covered by cones and terraces that have diameters of up to 1 km or more and sides up to several hundreds of meters high (Clague et al., 2002; Smith et al., 2002b). Many of the cones have craters in their tops that are 100s of meters in diameter.

Kilauea

East Rift Zone: Onshore section

Kilauea's rift zone system is one of the best studied in the world (Tilling and Dvorak, 1993, and references therein). The volcano is fed from a central magma chamber (or system of magma conduits) beneath the summit (Ryan et al., 1981; Ryan, 1988). Lava is erupted at the summit and/or one of the volcano's two rift zones, the Southwest Rift Zone (SWRZ) and the ERZ. Typically, the onset of a rift zone eruption is marked by seismicity that migrates from the summit region down a rift zone to the site of eruption. The early phase of eruption is normally through a fissure as long as several hundreds of meters (e.g., Klein et al., 1987; Wolfe et al., 1987). If fissure eruptions persist, they normally become confined to a single vent.

Since 1983, eruptions have occurred continuously along the ERZ centered at either the Pu'u 'O'O or Kupaianaha vents with more than 1 km^3 of lava being erupted since 1983 (Wolfe et al., 1987; Mangan et al., 1995). Surface deformation associated with seismic activity (Pollard et al., 1983), the fissure eruptions, and the observation of dikes within the eroded cores of Hawaiian volcanoes (Walker, 1988) indicate that rift zone eruptions are dike-fed and that the seismic activity is associated with magma moving through the underlying magma conduit system (Rubin and Gillard, 1998). The zone of eruptive fissures ranges in width from 1.5 to 3.0 km (Holcomb, 1987; Moore and Trusdell, 1991).

The ERZ of Kilauea extending from the southeast side of the summit caldera trends southeastward for ~15 km and then changes its trend towards the northeast (N65°E), an orientation that it maintains to the shoreline. The along-axis gradient is about 23 m/km, a uniform gradient that led Lonsdale (1989) to infer that as a result of high eruption rates the rift has been able to maintain a more or less constant, gentle profile. Superimposed on the long-axis profile are secondary features such as pit craters, cones, spatter ridges, low mounds, and phreatomagmatic ash near the rift's axial zone. Parfitt et al. (2002) classified eruptions along the ERZ into five categories: short-lived fissure eruptions that last less than five days and often less than one day, consisting of linear spatter ramparts and lava flows; longer-lived fissure eruptions of ~5 to 15 days duration, consisting of linear spatter ramparts and lava flows; fissure eruptions evolving into a central vent; episodic eruptions; long-lived steady state eruptions. They ascribe these different eruption types to differing thermal and driving pressure behavior in the feeder dikes. Epp et al. (1983) proposed that ERZ eruptions tend to drain the summit magma reservoir to pressure levels corresponding to the elevations of the eruptive vents.

East Rift Zone: Offshore section

The 70 km long Puna Ridge is the submarine extension of the onshore ERZ (Figures 15 16 and 17). Unlike

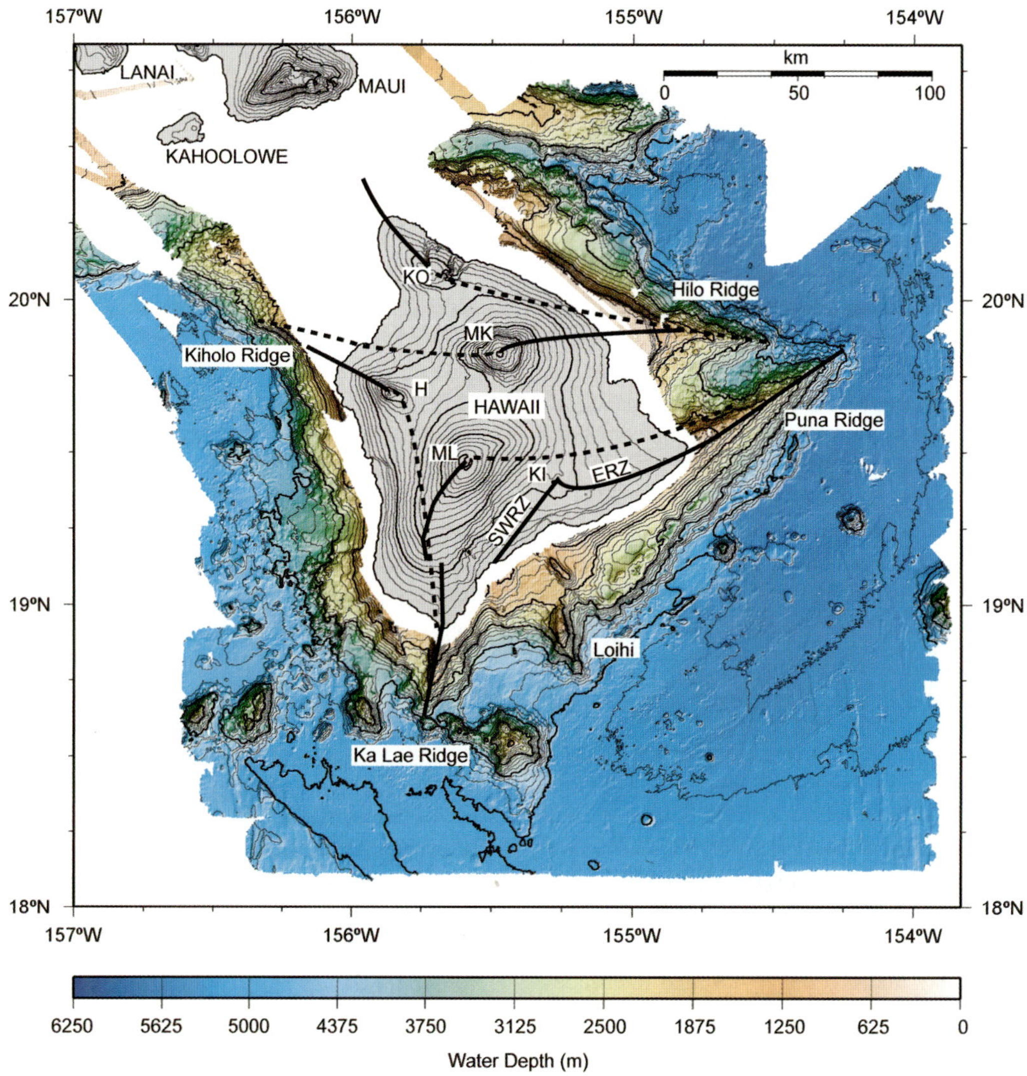

Figure 15. Topographic/bathymetric map of Hawaii and the surrounding seafloor, modified from Smith et al. (2002b). Volcanoes are labeled as: KO-Kohala; H-Hualalai; MK-Mauna Kea; ML-Mauna Loa; KI-Kilauea. Solid lines: distribution of rift zones from Fornari (1986). Dashed lines: distribution of rift zones from Holcomb et al. (2000). The East Rift Zone (ERZ) and Southwest Rift Zone (SWRZ) of Kilauea are labeled. Offshore extensions of the rift zones also are labeled. Loihi, the youngest Hawaiian volcano is located to the south of Kilauea.

the submarine extensions of the rift zones off the Canary Islands, the submarine rift zones of Hawaii have recently been the site of several high-resolution studies, and we are much more confident about the origin of submarine features here than at the submarine rift zones off shore La Palma and El Hierro. Multibeam bathymetry data for the entire Puna Ridge (Clague et al., 1994; Monterey Bay Aquarium Research Institute (MBARI), 2000), along with high resolution side-scan sonar images (Lonsdale, 1989; Smith et al., 2002a), photographic imagery (Moore and Fiske, 1969; Clague et al., 1988; Lonsdale, 1989; Smith et al., 2002a; Parfitt et al., 2002; Gregg and Smith, 2003), submersible dive observations (Fornari et al., 1978; Clague et al., 2000; Johnson et al., 2002), and GLORIA side-scan sonar data (Holcomb et al., 1988), verify that the Puna Ridge crest is the location of dike intrusions and fissure eruptions. Sea-surface magnetic data show a normally polarized anomaly centered over the ridge axis, consistent with the presence of a 11 km wide, 70 km long, nearly-vertical magnetic source, presumably representing a dike complex at depth (Malahoff and McCoy, 1967). High-resolution magnetic data show regional highs in the crustal magnetization along the ridge axis, perhaps

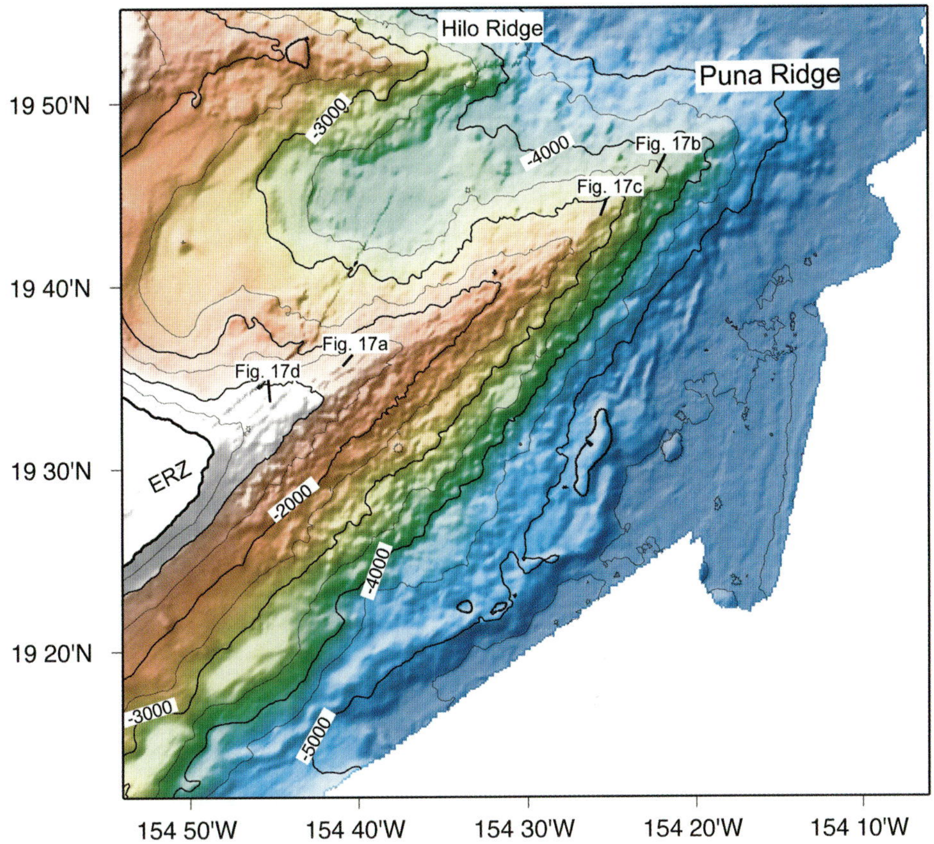

Figure 16. Color shaded relief bathymetry of the Puna and Hilo Ridges showing locations of the photomosaics in Figure 17. Contours in meters. The East Rift Zone (ERZ) is labeled. Modified from Smith et al. (2002a).

indicating areas of recent lava deposition (Smith et al., 2001).

Relatively few dike intrusions have been recorded from the summit of Kilauea to the Puna Ridge in historic time. Submarine lavas from the crest and flanks of the Puna Ridge appear to be older than most subaerial lavas. Holcomb (1987) estimated that 70% of the subaerial portion of Kilauea is younger than about 500 years. Based on palagonite thicknesses, Clague et al. (1995) estimated that dredged lavas from the Puna Ridge are 700–24000 years old, but mostly 2000–7000 years old. The most recent submarine eruptions are thought to have occurred in 1790, 1884 and 1924 (Stearns and Macdonald, 1946). The 1884 eruption was witnessed just offshore at 20 m water depth. In 1790 and 1924, magma withdrawal from the summit reservoir is inferred to have fed submarine eruptions on the Puna Ridge (Stearns and Macdonald, 1946). There has been a lack of seismicity along the ridge since 1960, suggesting recent inactivity (Klein et al., 1987).

The average along-axis slope of the Puna Ridge is more than three times as steep as that of the onshore section (~73 m/km vs. 23 m/km, respectively) (Figure 14).The flank slopes of the Puna Ridge range between 160–275 m/km (Fornari et al., 1978; Smith and Cann, 1999), also about three times steeper than on the onshore ERZ (~51 m/km). The slopes on the south flank of the Puna Ridge do not change significantly with distance from the shoreline until the crest is below 3500 m water depth. The slopes are steepest on the north flank between about 1200–2000 m water depth. Multibeam bathymetry data (e.g., Swanson et al., 1976; Moore and Chadwick, 1995) and high-resolution side-scan sonar imagery (Smith et al., 2001, 2002a) suggest a landslide scarp in this region.

Moore and Chadwick (1995) identified slump deposits at the southern base of the Puna Ridge. The deposits consist of anomalous terraced terrain that extends from about 4000 m water depth down to the base of the ridge where it meets the Hawaiian Deep at about 5500 m. They suggested that the terraces represent the

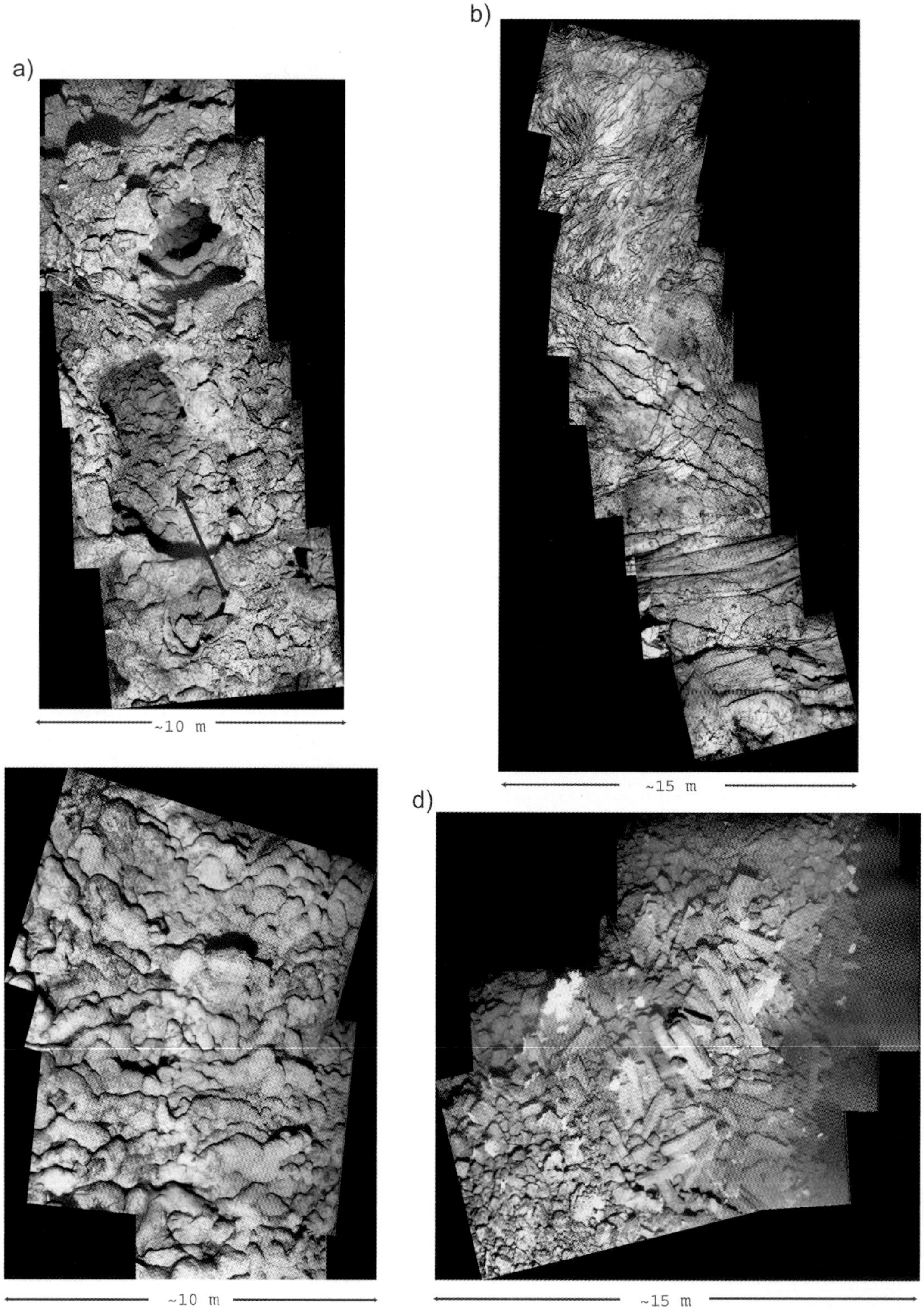

Figure 17. Photomosaics of bottom photographs obtained using a deeply towed photo-imagery system. Figure modified from Smith et al. (2002a). Locations of photomosaics are marked on Figure 16. (a) skylights in a lava tube, surface flows on top or a lava terrace. Black arrow marks a large skylight. (b) drapery folds in lava on the top of a terrace located at about 3300 m depth. The morphology is similar to that observed on the subaerial ERZ. (c) flanks of a pillow cone, covered with tubular pillows; downslope is to the right. (d) columnar joints that have tumbled out of a breach in the crater of a cone located at about 900 m water depth.

distal margin of a massive slump block that is now mostly covered by volcanic flows from the summit of the Puna Ridge. No slump deposits have been identified at the base of the north flank of the ridge (Moore and Chadwick, 1995) although they may be covered or modified by lavas from either the Puna Ridge or the submarine Hilo Ridge to the north (Moore and Chadwick, 1995).

From sea level to about 2100 m depth, the axis of the Puna Ridge trends 65° (Lonsdale, 1989), in line with the lower ERZ. The axis has a broad, 3-4 km wide crest, with a small (about 1 km) right-stepping offset near a water depth of 2000 m. Volcanic constructions, up to 140 m high, are scattered along the ridge crest, especially at the shallower end. From about 2100–3400 m depth, the axis of the Puna Ridge bends to 45° (Lonsdale, 1989), and the crest narrows. At about 2700 m depth, the axis has a right-stepping offset and the along-axis slope increases. Using a towed side-scan instrument, Lonsdale (1989) observed tightly clustered fissures and collapse pits concentrated in an inflated portion of the ridge above the offset, and suggested that intrusive rather than extrusive magmatism predominates here. Below 3400 m depth the ridge crest is poorly defined. The axis is composed of 'steps' that Lonsdale (1989) interpreted as flow fronts steepened by mass wasting.

The styles of volcanic features on the lateral slopes of the ERZ change significantly just as the crest of the rift dips below sea level. As mentioned above, lavas erupted from the onshore ERZ form smooth, low-angle slopes, except where interrupted by faults. Gently dipping low-relief lava flows mantle the slopes. In contrast, volcanic features on the flanks of the Puna Ridge are common and include flow fields, pillow ridges, cones, and large semicircular flat-topped features (terraces) that have diameters of 1 km or more and sides several hundreds of meters high (Smith et al., 2002a). These terraces, many with pits on their tops, often appear to form staircases, one on top of the next.

On a finer scale, three basic flow morphologies are observed: pillow, lobate, and sheet flows (Figure 17). All three are found in each of seven camera sites at the Puna Ridge at water depths ranging between 660–3700 m (Smith et al., 2002a). GLORIA images (Holcomb et al., 1988) and ground-truth sampling and photography (Clague et al., 1988) indicate that sheet flows cover large areas north and southeast of the ridge base.

Kilauea's Southwest flank

At Kilauea's ERZ diking is accommodated by seaward slipping of the south flank on a basal décollement (Dieterich, 1988; Delaney et al., 1990; 1993) that is located 7-9 km beneath the surface (Owen et al., 1995). The north flank of the ERZ is relatively stable and does not appear to respond to diking events, presumably because it is buttressed by Mauna Loa Volcano and/or because seaward displacement of the south flank is favored by the pre-existing gravitational stress field imposed by Mauna Loa (Swanson et al., 1976). The average "spreading rate" of the ERZ is estimated to be ~10 cm/yr, the rate that the south flank is slumping (Swanson et al., 1976). Between the SWRZ and ERZ of Kilauea are the Koae and Hilina Fault Systems. These fault systems are inferred to be associated with southward displacement of the south flank (e.g., Parfit and Peacock, 2001). Structurally the Hilina Fault System is comparable to the San Andrés Fault System between the South and Northeast rifts on El Hierro, Canary Islands. It is along this fault system that the Las Playas I and II slumps in the Las Playas Embayment became detached from the rest of the Northeast Rift.

Some workers have proposed that the southward displacement of the southeast flank of Kilauea is due to dike injection into the ERZ leading to compressive stresses in the rift zone flank that are relieved by seaward displacement (Duffield, 1975; Thurber and Gripp, 1988; Morgan et al., 2000). Others have inferred that the displacement of the southeast flank is due to gravitational spreading and forces exerted deep within the volcano by the accumulation of magma (Denlinger and Okubo, 1995). There is likely some balance reached between the horizontal stress generated by dike intrusion, the weight of the flank, and the friction on the basal décollement to produce the landslides and slumps that are observed.

Southwest Rift Zone

Kilauea's SWRZ extends 32 km from the summit caldera to the shoreline. It has been suggested that the SWRZ acts as a headwall for the south flank slump. The SWRZ is <1.5 km wide in its northern section. The rift zone changes trend from 230° to 195° ~15–20 km from the summit caldera. The southern section of the SWRZ is dominated by the Great Crack, a 15 km long pre-1790 AD fissure and graben system, and the site of an 1823 eruption (Holcomb, 1987).

West of the upper end of the crack are two 30 m high normal fault scarps (Puu Nahaha Faults) mantled by lavas as young as 500 years old. South of the Nahaha Faults the rift zone widens to at least 5 km. According to Dzurisin et al. (1984) only 10 % of the magma supply to the summit of Kilauea from July 1956 to April 1983 was injected into the SWRZ, 55% was intruded into the ERZ, and 35% was erupted at the summit. The submarine extension of the SWRZ is poorly defined.

Mauna Loa

The Onshore and Offshore rift sections

Mauna Loa is an active volcano on the island of Hawaii. Mauna Loa rose above sea level about 300 Ka and covers the largest area of the island, about 5000 km^2 (Moore and Chadwick, 1995). At present Mauna Loa is erupting a greater portion of lava above sea level than on its submarine sections. Mauna Loa has two rift zones: the Southwest Rift Zone (SWRZ) and the Northeast Rift Zone (NERZ) (Figures 13 and 14). The 65 km long subaerial SWRZ extends another 35 km offshore (García et al., 1995). The submarine section of the SWRZ is called Ka Lae Ridge (Holcomb et al., 2000). The NERZ does not reach the shoreline, or has been covered by Kilauea lavas (e.g., Holcomb et al., 2000). Here we concentrate on the SWRZ.

The subaerial section of Mauna Loa's SWRZ is disrupted by a 40° change in trend near an elevation of 2400 m. This may be due to the interaction of the rift zone with the growing Kilauea edifice, which has inhibited the inflation of Mauna Loa (Lipman, 1980). Presumably, the presence of Kilauea prevents slumping of the southeast flank of Mauna Loa, causing the SWRZ to migrate westward, and leading to the bend in the middle section of the rift. South of the bend, the rift zone is defined by the fault line scarp Kahuku Pali. The fault has a relief of about 170 m subaerially and it increases dramatically to about 1800 m of relief offshore (Fornari et al., 1979). It extends offshore for about 50 km (Fornari, 1986). The fault cuts through the Ka Lae Ridge, which was created from the intrusion and eruption of dikes along the SWRZ. Submersible observations coupled with bathymetric data show that the top of the Ka Lae Ridge does not have any volcanic cones; this is unusual for submarine rift zones of Hawaii (Clague et al., 2000). In order to explain this observation and the characteristics of the magnetic dipole over the submarine ridge, García et al. (1995) have suggested that the scarp did not cut the ridge at its axis, but rather the scarp cut the eastern flank of the Ka Lae Ridge and its axis is offset to the west. Recently, Holcomb et al. (2000) suggested that Ka Lae Ridge is part of Hualalai Volcano rather than Mauna Loa, although this remains to be proven.

Kohala

The Offshore section of the Southeast Rift

The Kohala Volcano, at the northwest tip of Hawaii, completed its shield phase 245 Ka (Moore and Clague, 1992). The deeper sections of the submarine Hilo Ridge, extending from the east coast of Hawaii east of Mauna Kea, have recently been hypothesized to be the continuation of the southeast rift zone of Kohala (Holcomb et al., 2000; Kauahikaua et al., 2000). Previously, the entire Hilo Ridge had been interpreted as the extension of the east rift zone of Mauna Kea (e.g., Moore and Clague, 1992).

The volcanic morphology of the 60 km long Hilo Ridge varies along its axis. The proximal part of the ridge (out to about 20 km from the shoreline) is dominated by two submerged terraces. The leading edge of the shallow terrace is at a depth of about 400 m; the leading edge of the deeper terrace is at about 1100 m. The shallow terrace is well known and thought to be associated with Mauna Kea (Moore and Clague, 1992). The deeper terrace is thought to be related to a terrace at similar depths that extends around the northern section of Kohala Volcano and that we speculate to document the end of Kohala's shield phase at about 245 Ka.

If this correlation is correct then the segment of Hilo Ridge deeper than 1100 m is older than Mauna Kea (Holcomb et al., 2000). Such a correlation is supported by the isotopic composition of basalts from the terrace that are distinct from those from Mauna Kea, but are similar to those from Kohala. This conclusion also is supported by three-dimensional gravity modeling that shows a gravity high extending from Kohala, across the northeast slope of Mauna Kea to Hilo Ridge (Kauahikaua et al., 2000). The morphology of Hilo Ridge deeper than 1100 m is very similar to that of the Puna Ridge. Flat-topped volcanoes and terraces mark its crest and flanks (Clague et al., 2000). The average along-axis slope for the ridge deeper than 1100 m is 60 m/km. The flanks have slopes of ~200 m/km. Both of these values are similar to those at the Puna Ridge. It seems that both submarine ridges were built by the same means: dike injection and eruption in a narrow axial zone to build a distinctive volcanic ridge.

Discussion

Studies of the Hawaiian Islands have led to a clearer understanding of how oceanic islands evolve. These investigations have demonstrated that oceanic islands are constructed not only by summit eruptions from shallow magma reservoirs, but by dike intrusions and fissure eruptions along well-defined rift zones. Most oceanic island volcanoes have radial rift zones of varying lengths, some attaining lengths of over 100 km. Rift zones may develop along fractures resulting from the stress field generated by an upwelling plume (Wyss, 1980). Rift zone locations also are strongly influenced by the gravitational stress fields created within an individual growing volcano modified by the stress fields of neighboring volcanoes (Fiske and Jackson, 1972). Once a rift zone is initiated its development is encouraged by the concentration of tensional gravity stresses along the rift zone axis and by rift zone flanks spreading across the underlying oceanic crust (e.g., Borgia and Treves, 1992).

Several generalizations can be made about the morphologic similarities and differences between the volcanic rift zones of La Palma and El Hierro in the Canary Islands and those of the Island of Hawaii that may illuminate similarities and differences in magmatic processes between them. In general, the rift zones on La Palma and El Hierro are shorter (a few tens of kilometers in length) than those on Hawaii (ranging up to >100 km in length). One possible explanation for this is the composition of the magma. In Hawaii, the shield-phase lavas are tholeiitic and the post-erosional lavas are alkali basalts, whereas the shield and post erosional phase lavas in the Canary Islands are both alkali basalts (Carracedo et al., 2001). We infer that as alkali basalts are more viscous than tholeitic liquids, encourages building of the volcano upwards rather than outwards. An important consequence of this is that volcano flank slopes are steeper, which in turn leads to catastrophic collapses of the volcano. The dramatic embayments on La Palma and El Hierro are a result of such massive flank failures. Another possibility is that dikes are able to propagate further, for a number of reasons including the effects of pressure within the magma pressure and gravity acting downslope along the rift.

Many of the rift zones on Hawaii have well defined axial zones, both in the on- and offshore portions. Typically rift volcanism builds a ridge that has a relatively narrow (a few kilometers wide) crest characterized by structures associated with dike intrusion and eruption: faults, volcanic flows and edifices, and collapse pits. In contrast, volcanism along some of the rift zones we have considered in this paper on El Hierro and La Palma is more diffuse and often it is difficult to define the lava pathways that fed the scattered volcanic edifices. This is particularly true for the Northwest and Northeast Rifts of El Hierro (Figures 9 and 10).

Submarine rift zones of Hawaii show some evidence of widening at their distal ends, where the stress field within the rift may be insufficient to trap dikes within a narrow region. At these places, the axis of the rift zone is hard to define and volcanic lineations fan out (Lonsdale, 1989; Clague et al., 1995). The widening observed at the distal end of rift zones in Hawaii, may be a small-scale model for the fanning of the Northeast and Northwest Rifts on El Hierro. This fanning is so pronounced in these rifts, however, that they closely resemble deep-sea sedimentary submarine fans.

Another possibility is that rift zones fan out down slope because dike injection and volcanism has shifted laterally. Such lateral migration of volcanism may be an important process at the Northwest Rift of El Hierro. The more recent igneous activity along this rift is concentrated along its north side where it appears that lava has flowed onto an adjacent debris flow. In contrast, the south side of the rift has the appearance of a sediment apron, suggesting that either it has not been active recently or that its mode of emplacement differs from that on the north side.

Lénat et al. (1986) suggested that the same processes that form the narrower, ridge shaped submarine rift zones in Hawaii have constructed the wide submarine rift zones on Réunion Island. That is, dikes injected from the summit reservoir move along a narrow zone and erupt along linear vents. The difference in the morphology of the rift zones between those in Hawaii and those in Réunion may be explained, according to Lénat et al. (1986), by the more complex history of volcano building in Réunion than in the Hawaiian volcanoes. Lénat et al. suggested that the presence of an older volcano beneath a younger one in Réunion may lead to a more complex distribution of gravitational stresses within the volcanic edifice leading to a broader zone of dike injection. This may be the case on El Hierro as well where El Golfo Volcano has been constructed on top of El Tiñor Volcano.

In Hawaii, the stress field associated with a volcanic edifice may remain more or less constant throughout its history. If another volcano begins to form nearby, a different stress field is superimposed on the first that is related both to the new volcano and the older one. In La Palma and El Hierro, this scenario may be slightly different. A volcano with its associated stress field is established. That volcano is destroyed and another one is created over the remains of the older one with its stress field superimposed on the older one. The orientation and magnitude of the new stress field may not necessarily be the same as the previous one, and thus may produce the apparent widening of rift zones down slope. If this is the case, it is important to note that in some places the old and new stress fields must be similar to produce narrow, ridge-like rifts (e.g., the South Rift of El Hierro, and Cumbre Vieja on La Palma) in addition to the wide rifts.

Why are the volcanoes in the Canary Islands built one atop another rather than next to each other? We assume that this is related to slow absolute plate motion in the region. In the Canary Island region the rate of motion is on the order of 1.9 cm/year, whereas in the region of Hawaii the absolute plate motion is five times faster, on the order of 10 cm/year (Carracedo, 1999).

Magma supply may also play a role in the width of rift zones. As a result of a low rate of magma supply, the orientation at which dikes are injected into the country rock may become random and not concentrated along a narrow rift zone. The reservoir pressure created by magma-supply mechanisms may be incapable of overcoming the compressive stresses along a previously intruded rift. The magma is forced to find other avenues of intrusion and extrusion into the country rock. As this new route is then closed, it in turn forces the magma to find other pathways creating a dike swarm fanning out from the magma body. This process may be responsible for the widening of the Northeast and Northwest Rifts in El Hierro, but also in Mauna Loa, a volcano that lacks large rift-zone faults and fissures and contains numerous radial vents outside its two rift zones (Rubin, 1990).

As mentioned, submarine rift zones in La Palma and El Hierro have built narrow ridges comparable to those in Hawaii. The ridge-shaped rifts in La Palma and El Hierro, however, are both shorter and more irregular along-axis, displaying significant curvature in plan shape than is typical on the Island of Hawaii. In addition the rifts in the Canary ridges are often discontinuous along strike. These differences also may be related to magma supply rate. Apparently the magma supply rate on La Palma and El Hierro is too low to smooth out any topographic irregularities along the axis. For example, the South Rift offshore El Hierro is cut by a 400 m deep east-west saddle located at 27°30′ N. This suggests that dike injection and eruption may have varied through time. Gee et al. (2001) proposed that the section of the ridge south of the saddle must represent an early phase of rift zone construction. For comparison, the rifts associated with Nintoku Guyot in the Emperor Seamount Chain have uneven crests with major reversals of slope as a result of more than 50 m.y. of volcanic inactivity (Vogt and Smoot, 1984).

Another possibility to explain the along-axis irregularities is that they are a result of the juxtaposition of ridges associated with different volcanoes. The north-south trending rift in La Palma does appear to represent distinct eruption cycles from two different volcanoes superimposed on one another. The eruption cycles differ in age by hundreds of thousands of years. A similar condition exists along the Hilo Ridge where the section of the ridge below 1000 m is an extension of a rift zone from Kohala Volcano. The section shallower than 1000 m was buried by lavas extruded during the construction of Mauna Kea. In the case of the Hilo Ridge, however, eruptions were large enough to smooth out the morphologic transition between the two ridges.

Other considerations may contribute to shaping the morphology of rifts including large scale factors such as: (1) age of the oceanic crust that the volcanoes are built on, (2) orientation of the tectonic fabric of the surrounding seafloor, (3) pre-existing sediment thickness, and 4) proximity of a plate boundary/continental margin. All of these must play some role in controlling the construction and evolution of oceanic islands, but we do not think they are the most important in controlling the characteristics of individual volcanic rift zones. Instead, as discussed above we infer that rift zone morphology is controlled primarily by the rate and constancy of magma supply, the regional stress field that may have been influenced by older volcanoes, and the rate of absolute plate motion.

Conclusion

Rift zones on La Palma, El Hierro and Hawaii Island are distinct and similar in many of their topographic characteristics. The rift zones in Hawaii tend to consist mostly of linear ridges. In contrast, the rift zones on La Palma and El Hierro display various geometries ranging from linear ridges having smooth to irregular crests to structures displaying a fan-like morphology in plan view. Although the toes of some of the Hawaiian rifts do partially fan out, they do not to the extent displayed by the rifts offshore El Hierro. The pronounced fanning may be a reflection of lower reservoir pressure, which is insufficient to overcome the compressive stress in the previous pathway so that new pathways are created that will be randomly distributed away from the magma source. It may also be caused by lateral shifts in the location of the rift zone. Another significant factor that may influence rift morphology and fanning of the rift may be absolute plate motion. In the Canary Islands, because plate motion is so slow, the volcanoes are built nearly one atop of another, with the stress field of one structure nearly superimposed on the other. The interaction of the stress fields may cause both widening and narrowing of the rift zone.

Irregularities and curvature of the rift zones in plan view on La Palma and El Hierro may be a reflection of differences in the rate of magma production in the islands groups (86 km^3/years for Kilauea Volcano versus 0.40–50 km^3/1000 years for El Hierro). There appear to be insufficient volumes of lavas erupted to smooth out topographic irregularities in the Canary Islands volcanoes. The superposition of rifts from different volcanoes may also add to the topographic irregularities, especially if eruption rates are low. Finally, such factors as the thickness of sediment cover over oceanic basement, the positions of the volcanic islands relative to the continent-ocean boundary and composition of the discharged magma also may have played roles in creating the rift morphologies observed in the Canary Islands and Hawaii.

Acknowledgements

We are indebted to the officers and crew of the R/V Hespérides and Vizconde de Eza for their cooperation during the various cruises to the Canary Island. Financial for the different cruises to the Canary Islands came from the Oceanographic and Hydrographic Research of the Spanish Economic Exclusive Program. Thanks are also extended to P. Clift, E. de Reus, Donna Blackman and an anonymous referee who helped greatly improve the manuscript. D.K. Smith's work in this study was funded by National Science Foundation Grant OCE-9618226. Contribution number 11171 of the Woods Hole Oceanographic Institution.

References

Ancochea, E., Hernán, F., Cendrero, A., Cantagrel, J.M., Fúster, J.M., Ibarrola, E. and Coello, J., 1994, Constructive and destructive episodes in the building of a young oceanic island, La Palma, Canary Islands, and genesis of the Caldera de Taburiente, *J. Vol. Geotherm. Res.*, **60**: 243–262.

Anguita, F. and Hernan, F., 1975, A propogating fracture model versus a hotspot origin for the Canary Islands, *Earth Planet. Sci. Letts.*, **27**: 11–19.

Atwater, T. and Severinghaus, J., 1989, Tectonic maps of the northeast Pacific, in E.L.Winterer, D.M. Hussong, and R.W. Decker, R.W. (eds,), *The Eastern Pacific and Hawaii, The Geology of North America*, **N**, *Geol. Soc. America*, 15–20.

Borgia, A. and Treves, B., 1992, Volcanic plates overriding the oceanic crust: structure and dynamics of Hawaiian volcanoes, in: Parson, L.M., Murton, B.J. and Browning, P. (eds), *Geol. Soc. Lond. Spec. Publ.* **60**: 277–299.

Bravo, T., 1982, Formaciones geológicas en la isla de El Hierro, *Inst. Estud. Canarios Aulas de Cultura, Cabildo de Tenerife*, 85–99.

Canales, J.P. and Dañobeitia, J.J., 1998, The Canary Islands swell; a coherence analysis of bathymetry and gravity, *Geophys. J. Int.*, **132**: 479–488.

Carracedo, J.C., 1994, The Canary Islands: an example of structural control on the growth of large oceanic-island volcanoes, *J. Vol. Geoth. Res.* **60**: 225–241.

Carracedo, J.C., 1999, Growth, structure, instability and collapse of the Canarian volcanoes and comparison with Hawaiian volcanoes, *J. Vol. Geoth. Res.* **94**: 1–19.

Carracedo, J.C., 1996, A simple model for the génesis of large gravitational landslide hazards in the Canary Islands, in McGuire, W.J., Jones A.P., and Neuberg, J. (eds.), Volcano Instability on the Earth and Other Planets, *Geol. Soc. Lond. Spec. Publ.* **110**, 125–135.

Carracedo, J.C., Badiola, E.R., Guillou, H., de la Nuez, J. and Perez Torrado, F.J., 2001, Geology and Volcanology of La Palma and El Hierro, Western Canaries, *Estud. Geol.* **57**: 175–273.

Carracedo, J.C., Day, S.J., Guillou, H. and Gravestock, P., 1999a, Later stages of volcanic evolution of La Palma, Canary Islands: rift evolution, giant landslides, and the genesis of the Caldera de Taburiente, *Geol. Soc. Am. Bull.* **111**: 755–768.

Carracedo, J.C., Day, S.J., Gillou, H. and Pérez–Torrado, F.J., 1999b, Giant Quaternary landslides of La Palma and El Hierro, Canary Islands, *J. Vol. Geoth. Res.* **94**: 169–190.

Cendrero, A., 1970, Report on the International symposium on volcanology (I.A.V.C.E.I.) held in Spain (Canary Islands) in September 1968, *Earth Extraterr. Sci.* **1**: 79–88.

Clague, D.A. and Dalrymple, G.B., 1989, Tectonics, geochronology, and origin of the Hawaiian-Emperor volcanic chain, in E.L.Winterer, D.M.Hussong, and R.W. Decker, R.W. (eds,), *Eastern Pacific Ocean and Hawaii, The geology of North America*, **N**, *Geol. Soc. America*, 188–217.

Clague, D.A., Hon, K.A. Anderson, J.L., Chadwick, W.W. and Fox, C.G., 1994, Bathymetry of Puna Ridge, Kilauea Volcano, Hawaii, *U.S. Geol. Survey. Misc. Invest. Ser.* **Map MF-2237**.

Clague, D.A., Moore, J.G., Dixon, J.E. and Friesen, W.E., 1995, Petrology of submarine lavas from Kilauea's Puna Ridge, *J. Petrol.* **36**: 299–349.

Clague, D.A., Moore, J.G. and Reynolds, J.R., 2000, Formation of submarine flat-topped volcanic cones in Hawaii, *Bull. Vol.* **62**: 214–233.

Clague, D.A., Moore, J.G., Torresan, M.E., Holcomb, R.T. and Lipman, P.W., 1988, Shipboard report for Hawaii GLORIA ground-truth cruise11-88-HW, *U.S. Geol. Survey Open File Rep.* **88–0292**: 54 pp.

Clague, D.A., Uto, K., Satake, K. and Davis, A.S., 2002, Eruption style and high emplacement in the submarine north arch volcanic field, Hawaii, in E. Takahashi, P.W. Lipman, M.O. García, J. Naka and S. Aramaki, (eds.), *Hawaiian Volcanoes. Deep Water Perspectives, Am. Geophys. Union Monogr.* **128**: 65–84.

Coello, J., Cantagrel, J.M., Hernán, F., Fúster, J.M., Ibarrola, E., Ancochea, E., Casquet, C., Jamond, C., Diaz de Téran, J.R. and Cendredo, A., 1992, Evolution of the eastern volcanic ridge of the Canary Islands based on new K-Ar data, *J. Vol. Geoth. Res.* **53**: 251–274.

Day, S.J., Carracedo, J.C., Guillou, H. and Gravestock, P., 1999, Recent structural evolution of the Cumbre Vieja volcano, La Palma, Canary Islands: volcanic rift zone reconfiguration as a precursor to volcanic flank instablity, *J. Vol. Geoth. Res.* **94**: 135–167.

Decker, R.W., 1989, Geologic map of the state of Hawaii, in E.L.Winterer, D.M. Hussong, and R.W. Decker, R.W. (eds,), *The Eastern Pacific and Hawaii, The geology of North America*, textbfN, *Geol. Soc. America.*

Delaney, P.T., Fiske, R.S., Miklius, A., Okamura, A.T. and Sako, M., 1990, Deep magma body beneath the summit and rift zones of Kilauea Volcano. Hawaii, *Science* **247**: 1311–1316.

Delaney, P.T., Miklius, A., Arnadottir, T., Okamura, A.T. and Sako, M.K., 1993, Motion of Kilauea Volcano during sustained eruption from the Pu'u 'O'o and Kupaianaha vents, 1983–1991, *J. Geophys. Res.* **98**: 17,801–17,820.

Denlinger, R.P. and Okubo, P., 1995, Structure of the mobile south flank of Kilauea Volcano, Hawaii, *J. Geophys. Res.* **100**: 24,499–24,507.

Dieterich, J.H., 1988, Growth and persistence of Hawaiian volcanic rift zones, *J. Geophys. Res.* **93**: 4258–4270.

Duffield, W.A., 1975, Structure and origin of the Koae Fault System, Kilauea Volcano, Hawaii, *U.S. Geol. Survey Prof. Paper*, **856**, 12 pp.

Dzurisin, D., Koyanagi, R.Y. and English, T.T., 1984, Magma supply and storage at Kilauea Volcano, Hawaii, *J. Vol. Geoth. Res.* **21**: 177–206.

Elsworth, D. and Day, S.J., 1999, Flank collapse triggered by intrusion: the Canarian and Cape Verde archipelago, *J. Vol. Geoth. Res.* **94**: 323–340.

Emery, K.O. and Uchupi, E., 1984, The Geology of the Atlantic Ocean, *Springer-Verlag*, New York, 1050 pp.

Epp, D., Decker, R.W. and Okamura, A.T., 1983, Relation of summit deformation to East rift zone, eruptions of Kilauea Volcano, Hawaii, *Geophys. Res. Lett.* **10**: 493–496.

Feraud, G., Giannerini, G., Campredon, R. and Stillman, C.J., 1985, Geochronology of some Canarian dike swarms: contribution to the volcano-tectonic evolution of the archipielago, *J. Vol. Geoth. Res.* **25**: 29–52.

Fiske, R.S. and Jackson, E.D., 1972, Orientation and growth of Hawaiian volcanic rifts: the effect of regional structure and gravitational stresses, *Proc. R. Soc. London* **A329**: 299–326.

Fornari, D.J., 1986, The geomorphic and structural evolution of Hawaiian submarine rifts, *U.S. Geol. Survey Prof. Paper* **1350**: 125–132.

Fornari, D.J. Campbell, J.F., 1986, Submarine topography around the Hawaiian Islands, *U.S. Geol. Survey Prof. Paper* **1350**: 109–123.

Fornari, D.J., Malahoff, A. and Heezen, B.C., 1978, Volcanic structure of the crest of the Puna Ridge, Hawaii: geophysical implications of submarine volcanic terrain, *Geol. Soc. Am. Bull.* **89**: 605–610.

Fornari, D.J., Malahoff, A. and Heezen, B.C., 1979, Submarine extension of the southwest rift zone of Mauna Loa Volcano, Hawaii: visual observations from U.S. Navy deep submergence vehicle DSV Sea Cliff, *Geol. Soc. Am. Bull.* **90**: 435–443.

Funck, T. and Schmincke, H.-U., 1998, Growth and destruction of Gran Canaria deduced from seismic reflection and bathymetric data, *J. Geophys. Res.* **103**: 15,393–15,407.

Fuster, J.M., Hernán, F., Cendrero, A., Coello, J., Cantagrel, J.M., Ancochea, E. and Ibarrola, E., 1993, Geocronología de la Isla de El Hierro (Islas Canarias), *Bol. R. Soc. Hist. Nat. (Sec. Geol.)* **88**: 85–97.

García, M., Hulsebosch, T. and Rhodes, J., 1995, Olivine-rich submarine basalts from the southwest rift zone of Mauna Loa Volcano: Implications for magmatic processes and geochemical evolution, in J.M. Rhodes and J.P. Lockwood (Eds.), *Mauna Loa revealed. Structure, composition, history and Hazards, Am. Geophys. Union Mon.* **92**: 219–239.

Gee, M.J.R., Masson, D.G., Watts, A.B. and Mitchell, N.C., 2001, Offshore continuation of volcanic rift zones, El Hierro, Canary Islands, *J. Vol. Geoth. Res.* **105**: 107–119.

Gregg, T.P.K. and Smith, D.K., 2003. Volcanic investigations of Puna Ridge, Hawaii: relations of lava flow morphologies and underlying slopes. *J. Vol. Getherm. Res.* **126**: 63–77.

Guillou, H., Carracedo, J.C. and Day, S.J., 1998, Dating of the upper Pleistocene-Holocene volcanic activity of La Palma using unspiked K-Ar technique, *J. Vol. Geoth. Res.* **86**: 137–149.

Guillou, H., Carracedo, J.C., Pérez, Torrado, F. and Rodríguez Badiola, E., 1996, K-Ar ages and magmatic stratigraphy of a hotspot-induced, fast grown oceanic island: El Hierro, Canary Islands, *J. Vol. Geoth. Res.* **73**: 141–155.

Hausen, H., 1959, On the geology of Lanzarote, Graciosa and the Isletas (Canarian archipelago), *Comm. Phys. Math.* **23**: 1–116.

Hausen, H., 1969, Some contributions to the geology of La Palma, *Comm. Phys. Math.* **35**: 1–140.

Hausen, H., 1973, Outlines of the geology of Hierro (Canary Islands), *Comm. Phys. Math.* **43**: 65–148.

Hernández Pacheco, A., 1982, Sobre una posible erupción en 1793 in la isla de El Hierro (Canarias), *Estud. Geol.* **38**: 15–25.

Hernández Pacheco, A. and Fernandez Santin, S., 1975, The submarine formation of the Caldera de Taburiente in La Palma (Canary Islands) and their metasomatic processes, IAVCEI Symposium, Santaigo de Chile, *Bull. Vol. Spec. V.*

Hills, D.J., Morgan, J.K., Moore, G.F. and Leslie, S.C., 2002, Structural variability along the submarine south flank of Kilauea Volcano, Hawaii, from a multichannel seismic reflection survey, in Takahashi, E., Lipman, P.W., García, M.O. and Aramaki, S. (eds.), *Hawaiian volcanoes. Deep Water Perspective, Am. Geophys. Union Monogr.* **128**: 105–124.

Hoernle, K., Tilton, G. and Schmincke, H.-U., 1991, Sr-Nd-Pb isotopic evolution of Gran Canaria: evidence for shallow enriched mantle beneath the Canary Islands, *Earth Planet. Sci. Letts.* **106**: 44–63.

Holcomb, R.T., 1987, Eruptive history and long-term behavior of Kilauea volcano, in R.W. Decker, T.L.Wright, T.L. and P.H. Stauffer (eds.), *Volcanism in Hawaii, U.S. Geol. Survey Prof. Paper* **1350**: 261–350.

Holcomb, R.T., Moore, J.G., Lipman, P.W. and Belderson, R.H., 1988, Voluminous submarine lava flows from Hawaiian volcanoes, *Geology* **16**: 400–404.

Holcomb, R.T., Nelson, B.K., Reiners, P.W. and Sawyer, N.L., 2000, Overlapping volcanoes: the origin of Hilo Ridge, *Geology* **28**: 547–550.

Holcomb, R.T. and Searle, R.G., 1991, Large landslides from oceanic volcanoes, *Mar. Geotechnol.* **10**: 19–32.

Holik, J.S., Rabinowitz, P.D. and Austin, J.A., 1991, Effects of Canary hotspot volcanism on structure of oceanic crust off Morocco, *J. Geophys. Res.* **96**: 12,039–12,067.

Johnson, K.T.M., Reynolds, J., Vonderhaar, D., Smith, D.K. and Kong, L.S.L., 2002, Petrological systematics of submarine lavas from the Puna Ridge, Hawaii: implications for rift zone plumbing and magmatic processes, in E. Takahishi, P.W. Lipman, M.O. García, J. Naka and S. Aramaki (eds.), *Hawaiian volcanoes. Deep underwater perspectives*, *Am. Geophys. Union Monogr.* **128**: 143–160.

Kauahikaua, J., Hildenbrand, T. and Webring, M., 2000, Deep magmatic structures of Hawaiian volcanoes, imaged by three-dimensional gravity models, *Geology* **28**: 883–886.

Klein, F.W., Koyanagi, R.Y., Nakata, J.S. and Tanigawa, W.R., 1987, The seismicity of Kilauea's magma system, *U.S. Geol. Survey Prof. Paper* **1350**: 1019–1185.

Klügel, A., 1997, Ascent rates of magmas and genesis, transport, and reactions of mantle and crustal xenoliths of the 1949 eruption on La Palma (Canary Islands), *Ph.D. thesis*, Christians-Albrechts-Universität, Kiel, 209 pp.

Langenheim, V.A.M. and Clague, D.A., 1987, The Hawaiian-Emperor volcanic Chain: Part II. Stratigraphic framework of volcanic rocks of the Hawaiian Islands, *U.S. Geol. Survey Prof. Paper* **1350**: 55–84.

Le Bas, M.J., Rex, D.C. and Stillman, C.J., 1986, The early magmatic chronology of Fuerteventura, Canary Islands, *Geol. Mag.* **123**: 287–298.

Lénat, J.F., Vincent, P. and Bachèlery, P., 1986, The off-shore continuation of an active basaltic volcano; Piton de la Fournaise (Réunion Island, Indian Ocean); structural and geomorphological interpretation from sea beam mapping, *J. Vol. Geoth. Res.* **36**: 1–36.

Lipman, P.W., 1980, The Southwest Rift Zone of Mauna Loa: implications for structural evolution of Hawaiin volcanoes, *Am. J. Sci.* **280-A**: 752–776.

Lipman, P.W., Clague, D.A., Moore, J.G. and Holcomb, R.T., 1989, South Arch volcanic field-newly identified young lava flows on the sea floor south of the Hawaiian Ridge, *Geology* **17**: 611–614.

Lonsdale. P., 1989, A geomorphological reconnaissance of the submarine part of the East rift zone of Kilauea Volcano, Hawaii, *Bull. Vol.* **51**: 123–144.

Malahoff, A. and McCoy, F., 1967, The geologic structure of the Puna Ridge, Hawaii, *J. Geophys. Res.* **72**: 541–548.

Mangan, M.T., Heliker, C.C., Mattox, T.N., Kauahikaua, J.P. and Helz, R.T., 1995, Episode 49 of the Pu'u 'O'o – Kupaianaha eruption of Kilauea volcano: breakdown of a steady-state eruptive era, *Bull. Vol.* **57**: 127–135.

Martí, J., Mitjavila, J. and Araña, V.,1994, Stratigraphy, structure and geochronology of the Las Cañadas Caldera (Tenerife, Canary Islands), *Geol. Mag.* **131**: 715–727.

Martí, J., Hurlimann, M., Ablay, G.J. and Gudmundsson, A., 1997, Vertical and lateral collapses on Tenerife (Canary Islands) and other volcanic ocean islands, *Geology* **25**: 879–882.

Masson, D.G., 1996, Catastrophic collapse of the volcanic island of Hierro 15 Ka ago and the history of landslides in the Canary Islands, *Geology* **24**: 231–234.

Masson, D.G., Watts, A.B., Gee, M.J.R., Urgeles, R., Mitchell, N.C., Le Bas, T.P. and Canals, M., 2002, Slope failures on the flanks of the Canary Islands, *Earth-Sci. Rev.* **57**: 1–35.

McDougall, I., 1979, Age of shield-building volcanism in Kauai and linear volcanism in Hawaiian Island chain, *Earth Planet. Sci. Letts.* **46**: 31–42.

Mehl, K.W. and Schmincke, H.-U., 1999, Structure and emplacement of the Pliocene Roque Nublo debris avalanche deposit, Gran Canaria, Spain, *J. Vol. Geoth. Res.* **94**: 105–134.

Mitchell, N.C., Masson, D.G., Watts, A.B., Gee, M.J.R. and Urgeles, R., 2002, The morphology of the flanks of volcanic ocean islands: A comparative study of the Canary and Hawaiian hotspot islands, *J. Vol. Geoth. Res.* **115**: 83–107.

Monterey Bay Aquirium research Institute (MBARI), 2000, Monterey Bay Aquarium Research Institute Hawaii multibeam survey, *Digital Series* **2**, Moss Landing CA.

Moore, J.G. and Campbell, J.F., 1987, Age of tilted reefs, Hawaii, *J. Geophys. Res.* **92**: 752–759.

Moore, J.G. and Chadwick, Jr., W.W., 1995, Offshore geology of Mauna Loa and adjacent areas, Hawaii, in Rhodes, J.M. and Lockwood, J.P. (eds.), *Mauna Loa Revealed. Structure, Composition, History and Hazards, Am. Geophys. Union Geophys. Mon.* **52**: 21–44.

Moore, J.G. and Clague, D.A., 1992, Volcano growth and evolution of the island of Hawaii, *Geol. Soc. Am. Bull.* **104**: 1471–1484.

Moore, J.G., Clague, D.A., Holcomb, R.T., Lipman, P.W., Normark, W.R. and Torresan, M.E., 1989, Prodigious submarine landslides on the Hawaiian Ridge, *J. Geophys. Res.* **94**: 17,465–17,484.

Moore, J.G. and Fiske, R.S., 1969, Volcanic substructure inferred from dredge samples and ocean-bottom photographs, Hawaii, *Geol. Soc. Am. Bull.* **80**: 1191–1201.

Moore, J.G. and Fornari, D.J., 1984, Drowned reefs as indicators of rate of subsidence of the island of Hawaii, *J. Geology* **92**: 752–759.

Moore, J.G., Phillips, R.L., Grigg, R.W., Peterson, D.W. and Swanson, D.A., 1973, Flow of lava into the sea, 1969–1971, Kilauea Volcano, Hawaii, *Geol. Soc. Am. Bull.* **84**: 537–546.

Moore, R.B. and Trusdell, F.A., 1991, Geologic map of the lower east rift zone of Kilauea Volcano, Hawaii, *U.S. Geol. Survey Misc. Invest. Ser.*, **Map I-2225.**

Morgan, W.J., 1970, Plate motions and deep mantle convection, in R. Shagan et al. (eds), *Studies in Earth and Space Sciences. Geol. Soc. America Memoir*, **132**: 7–22.

Morgan, W.J., 1972, Deep mantle convection plumes and plate motions. *Am. Assoc. Pet. Geol. Bull.* **56**: 203–213.

Morgan, J.K., Moore, V.G.F., Hills, D.J. and Leslie, S., 2000, Overthrust and sediment accretion along Kilauea's mobile south flank, Hawaii: evidence for volcanic spreading from marine seismic reflection data, *Geology* **28**: 667–670.

Moss, J.L., McGuire, W.J. and Page, D., 1999, Ground deformation monitoring of a potential landslide at La Palma, Canary Island, *J. Vol. Geoth. Res.* **94**: 251–260.

Owen, S., Segall, P., Freymueller, J., Miklius, A., Delinger, R., Arnadottir, T., Sako, M. and Burgmann, R., 1995, Rapid deformation of the south flanks of Kilauea Volcano, Hawaii, *Science* **267**: 1328–1332.

Parfitt, E.A., Gregg, T.K.P. and Smith, D.K., 2002, A comparison between subaerial and submarine eruptions at Kilauea Volcano, Hawaii, *J. Vol. Geoth. Res.* **113**: 213–242.

Parfitt, E.A. and Peacock, D.C.P., 2001, Faulting in the south flank of Kilauea Volcano, Hawaii, *J. Vol. Geoth. Res.* **106**: 265–284.

Peterson, D.W. and Moore, R.B., 1987, Geologic history and evolution of geologic concepts, island of Hawaii, *U. S. Geol. Survey Prof. Paper* **1350**, 149–189.

Pollard, D.D., Delaney, P.T., Duffield, W.A., Endo, E.T. and Okamura, A.T., 1983, Surface deformation in volcanic rift zones, *Tectonophysics* **94**: 541–584.

Price, I., 1980, Gravity tectonics on a passive margin.: Deep-Sea Drilling Project site 415 in the light of regional seismic data, in

Y. Lancelot, E.L. Winterer et al., *Init. Repts. of the Deep Sea Drill. Proj.* **50**: 759–771.

Ridley, W.I., 1971, The origin of some collapse structures in the Canary Islands, *Geol. Mag.* **108**: 477–484.

Rubin, A.M., 1990, A comparison of rift-zone tectonics in Iceland and Hawaii, *Bull. Vol.* **52**, 302–319.

Rubin, A.M. and Gillard, D., 1998, A reinterpretation of seismicity associated with the January 1983 dike intrusion at Kilauea Volcano, Hawaii, *J. Geophys. Res.* **103**:10,003–10,015.

Ryan, M.P., 1988, The mechanics and three-dimensional internal structure of active magmatic systems: Kilauea Volcano, Hawaii, *J. Geophys. Res.* **93**: 4213–4248.

Ryan, M.P., Koyanagi, R.Y. and Fiske, R.S., 1981, Modelling of the three dimensional structure of macroscopic magma transport systems: application to the Kilauea Volcano. Hawaii, *J. Geophys. Res.* **86**: 7111–7129.

Smith, D.K. and Cann, J.R., 1999, Constructing the upper crust of the Mid-Atlantic Ridge: a reinterpretation based on the Puna Ridge, Kilauea Volcano, *J. Geophys. Res.* **97**: 25,379–25,399.

Smith, D.K., Kong, L.S., Johnson, K.T.M. and Reynolds, J.R., 2002a, Volcanic morphology of the submarine Puna Ridge, Kilauea Volcano, in Takahashi, E., Lipman, P.W., García, M.O. and Aramaki, S. (eds.), *Hawaiian volcanoes. Deep Water Perspective, Am. Geophys. Union Monogr.* **128**: 125–142.

Smith, D.K., Tivey, M.A., Gregg, P. and Kong, L.S., 2001,, Magnetic anomalies along the submarine Puna Ridge, Kilauea Volcano, Hawaii, *J. Geophys. Res.* **106**: 16,047–16,060.

Smith, J.R., Satake, K., Morgan, J.K. and Lipman, P.W., 2002b, Submarine landslides and volcanic features on Kohala and Mauna Kea volcanoes, in Takahashi, E., Lipman, P.W., García, M.O. and Aramaki, S. (eds.), Hawaiian volcanoes. Deep Water Perspective, *Am. Geophys. Union Monogr.*, **128**: 11–28.

Staudigel, H. and Smincke, H.-U., 1984, The Pliocene seamount series of La Palma/Canary Islands, *J. Geophys. Res.* **89**: 11,190–11,215.

Stearns, H.T. and Macdonald, G.A., 1946, Geology and ground-water resources of the Island of Hawaii, *Hawaii Div. Hydro. Bull.* **9**: 363 pp.

Stillman, C.J., 1999, Giant Miocene landslides and evolution of Fuerteventura, Canary Islands, *J. Vol. Geoth. Res.* **94**: 89–104.

Swanson, D.A., Jackson, D.B., Koyanagi, R.Y. and Wright, T.L., 1976, The February 1969 rift eruption of Kilauea Volcano, Hawaii, *U.S. Geol. Survey Prof. Paper* **891**: 30 pp.

Thurber, C.H. and Gripp, A.E., 1988, Growth and persistence of Hawaiian rift zones, *J. Geophys. Res.* **93**: 4271–4278.

Tilling, R.I. and Dvorak, J.J., 1993, Anotomy of a basaltic volcano, *Nature* **363**: 125–133.

Troll, V.R., Walter, T.R. and Schmincke, H-U., 2002, Cyclic caldera collapse: piston or piecemeal subsidenced? Field experiment evidence, *Geology* **30**: 135–138.

Uchupi, E., Emery, K.O., Bowin, C.O. and Phillips, J.D., 1976, Continental margin off western Africa: Senegal to Portugal, *Am. Assoc. Petrol. Geol. Bull.* **60**: 809–878.

Urgeles, R., Canals, M., Baraza, J., Alonso, B. and Masson, D.G., 1997, The most recent megaslides on the Canary Islands: the El Golfo debris avalanche and the Canary debris flow; west of El Hierro Island, *J. Geophys. Res.* **102**: 20,305–20,323.

Urgeles, R., Masson, D.G., Canals, M., Watts, A.B. and Le Bas, T., 1999, Recurrent large-scale landsliding on the west flank of La Palma, Canary Islands, *J. Geophys. Res.* **104**: 24,331–25,348.

Vogt, P.R. and Smoot, N.C. 1984, The Geisha Guyots: multi-beam bathymetry and morphometric interpretation, *J. Geophys. Res.* **89**: 11,085–11,107.

Walker, G.P.L., 1988, Three Hawaiian calderas: an origin through loading by shallow intrusions? *J. Geophys. Res.* **93**: 14,773–14,784.

Ward, S.N. and Day, S., 2001, Cumbre Vieja volcano – potential collapse and tsunami at La Palma, Canary Islands, *Geophys. Res. Lett.* **28**: 3397–3400.

Watts, A.B., Peirce, C., Collier, J., Dalwood, R., Canales, J.P. and Henstock, T.J., 1997, A seismic study of lithospheric flexure in the vicinity of Tenerife, Canary Islands, *Earth Planet. Sci. Letts.* **146**: 431–447.

Wolfe, E.W., García, M.O., Jackson, D.B., Koyanagi, R.Y., Neal, C.A. and Okamura, A.T., 1987, The Pu'u 'O'o eruption of Kilauea volcano, episodes 1–20, January 3, 1983, to June 8, 1984, *U.S. Geol. Survey Prof. Paper* **1350**: 471–508.

Wyss, M., 1980, Hawaiian rifts and recent Icelandic volcanism: expressions of plume generated radial stress fields, *J. Geophys. Res.* **47**: 19–22.

Zazo, C., Goy, J.L., Hillaire-Marcel, C., Gillot, P.Y., Soler, V., Gozález, J.A., Dabrio, C.J. and Ghaleb, G., 2002, Raised marine sequences of Lanzarote and Fuerteventura revisited-a reappraisal of relative sea-level changes and vertical movements in the eastern Canary Islands during the Quaternary, *Quat. Rev.* **21**: 18–19: 2019–2046.

Marine Geophysical Researches (2003) 24: 91–112
DOI 10.1007/s11001-004-1335-3

Morphological and structural analysis in the Anaga offshore massif, Canary Islands: fractures and debris avalanches relationships

P. Llanes[1,*], A. Muñoz[2], A. Muñoz-Martín[1], J. Acosta[2], P. Herranz[2], A. Carbó[1], C. Palomo[2] & ZEE Working Group[**]
[1]*Universidad Complutense de Madrid. Facultad C.C. Geológicas. Departamento de Geodinámica. 28040 Madrid, Spain*
[2]*Instituto Español de Oceanografía. Grupo de Cartografía Multihaz. C. de María n° 8. 28002 Madrid, Spain*
[*]*Corresponding Author (Phone: +34-913944834; Fax: +34-913944631; E-mail: pllanes@geo.ucm.es)*

Key words: Anaga massif, bathymetry, debris avalanche, fractures, Tenerife island

Abstract

As part of the 'National Hydrographic and Oceanographic Research Plan for the Spanish Exclusive Economic Zone', multibeam bathymetry and seismic reflection profiles were obtained in the Canary Islands aboard the R/V Hespérides. The submarine flanks of the Anaga offshore extension of Tenerife Island are here studied to analyze its geomorphology. In the north sector of the Anaga submarine massif, the extension of the Anaga Debris Avalanche has been mapped for the first time, and a volume of 36 km^3 was calculated. The relationship between the Anaga and Orotava Debris Avalanches is also described. Faulting has been recognized as a key process for the occurrence of debris avalanches and the growth of volcanic lineaments. Moreover, faulting affects previous structures and the channelling of debris flows. Structural analysis shows the typical radial pattern of an oceanic island. In addition, a NE-SW dominant direction of faulting was obtained, consistent with the Tenerife Island structural trend seen in the Anaga Massif and Cordillera Dorsal. NW-SE and E-W are two other main trends seen in the area. Special interest is manifest in two long faults: 'Santa Cruz Fault' bounds the southern edge of Anaga offshore Massif with a length of 50 km and a direction that changes from NE-SW to almost E-W. The Güimar Debris Avalanche was probably channeled by this fault. The 'Guayotá Fault' was recognized in several seismic profiles with a N-S direction that changes towards NW-SE at its southern end. This fault affects the more recent sediments with a vertical offset of 25–30 m, along 60 km. It has been interpreted as a transpressive strike-slip fault.

Introduction

Multibeam mapping of the seafloor around oceanic islands has significantly increased the understanding of the evolution of volcanic islands. That is because the submerged zones of oceanic islands are considerably bigger than the subaerial parts, and they represent the initial growth stages and some of the major volcanic risk.

A great part of the studies related to the submarine flanks of the Canary Islands have described and characterized the giant landslides, studied the processes and age of their formation and compared them with the Hawaiian landslides (Holcomb and Searle, 1991, Carracedo, 1994; Watts and Masson, 1995; Masson, 1996; Geisslinger et al., 1996; Teide Group, 1997; Funck and Schmincke, 1998; Urgeles et al., 1999; Stillman, 1999; Carracedo, 1999; Watts and Masson, 2001; Krastel et al., 2001; Masson et al., 2002). A few other studies have focussed on seismostratigraphy, submarine volcanism or deep-sea sediment physical properties (Roberts and Cramp, 1996; Urgeles et al., 1998; Gee et al., 2001; Urgeles et al., 2000).

[**]F. Carrillo[a], C. Maté[a], M. Ballesteros[b], M. Vaquero[b], J. Martín-Dávila[c], M. Catalán[c] and J.A. Marín[c].
[a]Instituto Hidrográfico de la Marina. Cádiz.
[b]Instituto Español de Oceanografía. Madrid.
[c]Real Observatorio de la Armada. S. Fernando, Cádiz.

A previous study (Mitchel et al., 2003) was carried out in the region of Anaga with multibeam data. It described erosion process and used a geomorphology method to calculate the erosion depth. From the comparison of the old and eroded flank of Anaga with the young, mostly unaltered flank of El Hierro, these authors inferred a mean depth of Anaga's submarine erosion of 100 m. They also mapped canyon chutes and topographic peaks in the area.

In this paper we use geophysical methods to examine both the morphology and structure of the seaward extension of the Anaga Massif (NE Tenerife Island) in order to define the geological processes that controlled the geological evolution of this area. We have mapped erosional, volcanic and structural features. Structural lineations provide directions where the erosive processes preferentially take place. Because of the possible relationship between structural lineations and erosional process we carry out a structural analysis. We also discuss and try to establish the relationships between the submarine and the subaerial processes in relation with the debris avalanches of this area.

Geological setting

The Canary Islands are an archipelago formed by seven volcanic islands located in the eastern Atlantic Ocean, at the border of the northwest African continental margin (Figure 1). These islands are sited on Upper Jurassic oceanic crust (Roest et al., 1992) of the African plate, between magnetic anomalies M21 (148 Ma) and S1, interpreted as true oceanic crust (Roest et al., 1992) and dated by Klitgord and Schouten (1986) as 175 Ma. The African plate has been more or less stationary during the subaerial growth stage of Canary Islands, with an average of movement in this region of about 10 mm/yr since 60 Ma (Duncan, 1981; Morgan, 1983). The Canary volcanism is much younger than the underlying oceanic crust. The oldest subaerial volcanic activity occurred on Fuerteventura at 20 Ma (Coello et al., 1992), whereas the youngest volcanic activity happened in La Palma in historical time.

The structural evolution of a volcanic oceanic island may result from a complex interaction of magmatism with the regional stress field and the local stress field generated during the growth of the island itself (Marioni and Pasquarè, 1994). Volcanic structures are characterized by steep and gravitationally unstable slopes, and are commonly associated with recent seismic activity that favours the initiation of catastrophic slope failures (Acosta et al., 1997). Landslides are considered key processes in the evolution of oceanic islands, because they are a response to the volcanic edifice exceeding the mass that the island can support (Masson et al., 2002).

Tenerife island has a triangular shape and is the largest (2058 km^2) and highest (3718 m) of the Canaries, occupying the central position in the archipelago. This island is considered a triple rift zone, with a more complex scheme than others such as El Hierro, because it has a stratovolcano-type central volcano of differentiated (trachytic-phonolitic) magmas (Teide-Pico Viejo Volcanic Complex; Carracedo, 1994) (Figure 2).

The subaerial part of Tenerife was originally constructed by fissure eruptions of ankaramite, basanite and alkali basalts, that occurred between 12 and 3.3 Ma (Ancochea et al., 1990). According to Hausen (1956), the 'basaltic tableland series' products of these eruptions, are crossed by numerous dikes and outcrop at the three corners of the island, Anaga, Teno and Roque del Conde (Figure 2). They are known as the Old Basaltic Series.

In the recent past, it appears that Tenerife, together with El Hierro and La Palma, has been the most active of the Canary Islands, in terms of both volcanic and landslide activity (Urgeles et al., 1997). Such voluminous Quaternary volcanic activity was focussed on the island's centre (Cañadas Volcano, 1.9–0.2 Ma) and on a chain of basaltic eruption centres spreading from the central volcano to the northeast (Cordillera Dorsal, peak activity, 0.8 Ma) (Ancochea et al., 1990).

Debris avalanche deposits were first recognised offshore Tenerife by Watts and Masson (1995) and subsequent studies mapped at least four avalanche events offshore the north coast (Teide Group, 1997; Watts and Masson, 1998; Watts and Masson, 2001) and five in the south-eastern flank of Tenerife (Krastel et al., 2001). The valley of Güimar and shortly afterwards the Orotava valley were formed by large landslides after 0.8 Ma, and probably before 0.6 Ma. The Cañadas caldera may have had several collapse phases and the present caldera was formed by a landslide since 0.2 Ma (Ancochea et al., 1990). However, the key aspects of the history of the flanks of Tenerife are controversial, including the number of landslide events, their timing and nature and their possible relationship with onshore structures.

Anaga Peninsula, the northeast appendix of Tenerife, is 13 km wide and 26 km long and reaches 1024 m

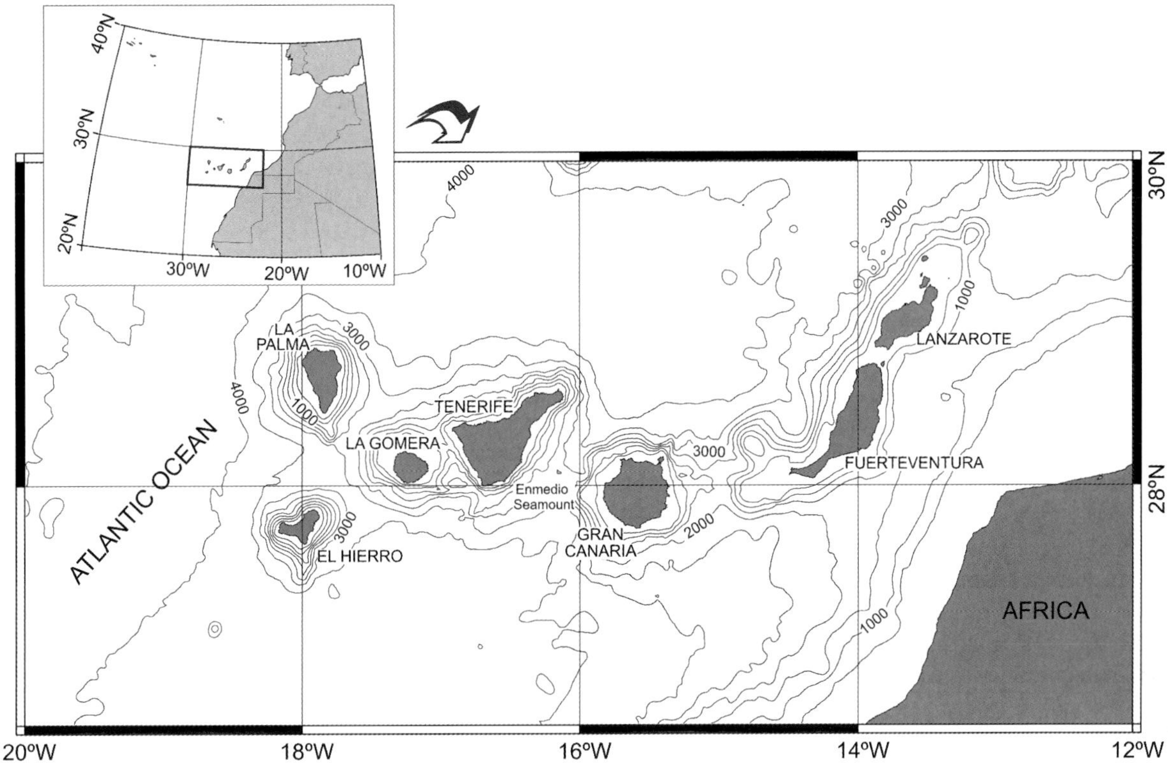

Figure 1. Location map of the Canary Archipelago. Depth in meters.

in height above sea level. It is incised by narrow valleys up to 500 m deep, mostly trending N-S to NNW-SSE. Anaga grew between 6.5 and 3.6 Ma, with major activity occurring between 6 and 4.5 Ma. It is formed by a complex sequence of alkali basaltic lava flows and volcanoclastic layers, intruded by subvolcanic bodies like dykes, domes and laccoliths of basalts, trachybasalts, trachytes and phonolites (Ancochea et al., 1990). Anaga Peninsula is linked to the central volcano of Tenerife (Cañadas Volcano) by a chain of basaltic centres, the Cordillera Dorsal (Figure 2).

Methods

The present study is part of the National Hydrographic and Oceanographic Research Plan for the Spanish Exclusive Economic Zone (ZEE), and was carried out aboard the Spanish R/V Hespérides.

The data we present here were collected in the Canarian Archipelago in June 1999, during the ZEE-99 cruise, in an area that covers 7050 km^2 of the Tenerife submarine flank, where we obtained bathymetric data and ultra-high resolution seismic profiles along 30 track lines (Figure 3). Five additional track lines of Teide 95 cruise were used to complete survey of the west side of Anaga offshore Massif.

The bathymetric data were used to identify the various morphologic features present on the Anaga insular slopes and to map their distribution, whereas the seismic reflection profiles were used to determine their internal structure.

Swath bathymetric data were obtained with a Simrad EM-12s System, giving 100% coverage of the seafloor. The EM-12s is a multibeam echosounder which transmits 81 beams across a total swath opening of 120°, producing a swath width equal to 3.5 times the water depth. The system works at a frequency of 12.5 kHz and is able to resolve depths of a few metres. These data were logged using Simrad's Mermaid and Merlin software and subsequently processed using Neptune and C-floor software. This allows the creation not only of bathymetric charts but also Digital Terrain Models and 3D block diagrams of the surveyed area. We have attempted to remove all artifacts on the maps, but a few remain. The more significant ones were

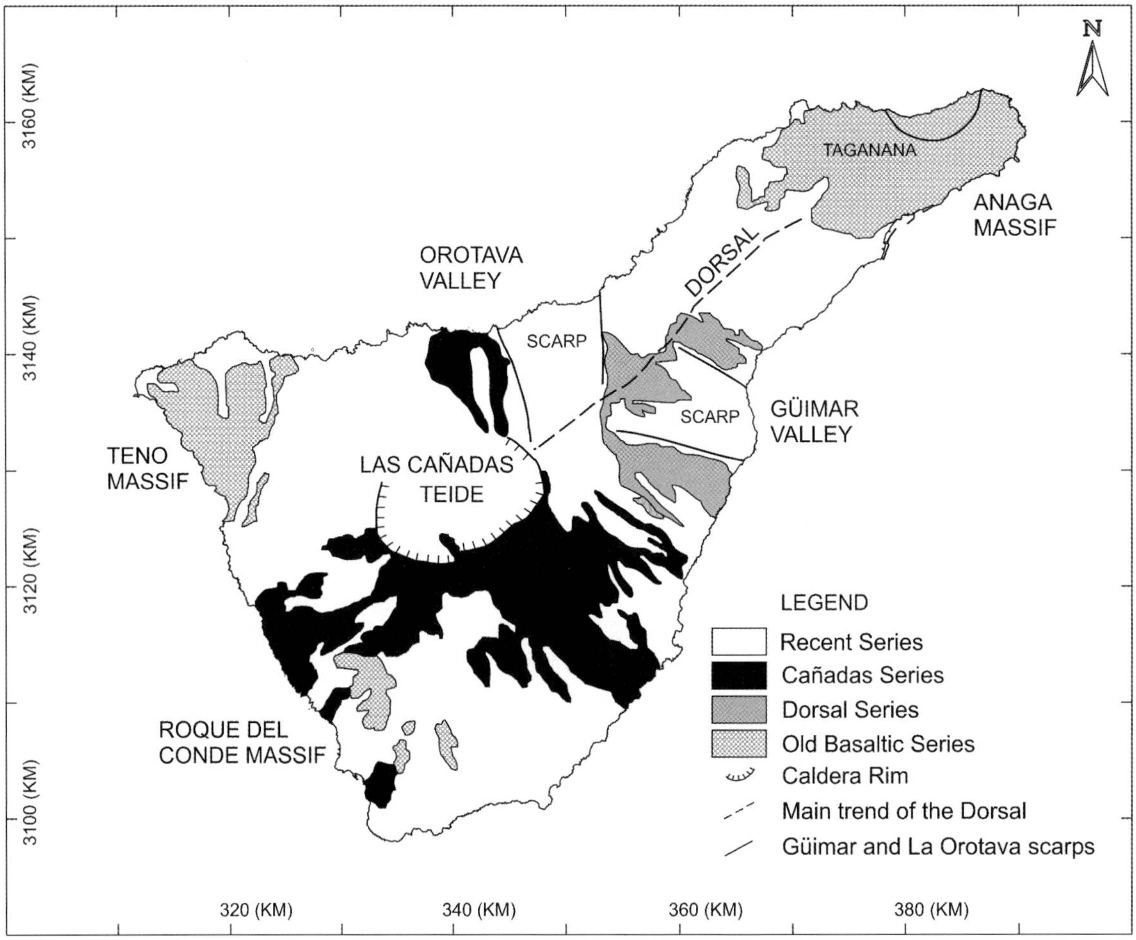

Figure 2. Simplified geologic map of Tenerife island. Modified from Ancochea et al. (1990). The coordinates are Universal Transverse Mercator (UTM) Zone 28 distances in km.

found in the flat seafloor of the abyssal plain in the NE area (starting at the 3625 m contour) where an E-W fabric is seen parallel to the ship tracks. Other artifacts with a N-S fabric are seen on the island flanks.

Three morphologic maps were then produced from bathymetry data, including a contour map (Figure 4), a slope map (Figure 5) and a shaded relief map (Figure 6). Ship's navigation, which is critical for quality control of the multibeam data, was by means of two simultaneous Differential GPS systems, integrated into R/V Hespérides central navigation system which transmits at any given time the ship's position to all the scientific stations via ethernet. Our navigation data has a 5 m precision.

Seismic reflection profiles were recorded using a Simrad Bentech Topographic Parametric Sonar, 018 bottom parametric source. This source has two primary frequencies, 15 and 18 kHz, giving secondary frequencies of 0.5 and 5 kHz. Thereby, it is possible to obtain very high resolution profiles, although with a corresponding lack of penetration that is implicit in the high resolution system.

Anaga offshore proximal region

General structure

To study the proximal offshore Anaga region only the bathymetric data have been used because ultra high resolution seismic profiles have a poor quality in those areas where the slopes are significant.

The offshore extension of the Anaga Massif is a wide structure extending seaward from the coast. It shows an elongated shape, with a major axis of 50 km,

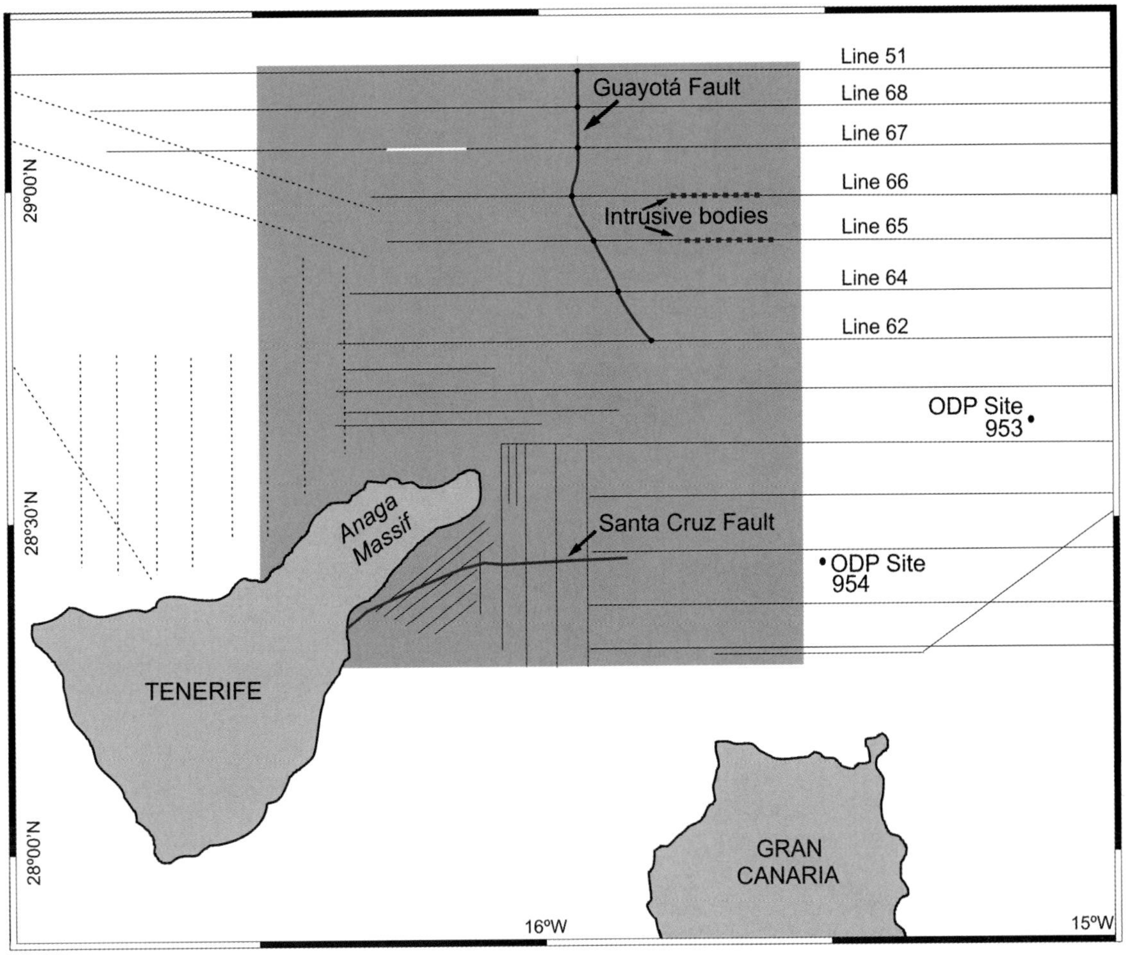

Figure 3. Track lines map. Geophysical data collected during ZEE-99 cruise is shown as black lines and survey lines collected during Teide cruise is shown as black dashed lines. Grey square box is the location of studied area. Circles mark the locations of Ocean Drilling Program (ODP Sites 953 and 954). Grey line mark situation of 'Guayotá Fault' and 'Santa Cruz Fault' and grey dashed line mark situation of intrusive bodies (see text, section 5). White line mark situation of the seismic segment display in Figure 13.

oriented NE-SW. The offshore structure starts with a relatively flat submarine shelf (labelled S in Figure 7) with slope gradients less than 1° (Figure 5). This is probably the result of wave erosion during Pleistocene glacially induced regressions (Teide Group, 1997). Mitchell et al. (2003) pay attention to the asymmetric effect of subaerial and submarine erosion of the Anaga Massif, finding that the north subaerial slopes of the island are steeper than those to the south, implying that the drainage divide lies north of the massif's center. From the analysis of trade winds data Mitchell et al. (2003) deduced that a persistently greater precipitation to the north and a more powerful sea leads to a greater width of the abrasion platform in the north of Anaga.

The shelf break lies at 125 m depth, where the slope gradient locally increases to more than 30°. The shelf break is reached at different distances from the coast, from 500 m in the south to more than 6000 m at the NE (Figure 4) where the shelf reaches its maximum development. Northeastwards, the Anaga offshore massif has its maximum development reaching the 3500 m depth contour. In this area, there are numerous pinnacles recognized in the bathymetric map (Figure 4). They are sometimes elongated or aligned downslope in groups of two or three, suggesting that they could be related to dyke activity radiating from a central volcanic zone. The pinnacles in the studied area are 1–3 km across and 100–200 m high. Other

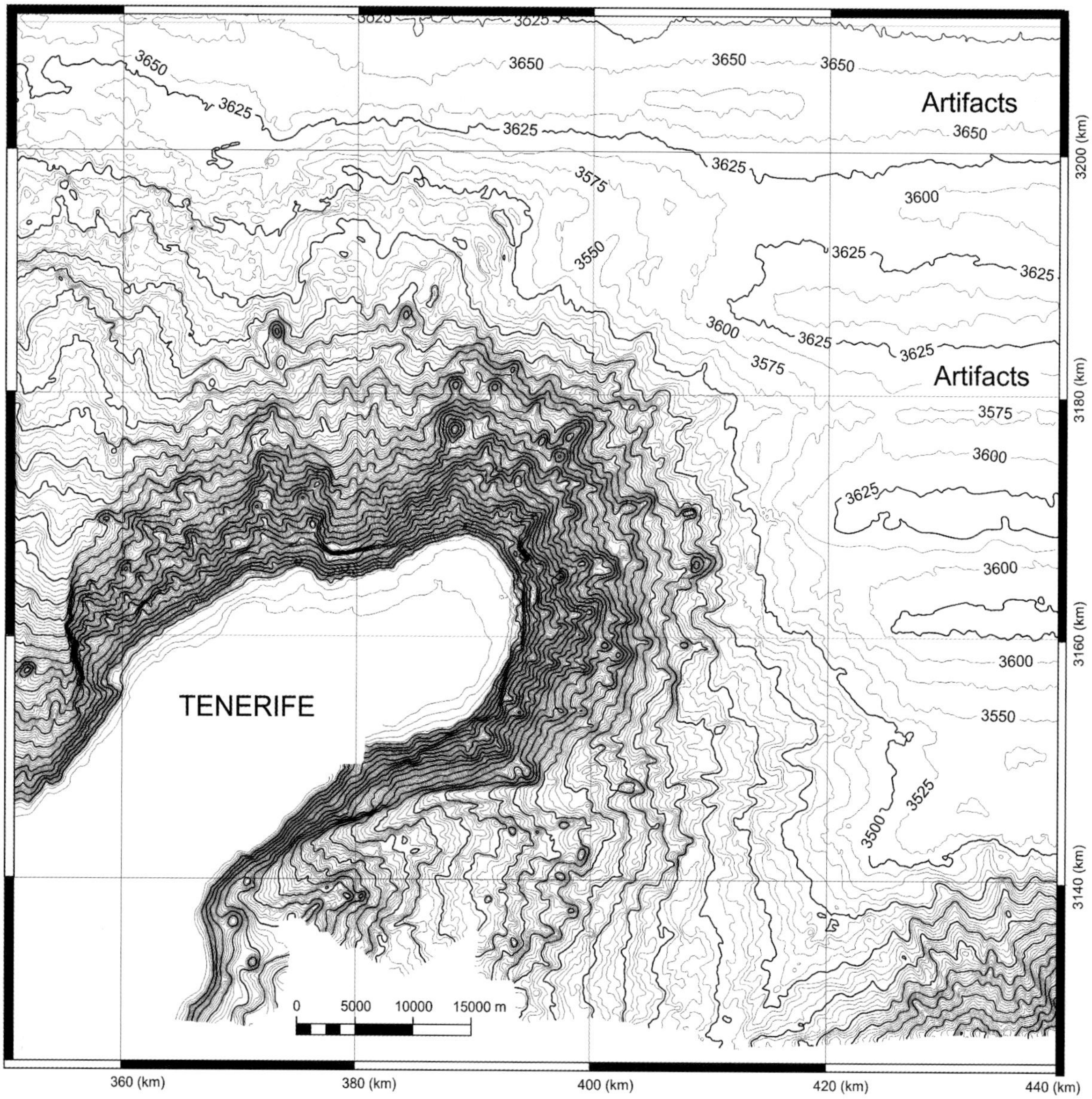

Figure 4. Bathymetric map of NE Tenerife. Multibeam coverage of 100%. The contour interval is 25 m. The coordinates are Universal Transverse Mercator (UTM) Zone 28 distances in km. Water depth in meters.

pinnacles, similar to these have been recognized and studied in the offshore continuation of volcanic rift zones on El Hierro (Gee et al., 2001; Mitchell et al., 2003) and Tenerife Island (Mitchell et al., 2003).

Dredged rocks in this area have been studied by Abratis et al. (2002), whose results indicate that igneous compositions correspond to alkali-rich ankaramites and basanites. The geochemical/isotopic signature of the Anaga Massif is apparently not restricted to the northeast part of the island itself, but is also characteristic of the Enmedio Seamount, located in the channel between Tenerife and Gran Canaria islands (Figure 1). Enmedio Volcano is the biggest submarine volcano in the surrounding area of Tenerife, reaching 660 m from a depth of 3300 m with a conical shape. It was first discovered by Schmincke and Rhim (1994) and later mapped in official bathymetric charts by Palomo et al. (1998). Its composition was studied by García-Cacho et al. (2000) and Abratis et al.

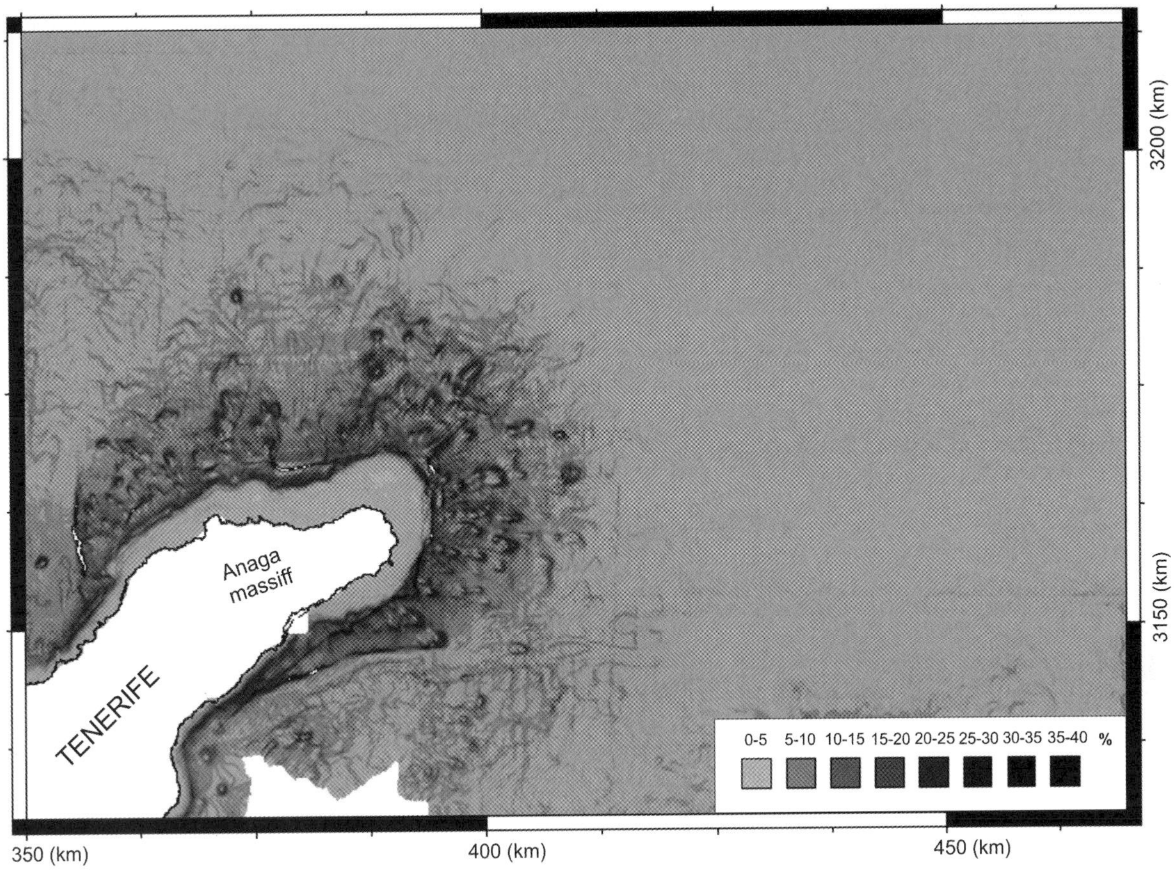

Figure 5. Slope map of NE Tenerife submarine region; dark areas are steepest slopes. Coastline is shown by solid black line. Image derived from 100-m gridded data. The coordinates are Universal Transverse Mercator (UTM) Zone 28 distances in km. For each particular point the slope is based on the direction of steepest descent or ascent at that point in a way that across the surface the gradient direction can change (the operation is similar to the way a First Directional Derivative is calculated but it defines the gradient direction at each point on the map automatically).

(2002), who concluded that Enmedio Volcano was comprised of benmoreites (Na-rich trachyandesites).

North sector

In the north of Anaga offshore massif area, we identify the result of mass wasting movements associated with slope failures, as demonstrated by the multi-beam bathymetry (Figure 4). Debris avalanche deposits were first recognised offshore Tenerife by Watts and Masson (1995) who noted that these deposits covered much of the submarine northern flank of the island. Their volume was estimated at 1000 km^3 by the Teide Group (1997). Our data show the presence of two fan-like structures where the bathymetric contours bulge seaward (Figure 4):

One of these aprons is located to the NW of the massif, at the end of a channel that limits the Anaga offshore structure to the west and has been recognised and studied in previous works as the 'Orotava Debris Avalanche' (ODA) (Teide Group, 1997; Watts and Masson, 2001; Masson et al., 2002).

The other lobe, called 'Anaga Debris Avalanche' (ADA), is the easternmost collapse deposit of Tenerife and was identified by Masson et al. (2002) based on the general bathymetric map, although its morphology was neither mapped nor completely studied due to the lack of detailed data in this area.

Orotava Debris Avalanche (ODA)

The Orotava Debris Avalanche is one of the largest landslides that has occurred in the Canary Islands (Masson et al., 2002) and can be traced upslope towards the Orotava Valley on Tenerife, where a flat floor flanked by steep scarps is seen (Palacios, 1994). This is supposed to be the product of lateral collapse

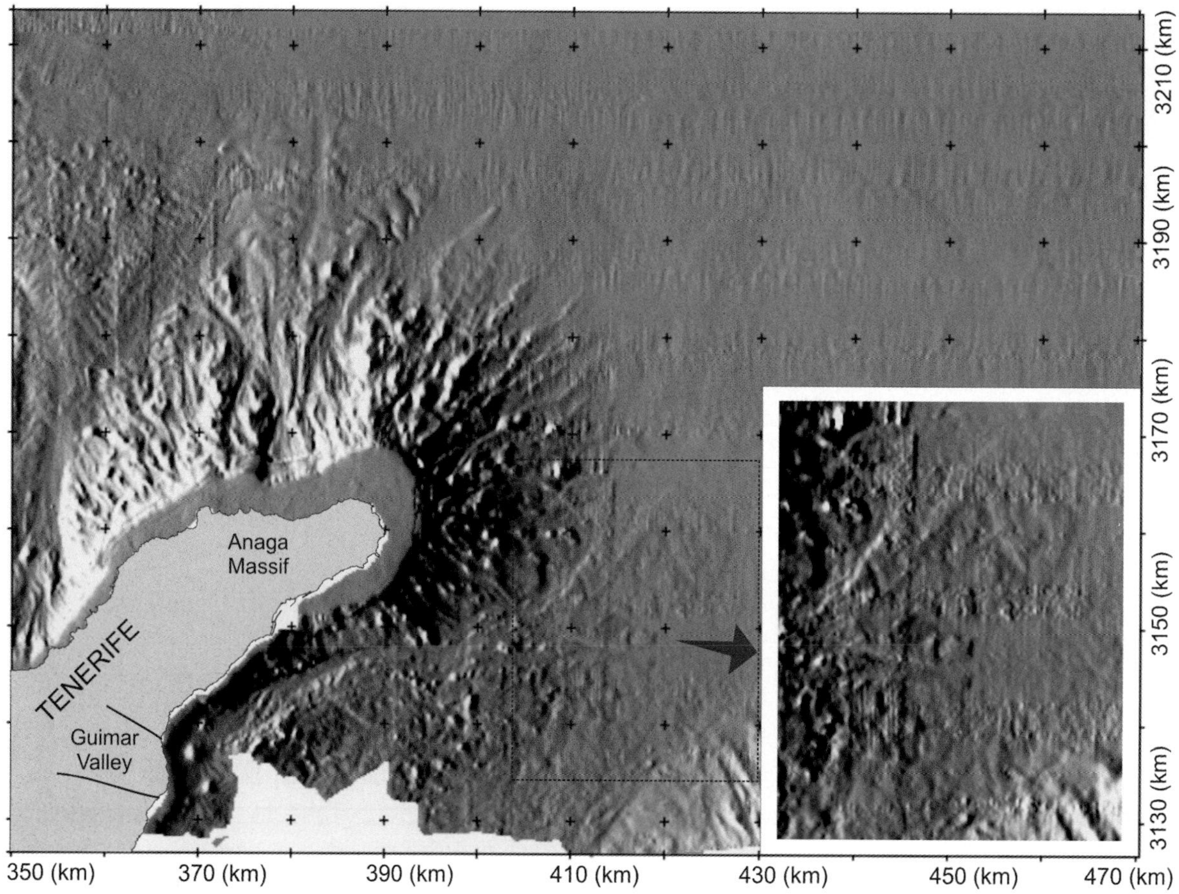

Figure 6. Shaded relief bathymetric map of NE Tenerife; apparent illumination from the northeast. Image derived from 100-m gridded data. The coordinates are Universal Transverse Mercator (UTM) Zone 28 distances in km.

of the Cañadas Volcano, estimated between 540 and 690 ka by the K/Ar dating of basaltic lavas in the upper part of the landslide scarp (Cantagrel et al., 1999).

The new data show that there is an abrupt change in the morphology of the area that limits the Anaga Massif structure offshore to the west. It can be deduced from the detailed observations based on bathymetry data that this limit has an original structural control and that faulting processes caused the abrupt change limiting the Anaga Massif in that area. This should have been favoured by the fact that Anaga Massif grew as an independent volcano that later volcanism linked to the dorsal series (Ancochea et al., 1990). The ODA was confined by this limit and the sediment apron also changed its morphology across this structure (Figure 6). However, it can be argued whether the avalanche was confined by the fault or if the scarp was created during the avalanche, indicating a purely morphological feature.

The roughness of the lobe is due to the presence of many small topographic highs between 30 and 110 m in height and between 50 and 1140 m in diameter. Limitation in the resolution of the bathymetric systems does not allow us to image highs less than 50 m diameter. Most of them can be interpreted as exotic blocks resulting from transport by the flow processes that formed the debris apron. However, some others seem to be pinnacles related to volcanic activity, but the resolution of the data is insufficient to determine the details of the morphology.

In the northeast area of the ODA, longitudinal pressure ridges have been identified (labelled P, Figure 7). These typical flow structures show the downslope movement of fan sediment due to gravitational processes. Pressure ridges were formed after the

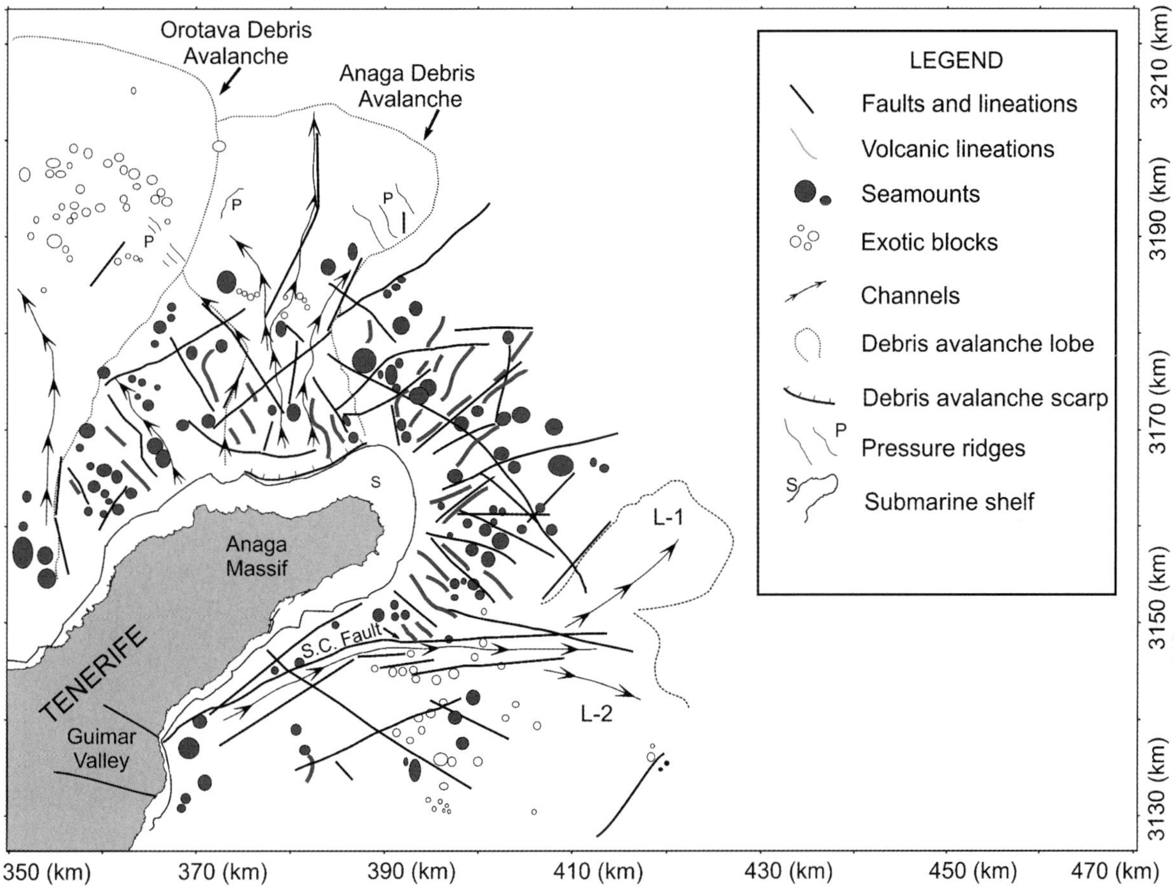

Figure 7. Morphological interpretation of Anaga surrounding flanks based on bathymetry data (bathymetry contour maps, slope maps, shaded relief maps and 3D images of bathymetry). The coordinates are Universal Transverse Mercator (UTM) Zone 28 distances in km. L-1 and L-2 show the locations of two lobes of the Güimar Debris Avalanche present on the study area. P shows location of shear and pressure ridges. S show location of the submarine erosive rocky shelf.

emplacement of the ODA, since they affect unconsolidated sediments. The length of these features varies between 27 and 182 m, with a relief that varies from 5 to 30 m and a separation between them ranging between 830 and 1180 m.

Anaga Debris Avalanche (ADA)

The new data have allowed us to map the real extension of the Anaga Debris Avalanche (ADA) (Figure 7). It is possible to differentiate two sectors of morphosedimentary significance associated to the ADA:

The proximal area of the ADA, where the collapse of the subaerial/submerged flank took place, starts with the Anaga landslide escarpment. This is not simple, being constituted by at least two scarps, which are identified in the 3D map (Figure 8). A broad entrant of 10 km length is recognised at 125 m relief, on the submarine shelf, that can reach 25–30° of slope. At 1250 m depth there is another entrant of 3.75 km length with a slope of 20°.

The sector extending from the abrupt area of scars to about 3000 m depth has a slope up to 10°. This sector represents the main channel of the debris avalanche. The width of the channel is 7000 to 10,000 m, but its relief is difficult of evaluate, because it is very modified. We estimate relief between 25 and 115 m. Mitchell et al. (2003) described how V-shaped channels can be seen formed around local highs with greater vertical relief in the flanks than lower down, which they consider consistent with the greater flow power of turbidity currents associated with the steeper gradient. These authors consider than the channels may have been active earlier in Anaga's history, and now the edifice base is an area of deposition rather than erosion. This is confirmed by low acoustic backscatter in sonar data (Masson et al., 2002). As the shelf

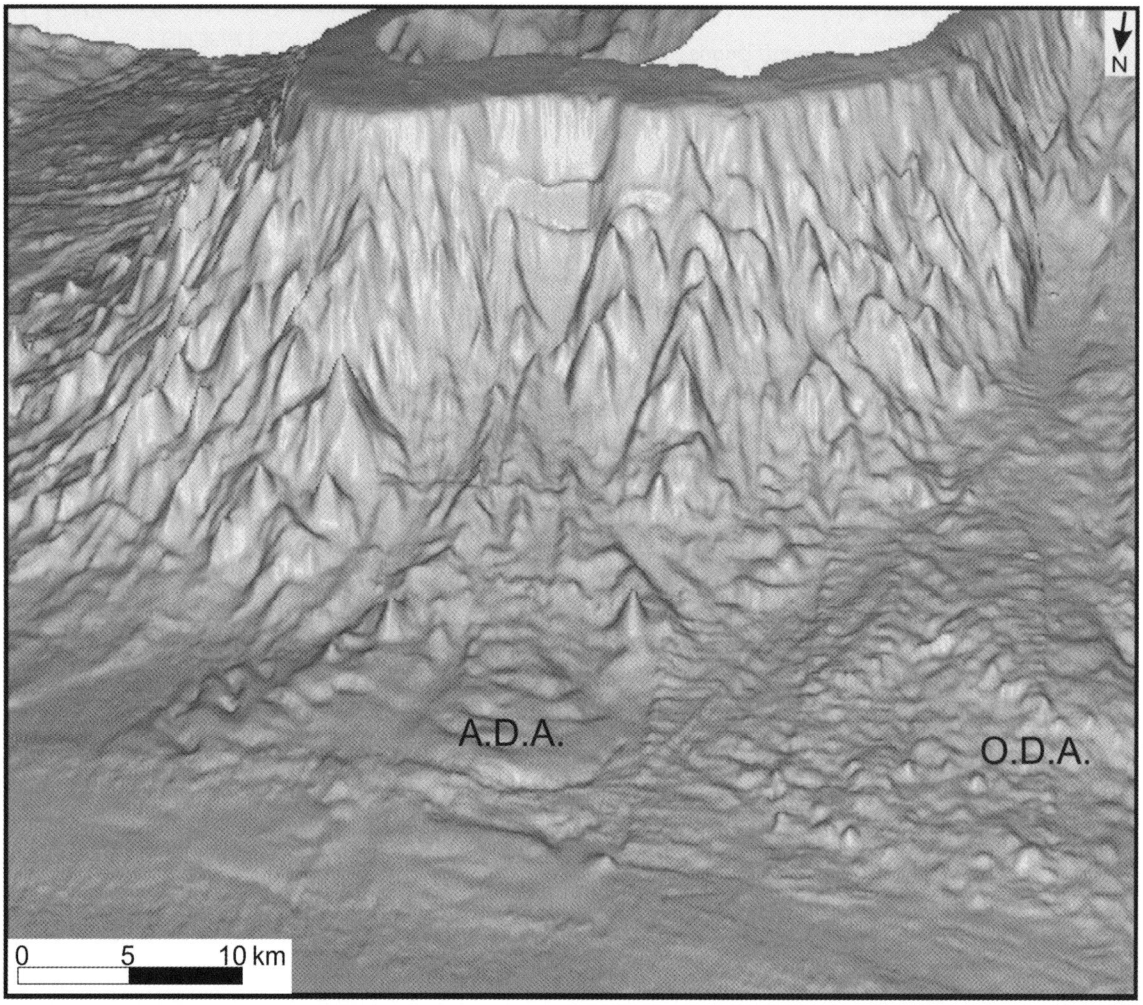

Figure 8. 3D image of Anaga north submarine flank showing the Anaga Debris Avalanche (ADA) and the Orotava Debris Avalanche (ODA). Scars of ADA are also recognised. View is from the NNW. Vertical exaggeration x10.

grew by coastal abrasion and subaerial erosion, the submarine canyons would have been progressively disconnected from their sources of hyperpycnal flows (Mitchell et al., 2003). Mitchell et al. (2002) studied the volcanic flanks of Tenerife, La Palma and El Hierro and found the bathymetric contours around Anaga more crenellated, its flank appearing more deeply gullied by erosion than the constructional flanks of El Hierro.

From our observations we found that these channels are frequently very linear and cut previous structures. We interpret many of them to be eroding tectonic lineaments.

The distal area of the ADA is characterized by the presence of the debris avalanche sediment apron. Its deposits could be recognised to a maximum water depth of 3650 m, 45 km from the coast line. The deposit is defined by the convex shape of the isobaths on the bathymetric map and by their morphology in the digital elevation model (Figures 4 and 6).

Accurate estimates of debris avalanche volumes are difficult. Urgeles et al. (1999) noted that the base of debris avalanche deposits can only rarely be determined from seismic reflection profiles. That is why volumes are usually approximated from thickness maps created by comparing failed slopes with adjacent unfailed regions or, more simply, by assuming a mean deposit thickness over the area of the deposit. We think it is more reliable to use sediment thickness maps than to assume a mean deposit thickness. To calculate the

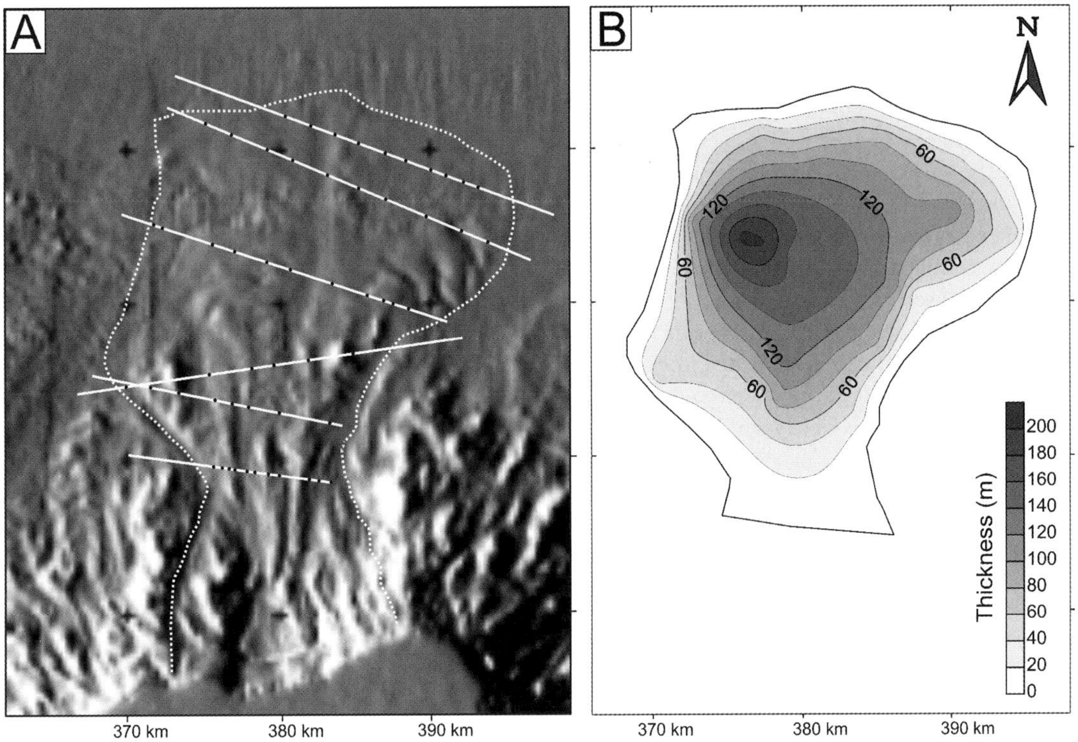

Figure 9. Panel A: Shaded relief bathymetric map of Anaga Debris Avalanche (ADA) area. White lines indicate the position of bathymetry profiles used for the thickness estimation. Black dots on the white lines are the points where the thickness of the avalanche has been estimated. Panel B: Thickness map of ADA. Thickness ranges between 0 and 220 m. An area of 502 km^2 for ADA deposits and a volume of 36 km^3 have been estimated. The coordinates are Universal Transverse Mercator (UTM) Zone 28 distances in km.

ADA volume, a thickness map was constructed using bathymetric profiles where the avalanche thickness was estimated in several points by comparing failed and unfailed slopes (Figure 9). A total volume of 36 km^3 was obtained for the ADA, whose deposits are extended over an area of 502 km^2. To establish this volume we did not consider the sediments placed at the avalanche's scarps or close to them, where erosion should be predominant. If there is sedimentation, it is difficult to quantify with geophysical methods due to the steep slope. At the northeast of the fan structure we have observed longitudinal pressure ridges (labelled P, Figure 7). As mentioned above, these flow structures show the downslope movement of fan sediment due to gravitational processes and are formed after the emplacement of the ADA. The length of these features ranges between 3400 and 5000 m. They have a relief that varies with a non-regular pattern from 3 to 38 m for the southern ones, and from 5 to 67 m for the northern ones. The separation between both sides is in the range of 1800 and 2800 m.

Hernández Pacheco and Rodríguez Losada (1996) studied in detail the geology of the Taganana Arc onshore Tenerife, located to the north of Anaga Massif. The Taganana Arc, formed by an old dyke-swarm, which cut breccias and massive materials, is curved and concave to the north. These authors suggested that Taganana Arc could be part of an edifice extending in the past to the north of the present-day coast. They proposed that the activity of a major fault, possibly located to the north of the Taganana margin, was the origin of the dip of the old dyke-swarm formed by dragging and rotating the northern side of the Anaga Block. They observed that the flexure the dykes have suffered was caused by fault drag and related them with the collapse of a volcanic edifice that extended in the past to the north of the present-day coast. They also suggested the possible existence of a big normal fault (not-mapped) to the north of Anaga that lead to this collapse. Finally, Hernández Pacheco and Rodríguez Losada (1996) suggested that the collapse to the north of Anaga is still active today, as could be seen from the existence of modern normal faults, small slides, scree and other recent avalanche deposits.

Our data appear to support the idea suggested by Hernández Pacheco and Rodríguez Losada (1996) that Anaga has suffered one or more collapses to the north. If we look at the scar described above, it seems possible than more than one landslide happened in this area. On other hand, the fan structure of the ADA seems to be the result of a single event. It is also likely that different collapses happened to the north, the last being the bigger, burying older events. We do not have evidence of how many collapses happened in this area with our detailed bathymetric analyses, so that more detailed studies with low frequency seismic reflection in the area of the scars will be needed to resolve this question.

Comparison between ODA and ADA

Our data confirms that the western boundary of the lobate deposits has been partially overlapped at its distal area by the ODA, as was suggested in previous works (Masson et al., 2002), which estimated an age of at least 600 ka for the ADA.

Comparing both avalanches (ODA and ADA) we note that ADA has some different characteristics:

- The surface of the ADA sediment lobe presents less exotic blocks. This deficiency of blocks could indicate that they had been buried by younger sediments, but it also can be attributed to a longer exposure to erosion, to the kind of material involved in the event, or to other flow characteristics.
- Both the channel and the fan structure of ADA have a larger degree of modifications or erosion compared to the ODA. Also, in ADA we have recognized at least four isolated seamounts, with diameters 800 and 1700 m, and heights between 140 and 350 m. Their size and shape in the resolution of our data is enough to be differentiated from exotic blocks. Other authors studying the hummocky terrain in the channel between Gran Canaria and Tenerife (Krastel and Schmincke, 2002) interpreted the large, isolated blocks as volcanic morphologies and the large number of small highs concentrated in an area as exotic blocks. Our interpretation seems to fit well with the conclusions of Krastel and Schmincke. It is not so evident studying our data, if the blocks are intruding the avalanche, or if they were partially buried by it. The fact than the ADA has suffered more triggering processes and volcanic intrusions, seems to indicate an older age for the ADA, which subsequently has been affected by erosional processes for a long time.
- The pressures ridges found in ADA have maximum lengths of 5 km, maximum relief of 38 m and a separation between them in the range of 1800–2800 m; while the ones in ODA, having a similar depth (as much as 30 m) are less clearly seen due to their shorter length (less than 200 m).

South sector

On the southeast coast of Tenerife, the dominant morphologic feature is the 'Cordillera Dorsal' (Figure 2), composed of a series of northeast trending ridges made of two flow series and separated by an unconformity. At the Cordillera Dorsal an 8–9 km wide valley was formed, with an amphitheatre-shaped head, an area of 73 km^2 and prominent side walls. The origin of this Güimar Valley was interpreted by Navarro and Coello (1989) to be the result of landslide processes, the same origin as for the Orotava Valley (Figure 2). Geological studies of Güimar Valley showed that it was formed by a flank collapse younger than 0.83 Ma, the most recent age obtained for lava flows in the scarp (Ancochea et al., 1990).

South of Anaga offshore massif, the most significant feature is a continuous scarp that produces a very significant slope break, that can be clearly identified in a three-dimensional representation of bathymetry (Figure 10) and whose trace is represented in Figure 7. The scarp is limiting the southern edge of Anaga offshore structure along at least 50 km, and its direction changes abruptly from N68°E to almost E-W. Between 600 and 3375 m water depth there is a clear change in all bathymetric contours, coinciding with the scarp (Figure 4). There are some differences between the areas at both sides of the scarp. The area at the north side of the scarp corresponds to the Anaga submarine massif. It has an irregular morphology characterized by numerous pinnacles, ridges and gullies that have been mapped in Figure 7. In some places, these forms are cut with a narrow and linear shape, indicating structural control. Submarine channels and canyons seem to exploit these tectonic directions. The area to the south side of the scarp has less slope, ranging from 0 to 5° (Figure 5) and is topographically smooth. The surface of this area has rough topography, similar to that of the ODA, with the presence of many small topographic highs, both pinnacles and perhaps exotic blocks. In the shallower part of the scarp along almost 30 km of the trace, there is a steep slope that varies

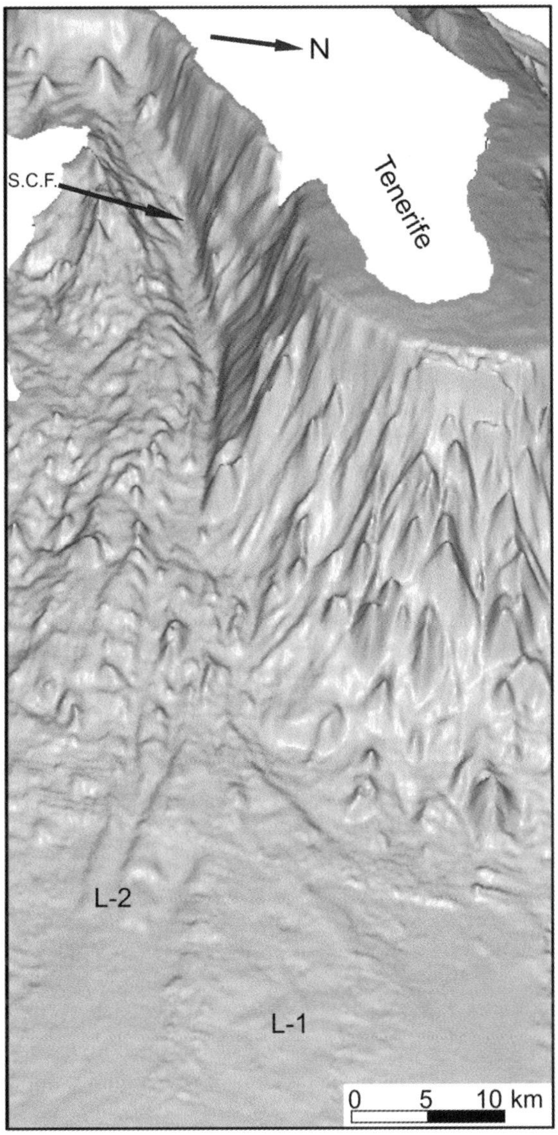

Figure 10. 3D image of Anaga SE submarine flank showing the Santa Cruz Fault and two lobes of Güimar Debris Avalanche (L1 and L2). View is from the ENE. Vertical exaggeration x10.

between 25° and 40°, but reaching slopes of 55° (Figure 4). In contrast, in the deepest part of the scarp, along 20 km, the slope decreases to 5–10°.

From the morphological observation of this scarp it is controversial whether it was created by sedimentary or tectonic processes or a combination of both. The origin of the steep escarpments bounding the seaward-facing embayment in the Canaries has been the subject of different interpretations, since its first association with gravitational processes by Bravo (1952). Nowadays it is accepted that giant landslides are a common feature in the Canary Islands and the relation between giant landslides and straight-walled valleys, calderas and wide coastal embayments, has been found onshore (Ancochea et al., 1990; Ancochea et al., 1994; Carracedo 1994; Guillou et al., 1996) and offshore (Watts and Masson, 1995; Teide Group, 1997; Masson, 1996; Urgeles et al., 1997; Carracedo, 1999, Mitchell et al., 2002; Mitchell et al., 2003, Masson et al., 2002 and Acosta et al., this issue). In Figure 7 the geographical position of the onshore Güimar's Valley scarp has been plotted, as well as the trace of the scarp found offshore. Looking at the coincidence of both features we could attribute the origin of the scarp entirely to the landslide process of the Güimar Debris Avalanche, being a continuation offshore of the eastern sidewall of this avalanche on land. Such an interpretation was suggested by Masson et al. (2002) for the sidewalls of El Golfo Debris Avalanche on El Hierro island. But a sedimentary origin for the submarine scarp without involving structural process can be argued. Observation of the relations between the amphitheatres and the scarps in other debris avalanches of the Canary Islands, such as the ones in El Hierro, La Palma, Tenerife and Gran Canaria islands, reveals that usually the head amphitheatre of the debris avalanche is aligned normal to the downslope direction of movement of the avalanche (for references see Masson et al., 2002; Acosta et al., this issue). Here the offshore scarp is not aligned with the direction of the onshore scarp, but instead is deviated from that direction by 55° to the north (Figure 7). Looking at the scarps of Güimar Valley on land, it seems obvious that if an avalanche happens in the absence of a tectonic stress, it would create an offshore scarp following more or less a straight direction with the onshore one, or having a gentle turn due to the regional bathymetry, as happened in some of the landslides in Hawaiian islands (Moore et al., 1989). The sharp angle of continuation between both scarps seems to indicate than the debris avalanche must have changed its direction sharply when it reached the coast of the island if the scarp was formed from the carving of the avalanche flow.

Faulting has been recognized as one of the factors that can directly cause or contribute to the occurrence of giant landslides. Day et al. (1999) described the well developed topographic San Andrés fault scarp on the flank of the steep-side NE rift of El Hierro, and related it to an aborted giant collapse; Vidal and Merle (2000) used a model to prove that the reactivation of a vertical fault in volcanic cones generates normal faults and an upturning of the layers that induces a flank collapse. In Hawaii, the Hilina slump has been recognized as an active landside that breaks the mobile southeast flank of Kilauea volcano and is headed on land by a system of seaward facing normal faults (Morgan et al., 2003). From the study of the characteristics of this scarp, its relation with the on-land features of Güimar Valley and the comparison with other avalanches in the Canaries, we suggest that a tectonic process seems to have created the scarp. Whether the fault scarp was created during on avalanche or before it remains unsolved, but the stress regimen in this area could have been the cause of the debris avalanche process on the buttressed flank of Tenerife. A submarine canyon has carved the sediments following the linear layout of the fault scarp. This canyon was also recognized by Krastel et al. (2001) and Krastel and Schmincke (2002), who mapped the Güimar Scarp on land but didn't interpret a debris avalanche scarp offshore.

In the absence of seismic lines transverse to the scarp, its origin remains unsolved and therefore we can interpret it as a fault. Future studies will be needed to solve the fact that Güimar Debris Avalanche seems to be a particular case among all others previously studied in the Canary Islands. Since it is the first time that this scarp is described and interpreted as a fault, we have called it 'Santa Cruz Fault Scarp', because the town of 'Santa Cruz de Tenerife' is the nearest main geographic place on land.

Güimar debris avalanche (GDA)

At a water depth of 3200–3300 m the sea-floor has a 0 to 5° slope. In this area there is an important accumulation of sediments that forms a deposit of 20 km length and 12 km width that has been mapped as L1 in Figure 7. Adjacent to this lobe to the south, another fan-shape structure has been recognised with a length of 8 km and a width that can not be exactly defined because it extends beyond the studied area limit (L2 in Figure 7). The thickness of these lobes varies between 5 and 15 m for L1 and 5 and 10 for L2. Hummocks can be recognized in the digital model elevation (Figure 6) and in the detailed bathymetric map. Because of the fan shape structure and the existence of the block field terrain, we interpret these deposits as products of a debris avalanche. Some of these topographically isolated highs seem to be volcanic cones rather than blocks transported by the debris avalanche, but distinction between them has not been possible, as there are no clear criteria to differentiate them. So, although it seems clear that this area is covered with the deposits left by a debris avalanche, some volcanic morphology may also be present. These fans can be seen in the digital model elevation (Figure 6) and the three-dimensional representation of bathymetry (Figure 10).

Our data indicate that the fans are the northern deposits of the Güimar Debris Avalanche. The total extension of the avalanche in Gran Canaria-Tenerife channel has been mapped by Krastel and Schmincke (2002) and by Acosta et al. (this issue). They interpreted Güimar Valley as the source area for the avalanche deposits of this area, and our results point to the same conclusion. These authors also recognized from backscatter data a large number of individual hummocks of different sizes and they discussed their nature, concluding that most of them are exotic blocks, although some, usually located close to the islands, are volcanic cones.

Anaga offshore distal region

Sedimentary characteristics

Topographic parametric sonar seismic profiles have been the basis for studying the Anaga offshore distal region, because bathymetric data in this area presents numerous artefacts (see Figure 4) that cannot be removed by standard processing.

The area beyond the 3500 m bathymetric contour is called the Anaga offshore distal region. From this contour, the slope stabilizes to less than 5°, resulting in a more or less flat seafloor, where depths range from 3500 to 3700 m, which are the maximum values northwest of Anaga. Seismic profiles show parallel and continuous reflectors (Figure 11) that characterize the distal region offshore Anaga Massif as a typical area of deep oceanic hemipelagic sedimentation. In the more distal region, the parametric system recorded a maximum of 0.12 sec (two way travel time). If we assume a velocity for marine sediments in Canarian Archipelago of 1.7–2.1 km/sec. (Dañobeitia and Canales, 2000), this implies a total thickness of 102–126 m.

In the studied seismic sequence we have selected four reflectors (named R1, R2, R3 and R4) to show the interpretations. We choose these because of their presence in most of seismic profiles. Although there were some stronger ones, they have less continuity along the profiles.

Taking into account the nearest deep ocean drill sites to our study area we can deduce a very similar

SEISMIC FACIES		DISTRIBUTION	INTERPRETATION
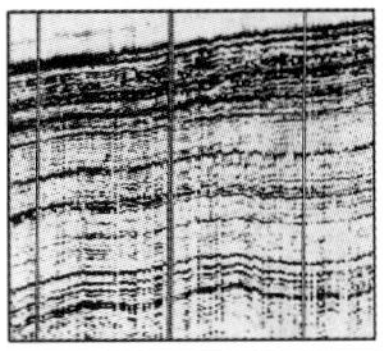	Parallel	Deep areas	Deep sea pelagic sedimentation
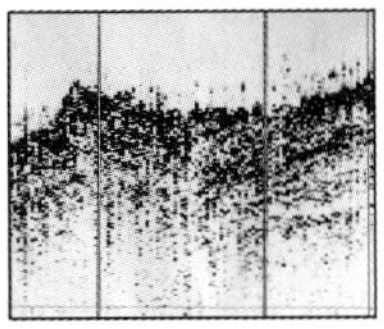	Chaotic	Ubiquitous	Large scale mass movement Small scale mass movement High quantity of sedimentation of island provenance Deformation caused by intrusive bodies
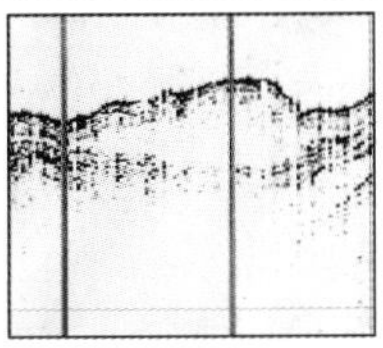	Transparent	Ubiquitous	Muddy strata seismically homogeneous Intrusive bodies
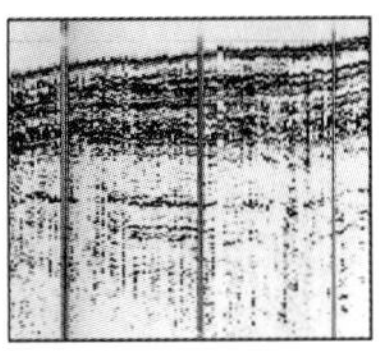	Convergent Divergent	Distal island areas	Lateral changes in the depositional rate
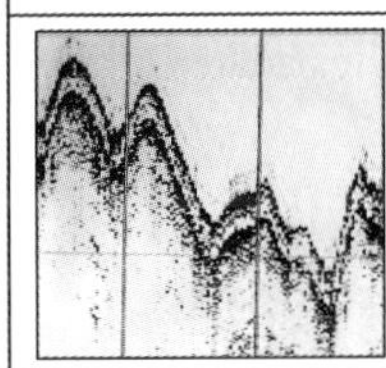	Undulating	High slopes areas Fault areas	Folding caused by sedimentary-gravitational processes Folding caused by tectonic processes

Figure 11. Seismic reflection facies defined in this study and their geologic interpretation.

composition for the material that forms the parallel and continuous reflector seismic sequence observed in profiles. Therefore, we used Ocean Drilling Program (ODP) Sites 953 and 954, located to the north and northeast of Gran Canaria (Figure 3), which reveal the lithological composition of the sediment in this area. In both sites, the first hundred metres corresponds to Unit I and are constituted by Holocene to upper Pliocene (0–3 Ma) sediments, consisting dominantly of pelagic clayey, nannofossil ooze and graded nannofossil clay-silt, with lesser amounts of foraminifer sands, lithic crystal sands, and silts. Coarse sands, rich in neritic biogenic material, recovered in the upper 100 m have been interpreted by Schmincke (1995) as turbidites possibly related to glacially controlled changes in sea level. The minor, thin fallout tephra layers at these sites were probably erupted on Tenerife and may represent the outer part of larger fallout fans (Schmincke et al., 1995). In this way, we can consider that the lithological composition in the Anaga offshore distal region may not differ too much of the one drilled on north margin offshore Gran Canaria.

Five distinct seismic facies are identified based on seismic reflection character and geometry. Environment interpretation of each seismic facies has been made based on Mitchum et al. (1977) (Figure 11). The detailed analysis of all the seismic profiles shows clearly the influence of Tenerife island. To the western edge of the seismic profiles the above mentioned reflectors are not recognized. The seismic sequence

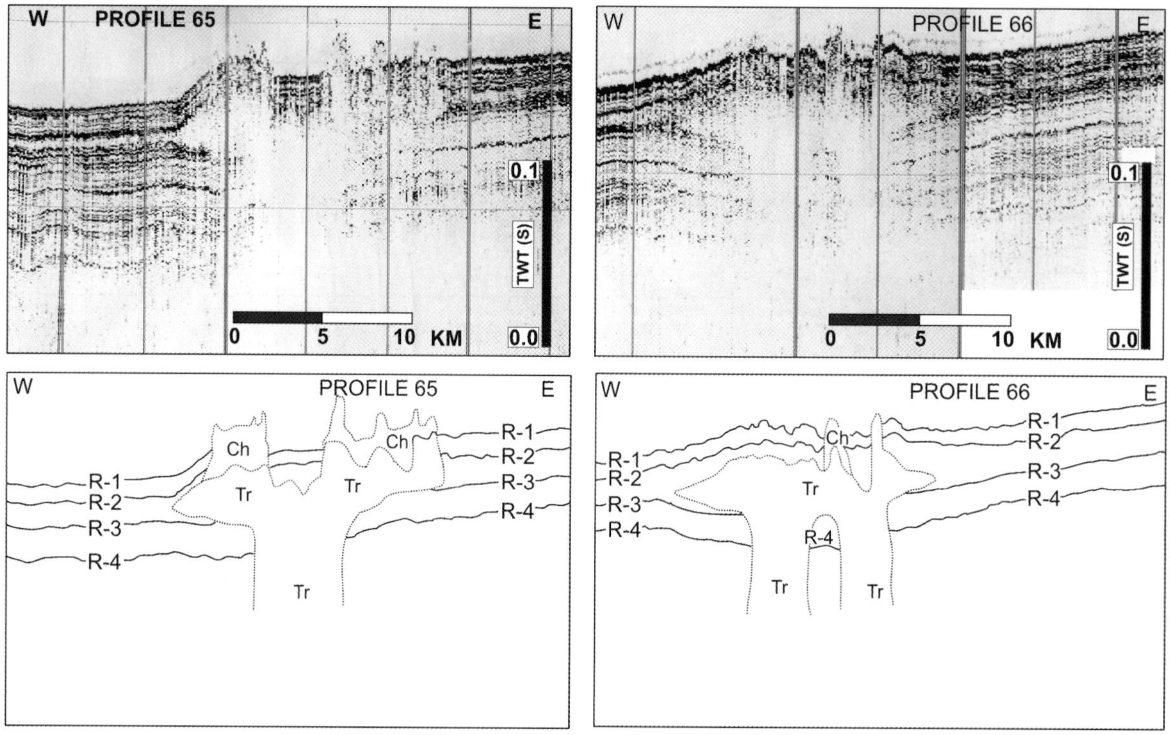

Figure 12. Interpreted parts of Profiles 65 and 66 showing two intrusive bodies of transparent facies (Tr) that cut parallel and continuous sequences. Location of the seismic profiles and the intrusive bodies are placed in Figure 3. Selected reflectors for this study are shown as R-1, R-2, R-3 and R-4 and chaotic facies as Ch. There is not enough evidence to interpret the volcanic or sedimentary origin of the intrusions.

loses its parallel and continuous aspect to show a more chaotic sequence nearer to the island. As the sequence is closer to the island, there is more volcanoclastic material mixed with the deep sea pelagic sediments. Bottom currents also affect the sediments, as has been reported in other areas of the Canaries, such as the channel between Gran Canaria and Fuerteventura, in which sediments are entirely modified by erosional currents (Funck and Lykke-Andersen, 1998), or the channel between Tenerife and Gran Canaria, where Teide Group (1997) recognized sediments whose internal morphology suggested deposition under the influence of bottom currents, probably the North Atlantic Deep Water at a depth of 1500–3900 m.

The Seismic sequence is interrupted in the distal area by an intrusive body. Continuity of the parallel facies of Profiles 65 and 66 ends sharply, as can be seen in Figure 12. Transparent facies material intrudes the sequence with one root in Profile 65 and two roots in Profile 66. These roots seem to be the conduits that fed the intrusive bodies. Both of them have a general oval shape, but in the shallow subsurface it turns more irregular with the intrusive material cutting the sediments that lie above. We have not enough criteria to distinguish if the transparent facies material is of volcanic or sedimentary origin.

Structural characteristics

TOPAS Profiles 51, 68, 67 and 66 show the structural complexity typical of an area proximal to a volcanic island. Numerous faults have been distinguished. In some areas these faults are separated from other tectonic structures by great distances, but there are also some areas where fracture density is higher and we found others faults every hundreds of metres or every few kilometres. This occurs along the western edge of Profile 67, where we recognized four faults over ten kilometres (Figure 13). The undulating reflector pattern in this region, indicating deformation, could be the result of the same tectonic process that generated the faults. The lack of penetration in this ultra-high resolution profile does not allow us to discern between these two possible mechanisms.

Among all the identified fractures, there is one that is noteworthy because of its particular characteristics, called the 'Guayotá Fault'. This fault has been recorded on seismic Profiles 51, 68, 67, 66, 65, 64 and

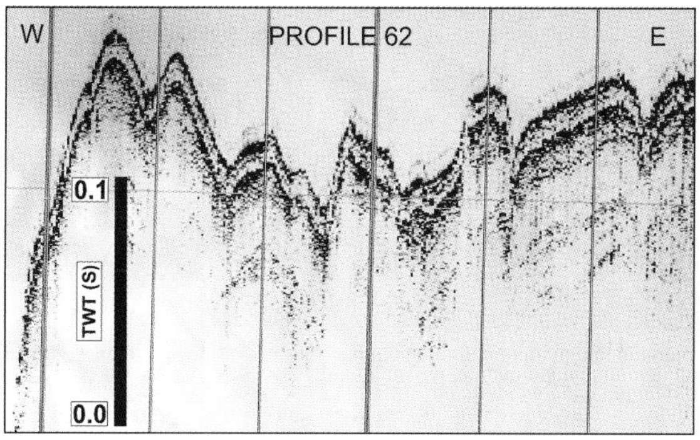

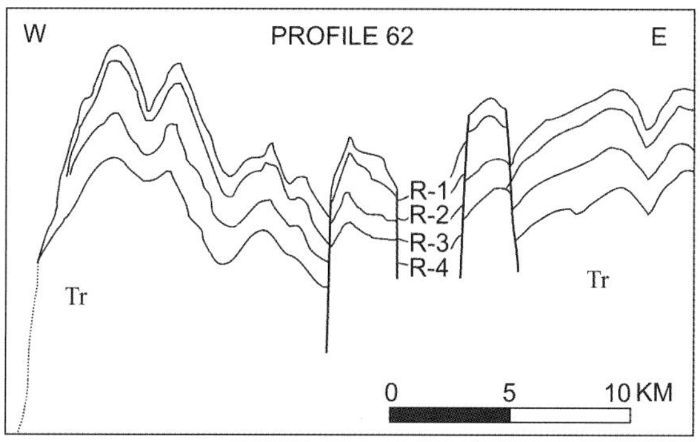

Figure 13. Interpreted segment of seismic Profile 67 showing a high fracture density. The undulated reflector pattern could be the result of the same tectonic process that generated the faults, or could be deformation caused by an igneous intrusion process, as transparent facies are seen below. Lack of penetration in this ultra-high resolution profiles does not allow us to discern between these two possible genesis mechanisms. Location of the seismic segment is shown in Figure 3.

62, in a north to south order, and as the profiles are separated by 10 km, its total length appears to be at least 60 km (Figure 3). It can be seen that the fault affects at least 100 metres of the more recent sediments, but as we do not have enough penetration, it is not possible to evaluate the actual depth extent of the fault. Additional seismic data is needed to provide a real estimate of the thickness of materials that have been affected by the fault and if they reach the basement or not. The Guayotá Fault is oriented N-S along its north edge, among seismic Profiles 51, 68, 67 and 66; and after an inflexion between profiles 66 and 65 it takes a direction of NW-SE, in profiles 65, 64 and 62 (Figure 3).

Figure 14 shows the segments of the seven profiles where the fault appears. The Guayotá Fault is a compressive structure with an offset varying between 25 and 30 metres. Because of the morphology seen in the profiles and the orientation that it presents we suggest that it is a contractional strike-slip fault.

Two of the profiles do not show a vertical offset. As they are the northern and southern ones and they probably indicate the limits of the fault. In Profile 51 the fault is recognized as a deformation area, but the vertical offset between the reflectors cannot be observed. In Profile 62 the fault is recognized by the abruptly disappearance of the reflectors, but again there is no offset, nor deformation in the upper area of the seismic sequence.

Structural analysis

Methodology of the structural analysis

Structural data are required to understand the complex regional context of the Canary Islands, but there are very few structural studies. Field structural data have been taken and interpreted on Fuerteventura by Stillman (1987) and Fernández et al. (1997), in Lanzarote by Marinoni and Pasquarè (1994) and in La Palma by Fernández et al. (2002).

Numerous faults have been observed in the Anaga offshore region (Figure 7). Interest in knowing which the prefered structural directions are, guided us to carry out a structural analysis. Previous analysis of the topography of the region (Mitchell et al., 2003) did not find faulting to be an important process in the evolution of this area. However we find that erosion is often taking place along existing structural directions.

Submarine faults were mapped by bathymetric maps contoured at different values. Slopes maps and digital elevation models have been used. The software we used has allowed us to do many tri-dimensional models that validate the interpretations. In the bathymetric maps, faults have been distinguished by the sharp inflexion produced in successive contours by faults. In the slope maps tectonic gradients have very clear slope contrasts. In the digital elevation models, faults are recognised as in aerial photographs by shadow changes, and also because they cut previous structures like volcanic lineations, channels and debris aprons. This kind of model helped us to distinguish faults that could not be identified in bathymetric and slope maps. The combined study of bathymetric maps, slope maps, three-dimensional models and digital elevation models is the best method to recognise submarine faults in a map.

To quantify the preferred fractured direction all the interpreted structures have been georeferenced, and

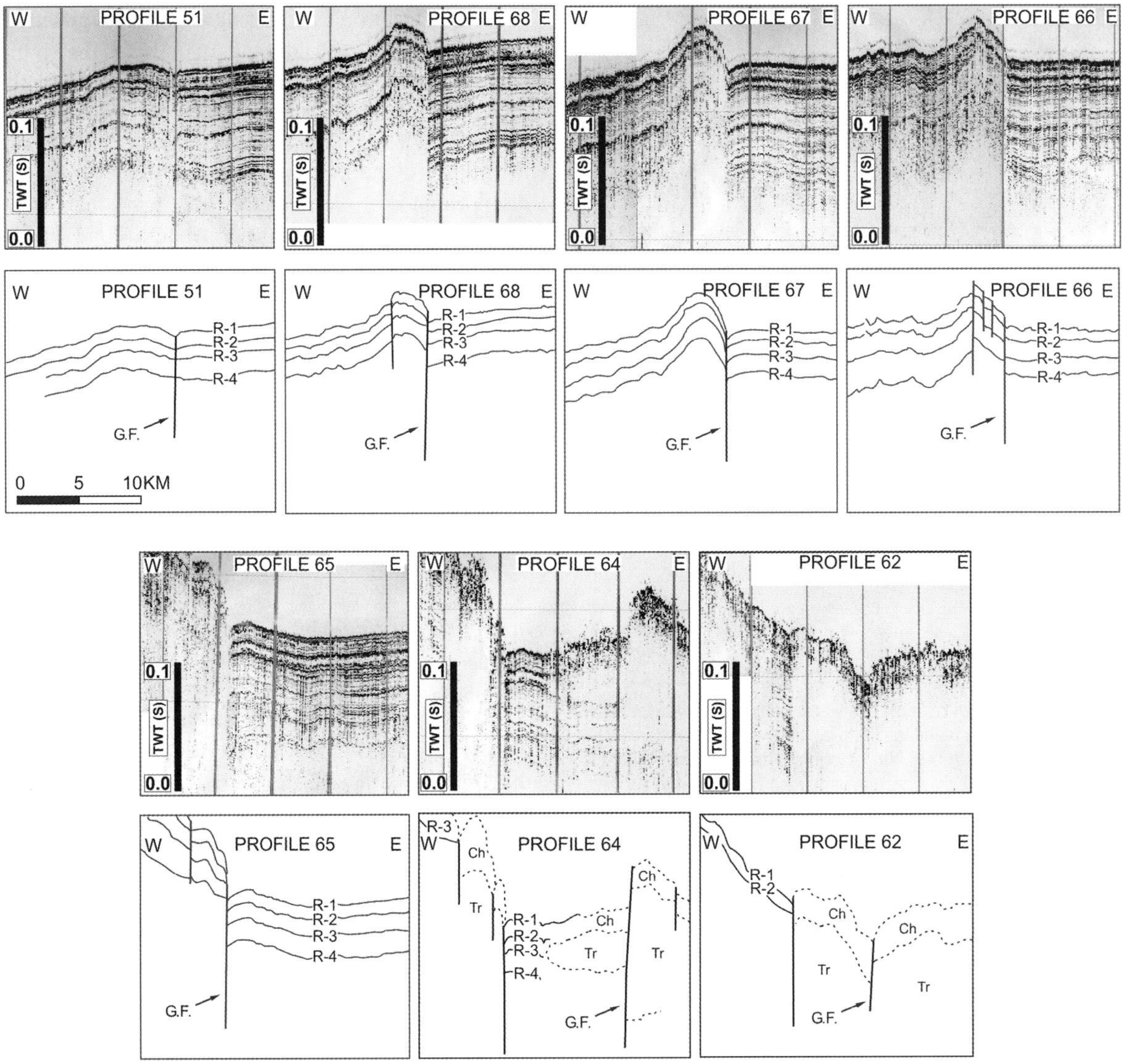

Figure 14. Interpreted seismic reflection Profiles 51, 68, 67, 66, 65, 64 and 62 showing the Guayotá Fault. Location of the seismic profiles and the interpreted fault are placed in Figure 3. The fault is affecting at least 100 metres depth of the more recent sediments. Its total length in a ground plan appears to be at least 60 km. Guayotá Fault direction is N-S in its north edge, among seismic Profiles 51, 68, 67 and 66; and after an inflexion between Profiles 66 and 65 it takes a direction of NW-SE in Profiles 65, 64 and 62. It is a compressive fault with an offset varying between 25 and 30 metres. We suggest that the Guayotá Fault is a contractional strike-slip fault.

when a fault changes its direction one or more times, we must consider these changes segmenting this lineation in more than one (Fault map in lower panel of Figure 15). For the structural analysis we assign a proportional weight to the fault trace length.

Results of the structural analysis

A total population of 63 fault segments is represented in a rose diagram with an interval class of 15° (Figure 15). Almost all directions are represented, showing a radial pattern typical of areas where extension of crustal layers occurs because of the intrusion of magma bodies or other kinds of domes. Faults are recognized radiating from the center, as they align in the direction perpendicular to the direction of maximum tension (Twiss and Moores, 1992; Ramsay and Lisle, 2000). Radial patterns have been previously identified in oceanic islands in fractures and dykes. Examples are found in the Canaries: like the ones

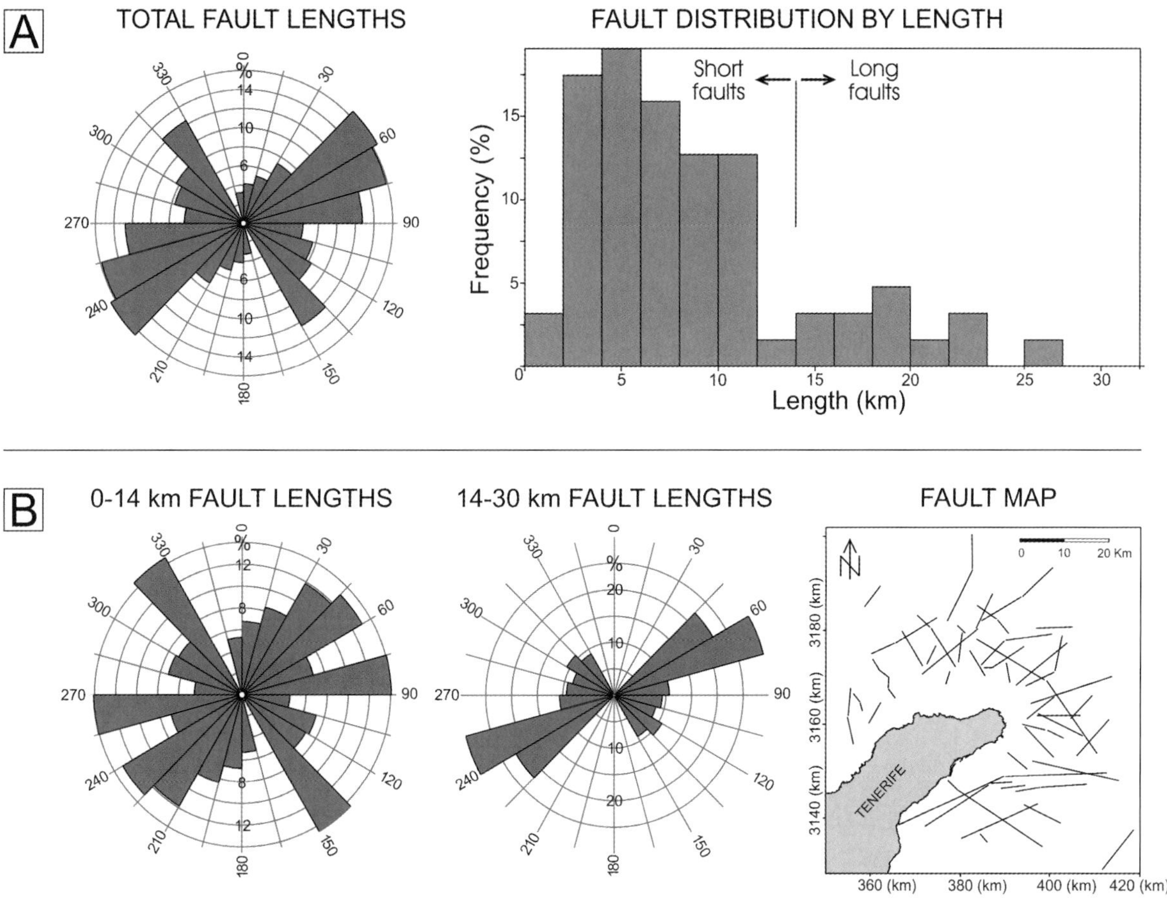

Figure 15. Panel A: Total fault length rose diagram and fault length distribution histogram. Panel B: Rose diagram for faults lengths between 0–14 km, rose diagram for faults lengths between 14–30 km, and the fault map used in the structural analysis. The coordinates are Universal Transverse Mercator (UTM) Zone 28 distances in km.

reported in Fuerteventura (Ancochea et al., 1996; Stillman, 1999), Tenerife (Ablay and Martí, 2000) or La Gomera (Ancochea et al., 2003); as well as in other archipelagos like Cape Verde (Day et al., 1999) or the Society Islands (Blais et al., 2000).

Further more, two major structural directions are recognized: The main one is NE-SW, with its conjugate at 90°, NW-SE, being the second in importance. The NE-SW submarine direction matches with the Tenerife direction in the Anaga Massif area and Cordillera Dorsal, clearly indicative of the importance that structural control has in the growth of oceanic island.

The interpreted faults have been analyzed as a function of length. Two modal values appear, allowing us to separate two populations of faults: The short ones less than 14 km, and the long ones for faults between 14 and 30 km. We can conclude that for the shorter faults the same radial pattern is recognised, although there are three preferred directions: Two of them were the same as seen in the general rose diagram, NE-SW and NW-SE, but an E-W direction also emerges. While in the shorter faults a radial pattern is recognized, when studying the longer faults the results point to a more defined pattern in a NE-SW direction.

Our results have been compared with structural field analysis undertaken on the Anaga Massif (Marinoni and Gudmundsson, 2000), where three main directions were obtained. Two of them, NW and NE, coincide roughly with the main axial trends of Tenerife, while another E-W trend does not correspond to the main axes of Tenerife, or with the rift-zone trends proposed by Carracedo (1994). Instead, it correlates with one of the major tectonic trends of the Canarian region, which consists of E-W structures (Banda et al., 1981: Marinoni and Pasquarè, 1994).

Structural analysis of the submarine fractures shows a good correlation with the main directions obtained in the Anaga onshore region.

Conclusions

The morphology of Anaga submarine slopes are partly a result of the volcanic terrain being reshaped by mass wasting and faulting. Bathymetric and TOPAS seismic data collected around the peninsula of Anaga have increased our understanding of the growth and destruction of Tenerife and oceanic islands in general.

1. The offshore Anaga Massif starts with a submarine shelf, a consequence of wave erosion during Pleistocene glacially induced regressions. Volcanic morphologies, faults that cut previous structures, and the results of mass wasting movements associated with slope failures can be recognized in the submarine Anaga structure. Below the 3500 m contour the slope stabilizes to less than 5° in a more or less flat topographic seafloor, being deepest NW of Anaga.
2. An area of 502 km^2 and a debris volume of 36 km^3 characterize in the ADA (Anaga Debris Avalanche). The presence of various scars shows that Anaga could suffer several collapses to the north, as also suggested by Hernández Pacheco and Rodríguez Losada (1996) studying the Taganana Arc geology onshore Tenerife. Our data confirm that the western boundary of the lobate deposits has been partially overlapped at its distal area by the ODA (Orotava Debris Avalanche), as suggested in previous works (Masson et al., 2002) which indicated an age of at least 600 ka for the ADA. Comparison between ODA and ADA also point to an older age for ADA, since more erosive processes and volcanic intrusions have affected both the channel and the fan structure. Moreover, ADA has less exotic blocks than ODA, due to burial by younger sediments.
3. Bathymetric data support the existence of many volcanic morphologies, between 1000–3000 m across and 100–200 m high around Anaga submarine massif. They could be related to dyke activity radiating from a central volcanic zone. In some areas distinction between volcanic morphologies and exotic blocks from avalanches is difficult and in the absence of other data there is no evidence to make a distinction.
4. An intrusive body of transparent facies has been seen in two seismic profiles interrupting the parallel facies of the seismic sequences in these areas. They have one and two roots respectively that acted as conduits feeding the intrusions. We do not have enough criteria to distinguish if the transparent facies material is of volcanic or sedimentary origin.
5. Numerous faults have been interpreted and represented from the new data. Most of them were recognized by the combined study of bathymetric maps, slope maps, tri-dimensional models and digital elevation models. They have been used to do a structural analysis. Two of the larger structures recognized have been called 'Santa Cruz Fault' and 'Guayotá Fault':
   - 'Santa Cruz Fault' was recognized in the bathymetric data limiting the southern edge of Anaga offshore structure along at least 50 km. Its direction changes sharply from N68°E to almost E-W.
   - 'Guayotá Fault' was recognized in seismic profiles at the NE distal area of Anaga along 60 km of strike. Guayotá Fault is a compressive structure that affects at least 100 m of sediments with a 25–30 m offset in a direction that changes N-S to NW-SE towards the south. We interpret it as a transpressional strike-slip fault.
6. From the analysis of all the submarine faults directions that have been interpreted, we found a radial pattern typical of oceanic islands. Besides this typical pattern, three main directions are emphasized in the rose diagrams: NE-SW, NW-SE and E-W. The main orientation, NE-SW, matches the Tenerife direction for the Anaga Massif and Cordillera Dorsal, and is also the one that controls the longer faults. There is good correlation with the main directions obtained in the Anaga onshore region by previous authors.
7. Faulting has been recognized as a key process for the occurrence of debris avalanches and the growth of volcanic lineaments. Moreover, faulting is modifying previous structures and channelling the flows. Submarine faults are more difficult to study compared with those onshore because of the environment in which they are located, but is important to quantify submarine faults as far as possible, as submarine flanks represent a major part of a given oceanic island.

Acknowledgements

This work was funded by the ZEE project (Oceanographic and Hydrographic Research of the Spanish Economic Exclusive Zone) and the 'Universidad Complutense de Madrid'. We are grateful for predoctoral grant to Pilar Llanes coming from the 'Conserjeía de Educación de la Comunidad de Madrid y Fondo Social Europeo'. We also thank the captain, officers and crew of B.I.O. Hespérides in cruise ZEE-99 for their cooperation at sea. Comments by G. De Vicente, from Universidad Complutense de Madrid, have greatly improved the structural analysis. Thoughtful reviews by Douglas G. Masson and Hans-Ulrich Schmincke were invaluable in preparing the final manuscript.

References

Ablay G.J., J. Martí, Stratigraphy, structure, and volcanic evolution of the Pico Teide-Pico Viejo formation, Tenerife, Canary Islands, *J. Volcanol. Geotherm. Res.*, 103, 175–208, 2000.

Abratis M., H.U. Schmincke, T.H. Hansteen, Composition and evolution of submarine volcanic rocks from the central and western Canary Islands, *Int. J. Earth Sci.*, 91, 562–582, 2002.

Acosta J., C. Palomo, E. Uchupi, A. Munoz, J. Escartin, P. Herranz and J.L. Sanz, Morphology and seismic character of north slope of Tenerife, Canary Island. Evidence for episodic massive landslides, *J. Geophys. Res.*, 102, 20325–20342, 1997.

Acosta J., E. Uchupi, A. Muñoz, P. Herranz, C. Palomo, M. Ballesteros and ZEE Working Group, Geologic evolution of the Older Canary Islands: Lanzarote, Fuerteventura, Gran Canaria and La Gomera, with a Brief description of the avalanches on the Younger Islands: Tenerife, La Palma and El Hierro, *Mar. Geophys. Res.*, This issue.

Ancochea E., J.M. Fúster, E. Ibarrola, A. Cendrero, J. Coello, F. Hernan, J.M. Cantagrel, and C. Jamond, Volcanic evolution of the island of Tenerife (Canary Islands) in the light of new K-Ar data, *J. Volcanol. Geotherm. Res.*, 44, 231–249, 1990.

Ancochea E., F. Hernán, A. Cendrero, J.M. Cantagrel, J.M. Foester, E. Ibarrola and J. Coello, Constructive and destructive episodes in the building of a young oceanic island, La Palma, Canary Islands, and genesis of the Caldera de Taburiente, *J. Volcanol. Geotherm. Res.*, 60 (3–4), 243–262, 1994.

Ancochea E., J.L. Brändle, C.R. Cubas, F. Hernán, M.J. Huertas, Volcanic complexes in the eastern ridge of the Canary Islands: the Miocene activity of the island of Fuerteventura, *J. Volcanol. Geotherm. Res.*, 70, 183–204, 1996.

Ancochea E., J.L. Brändle, M.J. Huertas, C.R. Cubas and F. Hernán, The felsic dikes of La Gomera (Canary Islands): identification of cone sheet and radial dike swarms, *J. Volcanol. Geotherm. Res.*, 120, 197–206, 2003.

Banda E., J.J. Danobeitia, E. Surinach and J. Ansorge, Features of crustal structure under the Canary Islands, *Earth Planet. Sci. Lett.*, 55, 11–24, 1981.

Blais S., G. Guille, H. Guillou, C. Chauvel, R.C. Maury and M. Caroff, Géologie, géochimie et geochronology de l'île de Bora Bora (Société, Polynésie francaise), Earth and Planetary Sciences 331, 579–585, 2000.

Bravo T., Aportación al studio geomorfológico y geológico de la costa de la fosa tectónica del valle de la Orotava, Bol. R. Soc. Esp. Hist. Nat., Tomo L, 1–30, 1952.

Cantagrel J.M., N.O. Arnaud, E Ancochea., J.M. Fúster, and M.J. Huertas, Repeated debris avalanches on Tenerife and genesis of Las Cañadas caldera wall (Canary Islands), *Geology*, 27 (8), 739–742, 1999.

Carracedo J. C., The Canary Islands: An example of structural control on the growth of large oceanic-island volcanoes, *J. Volcanol. Geotherm. Res.*, 60, 225–241, 1994.

Carracedo J.C., Growth, structure, instability and collapse of Canarian volcanoes and comparisons with Hawaiian volcanoes, *J. Volcanol. Geotherm. Res.*, 94, 1–19, 1999.

Coello J., J.M Cantagrel, F. Hernán, J.M. Fúster, E. Ibarrola, E. Ancochea, C. Casquet, C. Jamond, Díaz de Téran and A. Cendrero, Evolution of the eastern volcanic ridge of the Canary Islands based on new K-Ar data, *J. Volcanol. Geotherm. Res.*, 53, 251–274, 1992.

Dañobeitia J.J. and J.P. Canales, Magmatic underplating in the Canary Archipelago, *J. Volcanol. Geotherm. Res.*, 103, 27–41, 2000.

Day S.J., J.C. Carracedo, H. Guillou, Age and geometry of an aborted rift flank collapse: the San Andrés fault, El Hierro, Canary Island, Geol. Mag., 134, 4, 523–537, 1997.

Day S.J., S.I.N. Heleno da Silva, J.F.B.D. Fonseca, A past giant lateral collapse and present-day flank instability of Fogo, Cape Verde islands, *J. Volcanol. Geotherm. Res.*, 94, 191–218, 1999.

Duncan, R.A., Hotspots in the southern oceans – An absolute frame of reference for motion of the Gondwana continents, *Tectonophysics*, 74, 29–42, 1981.

Fernández C., R. Casillas, A. Ahijado, V. Perelló and A. Hernández-Pacheco, Shear zones as a result of intraplate tectonics in oceanic crust: An example of the Basal Complex of Fuerteventura (Canary Islands), *J. Struct. Geol.*, 19, 41–57, 1997.

Fernández C., De la Nuez J., Casillas R., and García Navarro E., Stress fields associated with the growth of a large shield volcano (La Palma, Canary Islands), *Tectonics*, 21, 4, 13–1 13–18, 2002.

Funck T. and H.-U. Schmincke, Growth and destruction of Gran Canaria deduced from seismic reflection and bathymetric data, *J. Geophys. Res.*, 103, 15393–15407, 1998.

Funck T. and H. Lykke-Andersen, Seismic structure of the volcanic apron north of Gran Canaria, Proceedings of the Ocean Drilling Program, Scientific Results, 157, 11–28, 1998.

García-Cacho L, Arana V., Romero C., López-González C.M., Palomo C., Acosta J., Herranz P., y Muñoz A., Submarine volcanism in the Canary Islands. Implications for tectonic controls, *2ª Asamblea Luso-Espanhola de Geodesia y Geofísica, Lagos (Portugal),* S15–11, 2000.

Gee M.J.R., D.G. Masson, A.B. Watts, N.C. Mitchell, Offshore continuation of volcanic rift zones, El Hierro, Canary Islands, *J. Volcanol. Geotherm. Res.*, 105, 107–119, 2001.

Geisslinger A., H.B. Hirschleber, M. Schnaubelt, J.J. Danobeitia and J. Gallart, Mapping of volcanic apron and the upper crust between Gran Canaria and Tenerife (Canary Islands) with seismic reflection profiling, *Geo-Marine Letters*, 16, 57–64, 1996.

Guillou H., J.C. Carracedo, F. Pérez Torrado, E. Rodríguez Badiola, K-Ar ages and magnetic stratigraphy of a hotspot-induced, fast grown oceanic island: El Hierro, Canary Islands, *J. Volcanol. Geotherm. Res.*, 1996, 73, 141–155, 1996.

Hausen, H., Contributions to the geology of Tenerife (Canary Islands), *Soc. Sci. Fennicae, Comm. Phys.-Mat.*, 18, 254 pp, 1956.

Hernández-Pacheco A. and J.A. Rodríguez Losada, Geología y estructura del Arco de Taganana (Tenerife, Canarias), *Rev. Soc. Geol. España*, 9 (3–4), 169–181, 1996.

Holcomb R.T. and R.C. Searle R.C., Large landslides from Oceanic Volcanoes, *Marine Geotechnology*, 10, 19–32, 1991.

Klitgord K.D. and Schouten H.S., Plate Kinematics of the Central Atlantic, Vogt P.R. and Tucholte B.E. eds, The Western North Atlantic Region: Boulder, Colorado, *The Geological Society of America, Geology of North America*, M, 351–378, 1986.

Krastel S., H.-U. Schmincke, C.L. Jaacobs, R. Rihm, T. P. Le Bas, and B. Alibés, Submarine landslides around the Canary Islands, *J. Geophys. Res*., 106, 3977–3997, 2001.

Krastel S. and H.-U. Schmincke, The channel between Gran Canaria and Tenerife: constructive processes and destructive events during the evolution of volcanic islands, *Int. J. Earth Sci.*, 91, 629–641, 2002.

Marinoni, L.B. and G. Pasquarè, Tectonic evolution of the emergent part of a volcanic ocean island: Lanzarote, Canary Islands, *Tectonophysics*, 239, 111–135, 1994.

Marinoni L.B. and Gudmundsson A., Dykes, faults and palaeostresses in the Teno and Anaga massifs of Tenerife (Canary Islands), *J. Volcano. Res.*, 103, 83–103, 2000.

Masson, D. G., Catastrophic collapse of the volcanic island of Hierro 15 ka ago, and the history of landslides, *Geology*, 24, 231–234, 1996.

Masson D.G., A.B. Watts, M.J.R. Gee, R. Urgeles, N.C. Mitchell, T.P. Le Bas, and M. Canals, Slope failures on the flanks of the western Canary Islands, *Earth-Science Reviews*, 57, 1–35, 2002.

Mitchell N.C., G. Douglas, Masson, A.B. Watts, M.J.R. Gee and R. Urgeles, The morphology of the submarine flanks of volcanic ocean islands, A comparative study of the Canary and Hawaiian hotspot islands, *J. Volcanol. Geotherm. Res.*, 115, 83–107, 2002.

Mitchell N.C., W.B. Dade and D.G. Masson, Erosion of the submarine flanks of the Canary Islands, *J. Geophys. Res*., 108, 6002, doi:10.1029/2002JF000003, 2003.

Mitchum R.M.J., P.R. Vail and J.B. Sangree, Seismic stratigraphy and global changes of sea level, Part 6: stratigraphic interpretation of seismic reflection patterns in depositional sequences, *Seismic Stratigraphy-Applications to Hydrocarbon Exploration AAPG*, Memoir 26, Eds: Payton C.E., 117–133, 1977.

Moore J.G., D.A. Clague, R.T. Holcomb, P.W. Lipman, W.R. Normark and M.E. Torresan, Prodigious submarine landslides on the Hawaiian Ridge, *J. Geophys. Res*., 94, B12, 17465–17484, 1989.

Morgan W.J., Hotspot tracks and the early rifting of the Atlantic, *Tectonophysics*, 94, 123–139, 1983.

Morgan J.K., G.F. Moore and D.A. Clague, Slope failure and volcanic spreading along the submarine south flank of Kilauea volcano, Hawaii, *J. Geophys. Res*., 108, B9, 2415, 2003.

Navarro Latorre J.M. and Coello J., Depressions originated by landslide processes in Tenerife, in *ESF Meeting on Canarian Volcanism*, Eur. Sci. Found., Strasbourg, France, 150–152, 1989.

Palacios D., The origin of certain wide valleys in the Canary Islands, Geomorphology, 9, 1–18, 1994.

Palomo, C., J. Acosta, A. Munoz, P. Herranz, J.L. Sanz, J. Molinero, M.A. Bécares, R. Gómez, Mapa batimorfológico del canal entre las islas de Tenerife y Gran Canaria, *Ed. Instituto Español de Oceanografía*, Dpto. de Geología y Geofísica, Madrid, 1998.

Ramsay J.G. and Lisle R.J., The techniques of modern structural geology, v. 3: Applications of continuum mechanics in structural geology, session 28, 2000.

Roberts J.A. and A. Cramp, Sediment stability on the western flanks of the Canary Islands, *Marine Geology*, 134, 13–30, 1996.

Roest, W.R., J.J. Dañobeitia, J. Verhoef and B.J. Colette, Magnetic anomalies in the Canary Basin and the Mesozoic evolution of the Central North Atlantic, *Mar. Geophys. Res.*, 14, 1–24, 1992.

Schmincke H.-U. and R. Rhim, Ozeanvulkan 1993, Cruise N° 24, Meteor-Berichte, Rep. 94–2; 88 pp., Univ. Hamburg, Hamburg, Germany, 1994.

Schmincke H.-U., P.P.E. Weaver and J. Firth, The Clastic Apron of Gran Canaria and the Madeira Abyssal Plain, *JOIDES Journal*, Science Operator Report Leg 157, 14–22, 1995.

Stillman C.J., A Canary Islands dyke swarm: Implications for the formation of oceanic islands by extensional fissural volcanism, Mafic Dyke Swarms, H.C. Halls and W.F. Fahring, Geol. Assoc., Can. Spec. Pap., 34, 243–255, 1987.

Stillman C.J., Giant Miocene Landslides and the evolution of Fuerteventura, Canary Islands, *J. Volcanol. Geother. Res.*, 94, 89–104, 1999.

Teide Group: C. Palomo, J. Acosta, J. L. Sanz, P. Herranz, A. Munoz, E. Uchupi, and J. Escartin, Morphometric interpretation of the northwest and southeast slopes of Tenerife, Canary Islands, *J. Geophys. Res.*, 102, 20325–20342, 1997.

Twiss R.J. and Moores E.M., Structural Geology, *Ed. W.H. Freeman and Company*, 1992.

Urgeles R., M. Canals, J. Baraza, B. Alonso, and D. Masson, The most recent megalandslides of the Canary Islands. El Golfo debris avalanche and Canary debris flow, west El Hierro Island, *Jour. Geophys. Res.*, 102, 20305–20323, 1997.

Urgeles R., M. Canals, J. Baraza, B. Alonso, Seismostratigraphy of the western flanks of El Hierro and La Palma (Canary Islands): a record of Canary Islands volcanism, *Marine Geology*, 146, 225–241, 1998.

Urgeles R., D.G. Masson, M. Canals, N.B. Watts and T. Le Bas, Recurrent large-scale landsliding on the west flank of La Palma, Canary Islands, *J. Geophys. Res.*, 104, B11, 25331–25348, 1999.

Urgeles R., M. Canals, J. Roberts and the SNV "Las Palmas" Shipboard Party, Fluid flow from pore pressure measurements off La Palma, Canary Islands, *J. Volcanol. and Geotherm. Res.*, 101, 253–271, 2000.

Vidal N. and O. Merle, Reactivation of basement faults beneath volcanoes: a new model of flank collapse, *J. Volcanol. and Geotherm. Res.*, 99, 9–26, 2000.

Watts A. B. and D.G. Masson, A giant landslide in the north flank of Tenerife, Canary Islands, *J. Geophys. Res.*, 100, 24487–24498, 1995.

Watts A.B. and D.G. Masson, Reply (to comment on 'A giant landslide on the north flank of Tenerife', Canary Islands by J. Marti). *J. Geophys. Res.*, 103, 9949–9952, 1998.

Watts A.B. and D.G. Masson, New sonar evidence for recent catastrophic collapses of the north flank of Tenerife, Canary Islands, *Bulletin of Volcanology*, 63, 8–19, 2001.

Marine Geophysical Researches (2003) 24: 113–127
DOI 10.1007/s11001-004-1336-2

Gravity analysis offshore the Canary Islands from a systematic survey

A. Carbó[1], A. Muñoz-Martín[1,*], P. Llanes[1], J. Álvarez[1] & EEZ Working Group[**]
[1]*Universidad Complutense de Madrid. Facultad de Ciencias Geológicas. Departamento de Geodinámica. 28040 Madrid (Spain) *Corresponding Author (Phone: +34-913944834; Fax: +34-913944631; E-mail: amunoz@geo.ucm.es)*

Key words: Bouguer anomaly, Canary Islands, free air anomaly, geodynamics, gradient zone, marine gravity

Abstract

As part of a systematic mapping program of the Hydrographic and Oceanographic Research Plan for the Spanish Exclusive Economic Zone (EEZ), gravity surveys were carried out offshore the Canary Islands. Using the gravity data collected during cruises between 1998 and 2000 aboard the RV Hesperides and satellite and land data, we construct free air and Bouguer anomaly maps and discuss the geodynamic implications.

Using maps of Bouguer anomaly, free air anomaly, vertical derivative, long wavelength Bouguer anomaly and short wavelength anomaly, a detailed description of the gravity characteristics of the archipelago is presented, describing gravity anomalies from a geologic point of view. The character of the crust throughout the studied area has been defined, as well as high gradient zones that limit crustal blocks of different density. High gradient zones have been mapped for the first time and interpreted as fracture zones, taking into account geophysical and geological information. Gravity highs and lows have been studied and related to crustal, mantle and volcanic effects.

Introduction

In 1995, the 'Instituto Español de Oceanografía' and the 'Instituto Hidrográfico de la Marina' undertook a geologic and oceanographic research study of the Spanish Exclusive Economic Zone (EEZ). The main objectives of this research were to map the EEZ in order to obtain bathymetric charts and to perform a geophysical survey, consisting of gravity, magnetic and other information compatible with a 10 knot ship velocity. The EEZ program involved the cooperation of the 'Real Observatorio de la Armada', which was responsible for magnetic data, and the 'Departamento de Geodinámica, Universidad Complutense de Madrid', responsible for gravity data. The investigations of the EEZ program were carried out offshore the Canary Archipelago (Figures 1 and 2), during cruises EEZ-1998, EEZ-1999 and EEZ-2000.

These cruises constituted the first systematic gravity survey performed in the Canary Islands region, providing data acquired using the same self-consistent acquisition system along quasi-regularly spaced survey tracks. The purpose of this paper is to present consistent, regional gravity maps of the Canary Islands region, and to interpret some of the geodynamic implications they provide. The publication of these maps is of great interest for tectonic and geodynamics studies. Among other applications, these maps, combined with other geological and geophysical information, provide constraints on the different tectonic models proposed to explain the origin of the Canary Islands (e.g., Anguita and Hernán, 2000; Carracedo et al., 1998), on the structure of the island crust and upper mantle (e.g. Watts et al., 1997), on the thermal and mechanical properties of the oceanic lithosphere underneath the archipelago (e.g. Canales and Dañobeitia, 1998), and on the nature of the ocean-continent transition in the region (Watts and Marr, 1995).

Geological setting and previous geophysical work

The Canary Islands are an intraplate volcanic archipelago formed of seven volcanic islands located

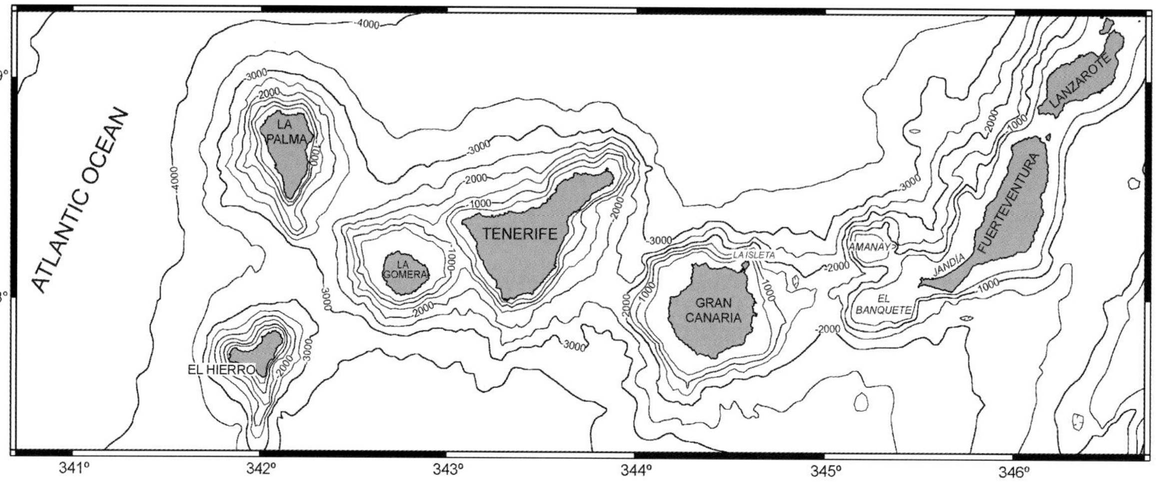

Figure 1. Location map of the Canary Islands and bathymetric map (depth in meters).

in the eastern Atlantic Ocean, close to the northwest African continental margin (Figure 1) and lying on Jurassic oceanic crust, between isochrons M21 and S1 (150 and 175 Ma, Roest et al., 1992). The easternmost island, Fuerteventura, is only 100 km from the African coast. Their proximity to the continent, together with the sparse data available, influenced discussions about the character of the crust under the Canaries. Hausen (1962) suggested a continental character for the islands while Rothe and Schmincke (1968) proposed a continental character for the eastern islands (Lanzarote and Fuerteventura) and an oceanic one for the western islands. Fuster et al. (1968) advocated an oceanic origin for all the Canary Islands. The oceanic character of the lithosphere under the Canaries has been confirmed by seismic multichannel reflection and refraction studies at both a crustal and a lithospheric scale (Bosshard and MacFarlane, 1970; Banda et al., 1981; Watts et al., 1997).

The first marine gravity study of the Canary Archipelago, together with some refraction seismic profiles, was published by Dash and Bosshard (1969) for the western islands. They obtained free air and Bouguer anomaly maps using a 50 mGal contour interval, described how free air anomalies followed the bathymetry quite closely, and noticed the absence of an offshore belt of negative anomalies associated with the islands, such as seen in Hawaii (Watts and Talwani, 1975). Based on Bouguer anomalies they interpreted all the islands as independent structural blocks, except Tenerife and La Gomera, which lie in the same block. They also interpreted two main faults, one between Gran Canaria and Tenerife trending NE-SW, and another through Tenerife, La Gomera and El Hierro trending in the same direction. Dash and Bosshard (1969) proposed oceanic crust for the west and a transitional type in the central area of the archipelago. Using the same data for the western island, Bosshard and MacFarlane (1970) constructed a more detailed Bouguer anomaly map and used the free air values to compute the depth of the Mohorovicic discontinuity and Bouguer values for structural models, but their general conclusions were the same as those of Dash and Bosshard (1969).

Regional and local gravity anomaly data were presented by MacFarlane and Ridley for the islands of Tenerife (1968) and Lanzarote (1969). Their results point to a gradational position of the islands between a truly oceanic and a truly continental environment. Bouguer anomaly values increase from a fairly typical continental value of 25 mGal over the continental shelf to over 250 mGal west of El Hierro. An interpretation of the regional and local gravity data was attempted and some broader genetic implications were briefly discussed. More recent local gravity studies have been done, basically to define the structure and volcanism in Tenerife and Lanzarote (Ablay and Kearey, 2000; Araña et al., 2000; Camacho et al., 2001; Vieira et al., 1986). On a regional scale the most recent studies focus on lithosphere analysis of the Canary Islands using both seismic and gravity data (Canales and Dañobeitia, 1998; Dañobeitia et al., 1994; Ranero et al., 1995; Watts, 1986).

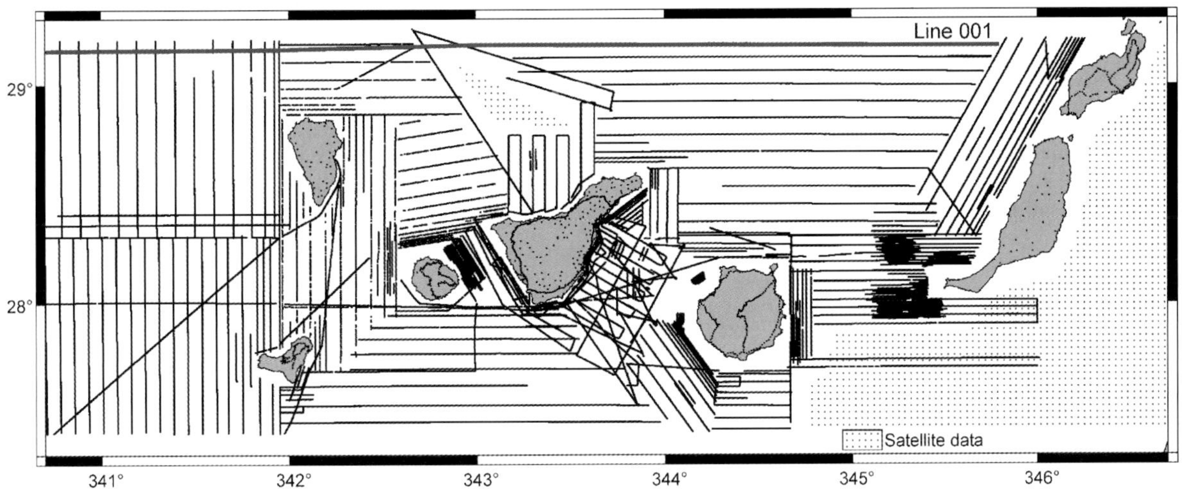

Figure 2. Track lines map. Line 1 is used in the spectral analysis of Figure 6.

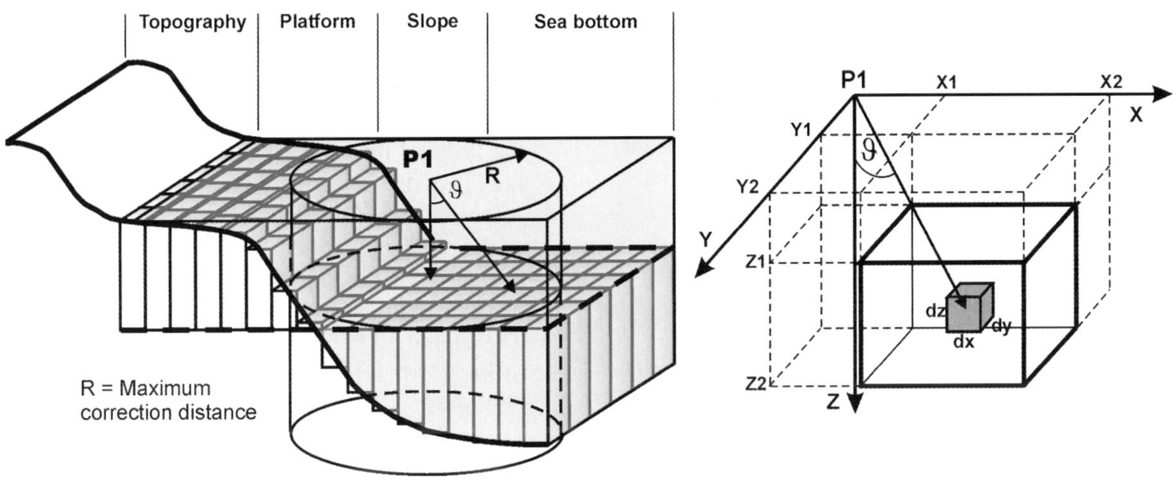

Figure 3. Block diagram that shows the parameters used in 'sea bottom corrections' (left) and variables used in calculations (right).

Data Collection

An area of more than 124 000 km^2 offshore the Canaries is studied here. The marine gravity data were collected during the EEZ cruises, where a total of 338,421 useful readings were obtained (Figure 2). Satellite-altimetry-derived gravity data (Sandwell and Smith, 1997) as well as data from a cruise previous to the EEZ project (Teide 95) were used to complete the study area. Land gravity data from the 'Instituto Geográfico Nacional' database, distributed over the seven major islands (I.G.N., 1996), were also used to avoid border errors inside the studied area. The locations of track lines together with land and satellite data are shown in Figure 2.

The acquisition of marine gravity data was carried out aboard the RV Hespérides using a ± 1 mGal accuracy Bell Aerospace BGM-3 gravity system, with an automatic pitch and roll elimination. The gravity meter used for the land link was a ± 0.01 mGal accuracy LaCoste & Romberg. The offshore gravity data set was corrected for instrumental drift and tied into the land network throughout the 'Instituto Geográfico Nacional' gravity bases of Cartagena (SE of Iberian Peninsula) and Las Palmas (Gran Canaria).

Ship navigation data was obtained by two simultaneous differential GPS (DGPS) systems integrated with RV Hespérides' central navigation system. A local DGPS base station and a second system based on a regional base station network were used, resulting in better than 5 m navigation accuracy. Water depth was recorded with both a multi-beam echosounder Simrad EM-12S and single beam echosounder Simrad EA-500.

Data Processing

Ship gravity data was merged with navigation data (date, time, ship's speed, water depth, course, latitude and longitude), and the Eötvos effect was corrected. Gravity data were reduced to the 1967 Geodetic Reference System (GRS), the water slab was corrected using a density of 1.03 g/cm^3 and Free-Air and Bouguer anomalies were calculated following the procedure of Nettleton (1976). Land gravity data were also reduced to the 1967 GRS using a density of 2.67 g/cm^3 for the Bouguer correction.

A total of 397 processed track lines, containing the data described above, were included in a georeferenced database. Spikes were corrected in each line and, due to the regional character of the map, high frequency anomalies were eliminated by a low pass filter (anomalies under 1.5 km of wavelength).

Sea bottom correction

Nettleton (1976, Chapter 5) described how to correct the effect produced by the water slab and irregularities of the ocean floor. This author calls the result of applying these corrections '*Bouguer anomalies*', though the corrections at sea are not exactly equal to those applied on land. It should be remembered that the correction for the water slab is done with a slab of water situated below the reference level. Also, the gravity effect produced by sea bottom irregularities, treated as positive and negative spikes above the lower limit of the slab of water, are computed at sea level and can be positive or negative. When working in areas close to the continental slope, or in archipelagos with large bathymetric variations, these corrections should be applied so that the map of gravity anomalies, obtained at sea level, reflects the distribution of the densities at the sea bottom and is not heavily affected by bathymetric irregularities.

The correction of these effects was performed for each point at which a reading was taken, using an original computer programme (LANZADAF). The programme can be explained as follows (Figure 3). The programme uses two matrixes: appearing in the first data matrix (DM) are the geographic coordinates of the reading points to be corrected, their bathymetry and the free air anomaly; in the second prism matrix (PM) are the geographic coordinates of each regular prism into which the study area has been divided and their average depths. It is first necessary to divide the sea bottom into regular prisms with the sides defined and the altitudes equal to the average bathymetry at the base of each prism.

The flux of the programme is carried out following three levels described as follows and by repeating the process for each reading point.

(1) Bouguer slab correction. This is done in the classic way, though the density under consideration is the negative contrast arising from the presence of seawater and the thickness of the slab is the depth of the reading point.

(2) Creation of a residual prism matrix (RPM) with the same geometrical disposition as in the PM. The new geographic coordinates and their altitude will be established by the difference between those of the PM and those of the point to be corrected. It is evident that the altitudes defined in this way will be positive when the bottom surface in the position of the prisms is greater than the lower limit of the slab of water used for the Bouguer correction and negative in the opposite case (Figure 3).

(3) The sum of all the effects generated by the prisms of the RPM above the reading point and the elaboration of a departure matrix in the appropriate format to execute the mapping of Bouguer anomalies. The operation will be broadened to a distance at which the calculated effects will be irrelevant. The relative X and Y coordinates, the slab correction, the depth correction, the free air anomaly and the Bouguer anomaly are included in the departure matrix. The gravity effect of a regular prism, with a density δ over a point (Figure 3), is expressed by:

$$g_z = \gamma \sigma \int_{x1}^{x2} dx \int_{y1}^{y2} dy \int_{z1}^{z2} \frac{z dz}{\left(x^2 + y^2 + z\right)^{\frac{3}{2}}}$$

where γ is the constant of universal gravitation and σ the density of the prisms.

As a base of calculation for programming, we have used the solution of this integral proposed by Nagy (1966), because of its simplicity and appropriateness for the programming of the problem, although there have been a great many further advances (e.g., Plouff, 1976; Danés, 1982, Ma and Watts, 1994). In

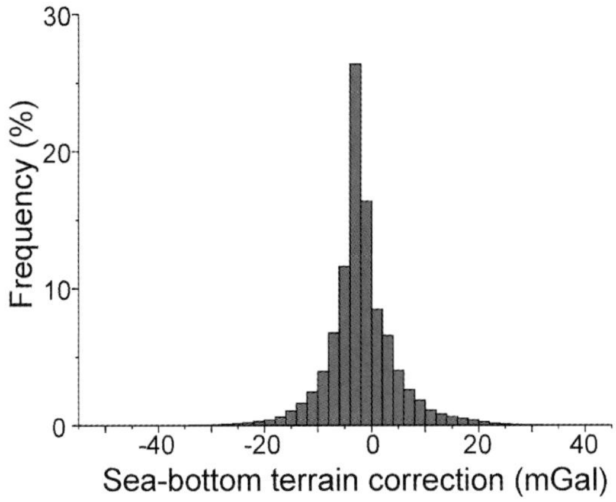

Figure 4. Histogram of sea bottom terrain values obtained for all the data.

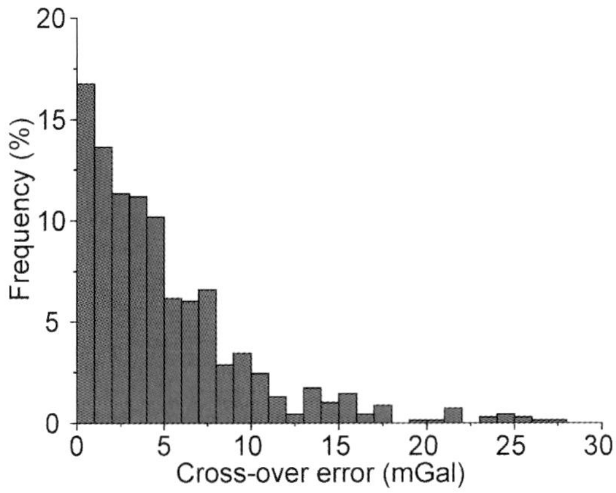

Figure 5. Histogram of differences between calculated Bouguer gravity anomalies at cross-over ship's tracks.

a complimentary way, Nagy (1966) also executes a classic topographical correction for nearby land elevations, breaking down the topography into straight prisms. This correction is necessary in cases like the Canary Islands, where altitudes of nearby emerged land reach 3500 meters above sea level.

The prism used was 2 km wide and the sea-bottom correction was extended to 22 km. The digital elevation model used included satellite sea-bottom data (Smith and Sandwell, 1997) and land data (ETOPO30). Statistical analysis of the sea-bottom corrected values showed a mean value of −1.92 mGal and a standard deviation of 6.64 mGal (Figure 4).

Levelling correction

To estimate the overall accuracy of the gravity measurements obtained during four different cruises, differences between gravity anomalies observed at intersecting ships' tracks were examined, since this is considered the best method to evaluate uncertainties in marine gravity data (Talwani, 1971; Wessel and Watts, 1988). The statistical cross-over analysis shows than 84% of the values have differences below 10 mGal and 64% of the cross-over errors are under 5 mGal (Figure 5).

The levelling correction method used was a statistical levelling that corrects errors by modelling a trend surface to the intersection errors and removing the trend from the tie line data. To level the tie lines we assumed that the difference between the tie line and all crossing survey lines can be used as an estimate of the error along the tie line. It is reasonable to assume that the tie line should, on average, match the crossing lines. Any single line may be out of level, but the average of all lines should follow the correct tie line. The statistical levelling was first applied to individual cruises and later to all the cruises together because there are some lines that cross groups of lines of other cruises.

Coherence analysis

Because interpolation of the anomaly maps included both ship gravity data and satellite gravity data (ERS-1, Geosat and Topex-Poseidon, from Sandwell and Smith, 1997) (Figure 6 a, b), a spectral analysis of the coherence between both kind of data was done. The objective of this analysis was to check the precision and the existence of a cut-off frequency that marked a lower limit in the satellite data accuracy. The longest ship track (495 km length), with an E-W direction, was selected for the analysis (Line 001, Figure 3). In that line, the auto- and cross-spectrum were calculated for free air anomaly values obtained by both ship and satellite (Figures 6a,b). Because the sample interval was different in each case, about 50 m for the ship and 3500 m for the satellite, the larger sample interval was chosen for the analysis. To avoid aliasing, the ship data were low-pass filtered before re-sampling to the coarser interval. For that interval, we have calculated the spectral density by Welch's method (1967) of ensemble averaging of multiple overlapped windows, using the standard error estimates of Bendat and Piersol (1986) and following the Wessel and Smith (1995) procedure. Finally, we have plotted the spectra for the two sources and the coherence between them. Power values are equivalent in both types of data (Figure 6c). The coherence has values near 1 for long wave-lengths >20 km (Figure 6d), while for

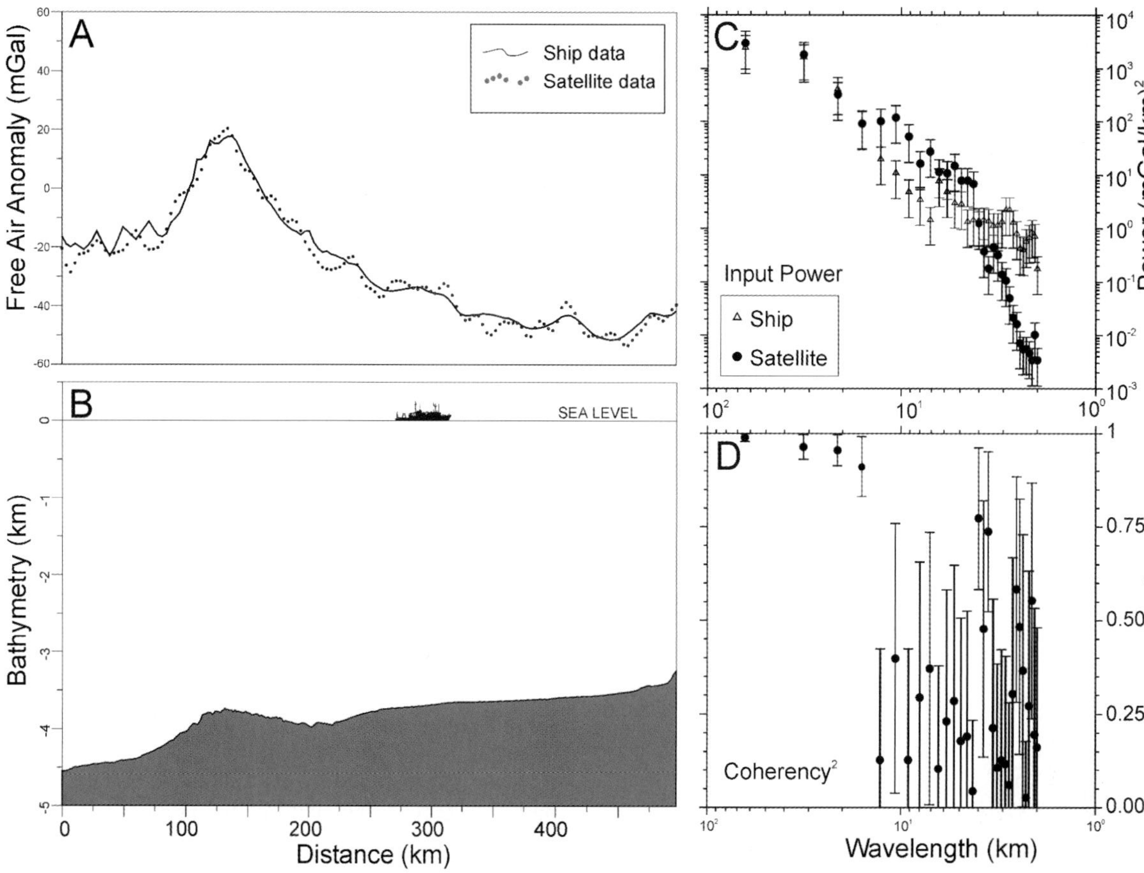

Figure 6. Coherence analysis of free air anomaly values. a) Representation of ship and satellite free air anomaly data for Line 001, sampling interval 3500 m. b) Ship bathymetry profile for Line 001 (located in Figure 2), sampling interval 50 m. c) Comparison of power spectrum of ship and satellite data along the line. d) Coherence analysis between ship and satellite gravity data along the line.

wavelengths shorter than 15 km, there is low coherence between both data sets, as also suggested by the power values (Figures 6c,d). These data agree with the resolution of satellite data established at 22–30 km in previous works (Yale et al., 1995). The good results of the coherence analysis may be due to the systematic character and quality of the gravity data used. The spectrum also shows that satellite data are sparse between 6 and 12 km wavelength. Below this value only the ship data have a significant signal level. Because of the regional character of the research area, it is possible to use both kinds of data to get a synthesized map, knowing that, for satellite data only wavelengths above 15 km are reliable. On the other hand, results of the spectral analysis of the coherence indicate that for high resolution studies the use of ship gravity data is essential.

Gravity maps

Map construction

All filtered offshore gravity data were merged with gravity data onshore to construct free air and Bouguer anomaly maps. The combined data were interpolated using a Kriging algorithm that divides the search area into six sectors. After trying different grid cell sizes and taking into account the results of the spectral analysis, a 6 km grid cell size was chosen as the most suitable for this study. The final maps include the anomaly contours over an illuminated coloured image computed from a directional derivative in the 310° direction and were drawn using GMT software (Wessel and Smith, 1995). The free air anomaly map was contoured at 10 mGal, while the Bouguer anomaly

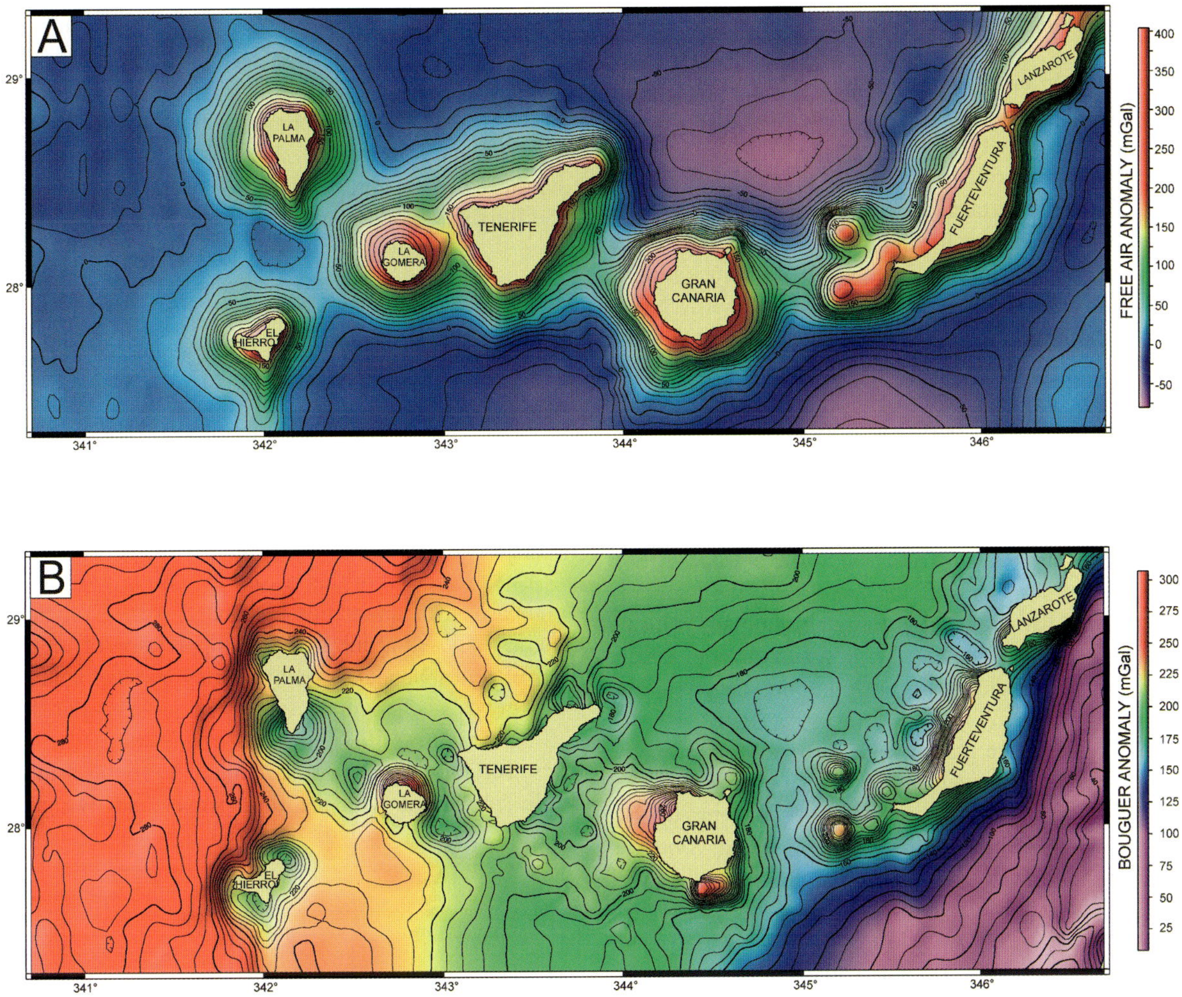

Figure 7. (a) Free air anomaly map (contour interval = 10 mGal). (b) Bouguer anomaly map (contour interval = 5 mGal, except to the east of Fuerteventura and Lanzarote where the contour interval = 10 mGal). The colour palette range has been emphasized for anomalies between 125–225 mGal.

map was contoured at 5 mGal, except in the area east of Fuerteventura and Lanzarote, where steep gravity gradients occur and a 10 mGal contour was chosen (Figures 7a and 7b).

Free air anomaly map

The free air anomaly map is shown in Figure 7a. Values above 100 mGal occur above sea level, reaching as much as +400 mGal on the highest island. Values between +100 and +50 mGal mark the submarine flanks of the islands with steep slopes. The 0 mGal represents deep seafloor with less island influence. Minimum values of −75 mGal appear east of Tenerife, NE of Gran Canaria and west of Fuerteventura. Looking at the gradual distribution, from positive values on the island to negative values in deep areas, there is a clear correlation with bathymetry. This is expected, since free air anomalies are dominated by the large density contrast at the seafloor (Lin et al., 1990). As free air anomaly values have a dependence on the topography and bathymetry, it is not possible to derive a correct interpretation, even qualitatively, without taking account of bathymetry. For geological interpretations Bouguer anomaly values are more useful, especially when a sea bottom terrain correction is done, because the anomalies can be interpreted in terms of density variations below the water slab and depth of the sources that generate these variations. This allows us to make an detailed description and to be able to discuss the gravity data based on the Bouguer anomaly map.

Bouguer anomaly map

The Bouguer anomaly map with both sea-bottom terrain and topography corrections is shown in Figure 7b. The Bouguer gravity values range from +315 mGal in the westernmost side to +30 mGal in the easternmost one. This change takes place in a steep and discontinuous way, shaped by three major elongated high gradient zones: near the eastern coast of Fuerteventura and Lanzarote, across Tenerife island (both with a NE-SW orientation) and just to the west of La Palma and El Hierro islands, with a N-S orientation. These high gradient zones separate four areas with different gravity responses. It is clearly observed that the Canary Islands occupy the two zones of intermediate Bouguer anomaly values (160–240 mGal). The mean value of the western islands is higher (220 mGal) than the eastern ones (180 mGal). In addition to the high gradient zones, there are other effects that disturb the general progressive variation in the Bouguer anomaly values from east to west, such as the presence of more local highs and lows and high gradient zones (Figures 7b and 8).

There are some gravity highs of special relevance because of their signal amplitude: Two of them are related to Gran Canaria and another is centred on La Gomera. Other gravity highs, located SW of Fuerteventura, have shorter wavelengths but considerable amplitude (40–50 mGal).

There are some gravity lows west of Lanzarote and Fuerteventura, and it is remarkable that the axis of minimum values has a N-S direction in the channel between Fuerteventura and Gran Canaria. The biggest gravity lows, because of their amplitude values, are related to La Palma and El Hierro.

The map shows different kinds of signals, due to the small grid side and the large size of the studied area. This characteristic makes it necessary to apply frequency separation techniques to analyse the different anomalies and to avoid possible distortion due to their superposition.

Gravity data filtering and transformation

The Bouguer anomaly map shows a combination of long- and short-wavelength features. In order to better determine gravity gradients that may indicate geological contacts between bodies of different density, and to distinguish between deeper and shallower sources, we have applied the following filtering and transformation techniques: A) vertical derivative maps, by taking the 2-D Forward Fast Fourier Transform; and B) filtering of the Bouguer anomaly map in different wavelengths by applying Fourier analysis (cut-off wavelength values have been obtained by the 2-D spectral analysis of the Bouguer anomaly map described later in the text).

First vertical derivative map

The first vertical derivative enhances shorter wavelengths, suppresses longer wavelengths and is accomplished by multiplying the Fourier transform of a potential field map by the wavenumber. The effect is to enhance anomalies caused by abrupt lateral changes in near-surface densities at the expense of broader anomalies caused by deeper or more gradual density changes. For this reason, the map is useful for comparison of anomalies associated with geological bodies exposed at the surface. Suppression of longer wavelengths also helps in comparing and distinguishing trends and anomaly fabrics in various domains (Simpson et al., 1986; Jachens et al., 1989). As the first vertical derivative map (Figure 9) has considerable similarities with the short wavelength Bouguer anomaly map, we will discuss both maps together below.

Bouguer anomaly filtered maps

The Fourier analysis was done with the regularly-spaced Bouguer anomaly grid shown in Figure 7b. Figure 10 represents the radial-averaged power spectrum of the Bouguer anomaly grid. It consists of a number of linear trends, which can be interpreted in terms of the mean depth to the major density contrasts within the lithosphere following the method of Karner and Watts (1983). We recognize two linear segments of differentiated slope. The steeper one is related to deep sources (mean depth = 65 km), responsible of long wavelength anomalies (between 200 and 40 km), while the other represents the shallower sources (mean depth = 20 km), responsible for short wavelength anomalies (between 40 and 12 km).

Using the spectral analysis results we have obtained two maps of filtered Bouguer anomalies, one of long wavelengths that show deep density variations and/or crustal flexures, and another of short wavelengths, from which the noise has been eliminated, that represents local density variations of crustal scale.

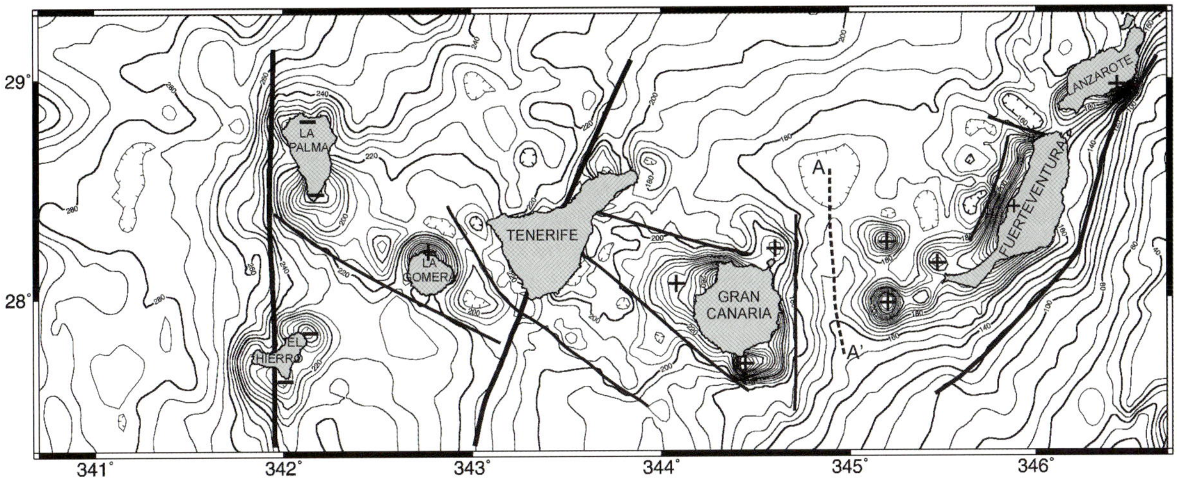

Figure 8. Interpreted Bouguer anomaly map (contour interval = 5 mGal, except to the east of Fuerteventura and Lanzarote where the contour interval = 10 mGal). Black solid lines show linear high gravity gradients. Gravity highs are represented with + symbols and gravity lows with – symbols. Dashed black line labelled as A-A′ represent an axis of gravity lows. See text for explanation.

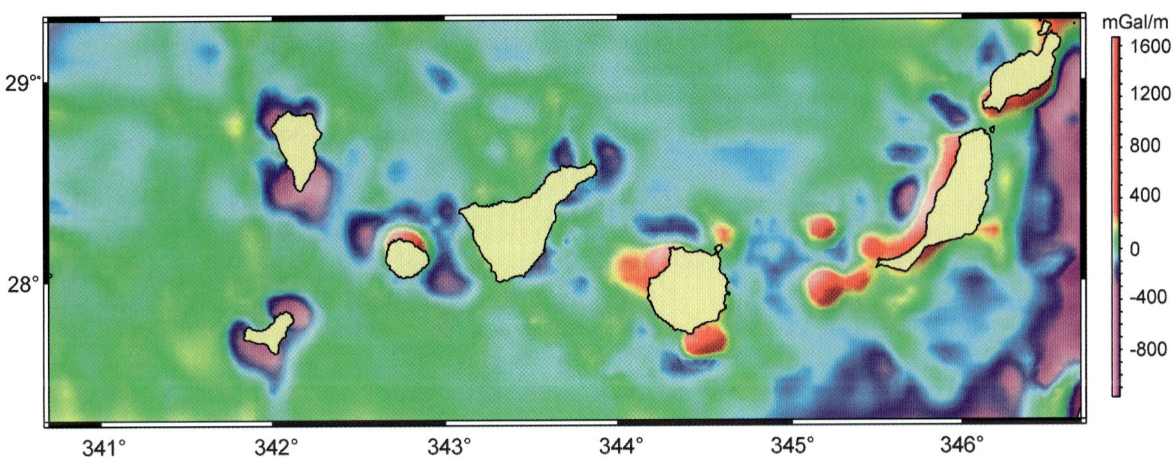

Figure 9. First vertical derivative of the Bouguer anomaly map.

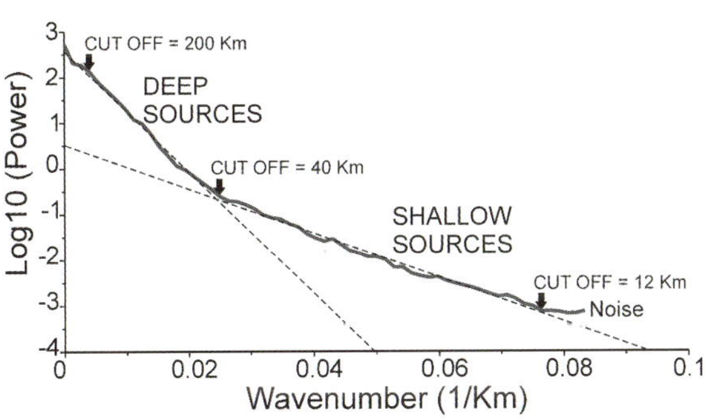

Figure 10. Radial-averaged power spectrum of Bouguer anomaly map and depth estimation for the density interfaces contributing to the long and short wavelengths.

Long wavelength Bouguer anomaly map

The description of the long wavelength Bouguer anomalies refers to the map of Figure 11a. It has some similarities with the Bouguer anomaly map (Figure 7b). As was mentioned previously, the change from low Bouguer anomalies in the east to high Bouguer anomalies in the west takes place in a steep and discontinuous way, allowing us to define four sectors separated by three major high gradient zones:

– Lanzarote – Fuerteventura Gradient Zone (LFGZ). The easternmost gradient zone is located to the east of Lanzarote and Fuerteventura. It runs in a general NE-SW direction, although near Fuerteventura it takes a NNE-SSW direction. The maximum gradient values are reached near Lanzarote (12 mGal/km).

– Tenerife Gradient Zone (TGZ). Crossing Tenerife in a NE-SW direction, there is a stronger gradient on the north side of the island than on the southern edge.
– La Palma - El Hierro Gradient Zone (PHGZ). This is the westernmost gradient trending N-S, and it passes west of La Palma and through El Hierro. In areas where it is not modified by wavelengths related to other anomalies, it reaches a 3.5 mGal/km.

Some differentiated anomalies seen in the general map appear to have an overlap effect as a unique long wavelength anomaly, while there are others whose effects seem to be absent.

The two independent gravity highs of Lanzarote and Fuerteventura that are seen in the unfiltered general map (Figure 7b) have a reflection as a unique and elongated gravity high (Figure 11a). This trends parallel to Fuerteventura's west coast in the long wavelength map (H1) and includes the Amanay and El Banquete gravity highs. The various individual gravity highs observed in the general map in relation to Gran Canaria have a unique reflection (H2). The long wavelength of this high suggests a deep source.

The gravity lows west of Lanzarote and Fuerteventura that appear in the general Bouguer map (Figure 7b) have a possible reflection as a single gravity low anomaly of low amplitude and a NE-SW orientation (L1), with its position displaced towards the northwest (Figure 11a).

The gravity lows located south of La Palma and east of El Hierro are clearly visible in the long wavelength map (L2 and L3, Figure 11a) as isolated gravity lows displaced eastwards respect to their position in the general map (Figure 7b).

Short wavelength Bouguer anomaly map

The description of the short wavelength Bouguer anomalies refers to the map of Figure 11b. In this map, as well as in the vertical derivative one (Figure 9), gravity anomalies caused by shallow bodies and high gradient zones are clearly visible. A detailed description of the most significant gradient zones, gravity highs and gravity lows is given here:
– Gradient Zone G1. This zone is an E-W trending gradient, located between Lanzarote and Fuerteventura. It clearly separates the gravity highs located off both islands. The associated gradient, of at least 50 mGal, seems to separate and displace both gravity highs in a dextral sense.
– Gradient Zone G2. This zone is located west of Fuerteventura, separating the high values of Fuerteventura's west flank of some gravity lows located to the west with a NE-SW direction. A change of 70 mGal is observed over a distance of less than 15 km.
– Gradient Zone G3. On the eastern flank of Gran Canaria there is a high gradient zone with N-S direction and a minimum Bouguer anomaly variation of 50 mGal.
– Gradient Zone G4. Zone G4 is located east of Tenerife and has an ENE-WSW orientation. Based on morphological criteria, a fault has been interpreted as occurring in the same place and having the same direction (Llanes et al., this issue). The fact that a density contrast is observed across the fault supports this interpretation.
– Gradient Zones G5, G6 and G7. There are three gradients with a NW-SE direction located between Tenerife and Gran Canaria. G4 and G5 are limited on the north and south, respectively, by a gravity high extending from Gran Canaria to Tenerife. Running between Tenerife and La Gomera there is a NW-SE axis of maximum values with a large longitudinal manifestation whose southern limit is the G7 gradient zone.
– Gradient Zone G8 is located south of La Palma and La Gomera with a NW-SE orientation. It separates a gravity high area to the south from an area of various gravity lows to the north with a change of 20 mGal in 10 km. In the northern area of gravity lows there is an isolated gravity high associated with La Gomera.
– Gravity highs H9 and H10. To the SE of Lanzarote there is a gravity high (H9) that has no continuation either towards the north or towards Fuerteventura. In contrast to the high located to the east, Fuerteventura has a gravity high on the west coast and its submarine flank (H10). Both highs are separated by the high gradient zone G1 described above.
– Gravity highs H3, H4 and H5. SW of Fuerteventura we find three closed gravity highs. H3 is located in the submarine flank of the Jandía Peninsula and the other two gravity highs (H4 and H5) are associated with the 'Amanay' and 'El Banquete' seamounts, respectively.
– Gravity highs H6, H7 and H8. There are three gravity highs associated with Gran Canaria. The larger one, at the NW of the island (H6), is limited by gradients G4 and G5, and shows attenuation towards Tenerife. Another gravity high is located at the submarine continuation of 'La Isleta' peninsula (H7) and extends in a NW-SE direction. The third one appears at the south of the island (H8) with a strong gradient that reaches 75 mGal in just 10 km.

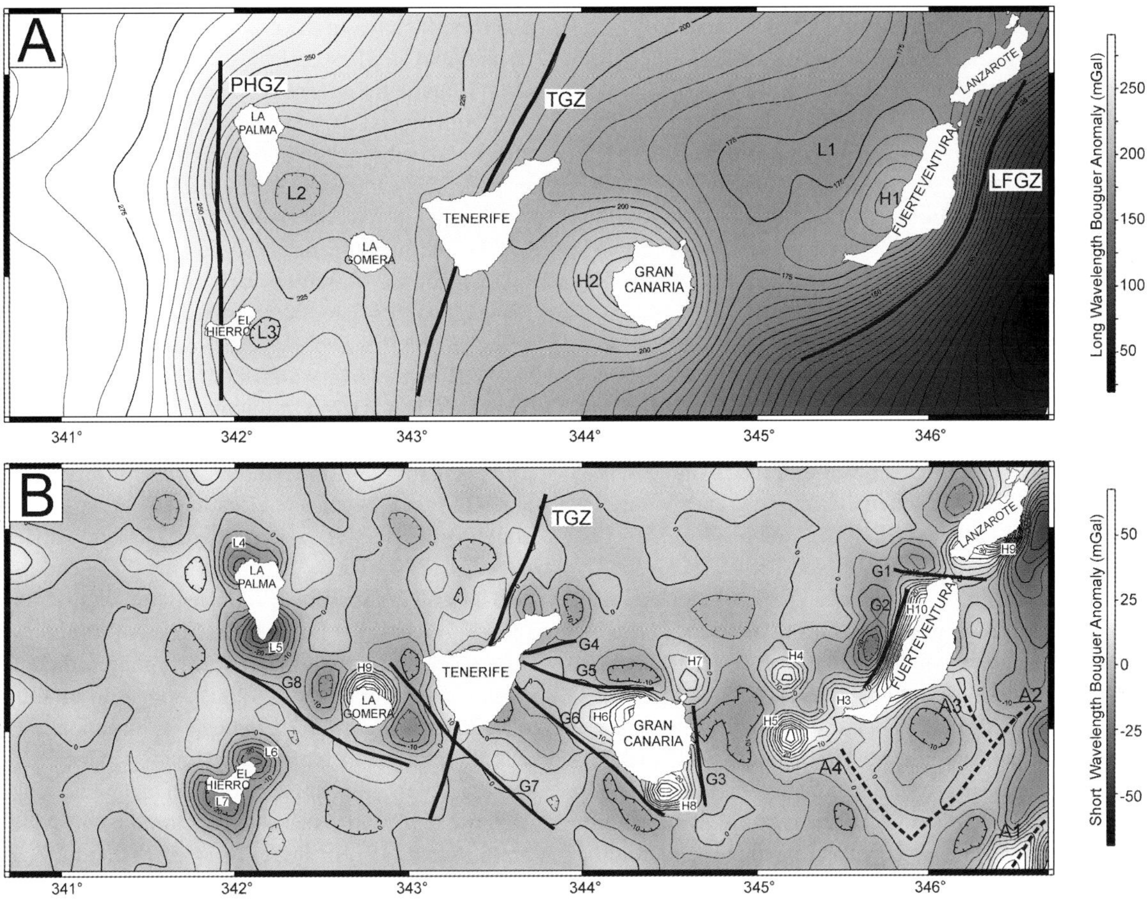

Figure 11. Low- and high-pass filtered maps. Black solid lines show linear high gravity gradients (LFGZ, TGZ and PHGZ). Gravity highs are marked as H and gravity lows as L. a) Long wavelength anomaly map ($\lambda > 40$ km). b) Short wavelength anomaly map (12 km $< \lambda <$ 40 km). Dashed black lines (labelled as A1 to A4) represents axes of gravity highs. See text for explanation.

– Gravity high H9. La Gomera has a very well defined gravity high that is located at the north side of the island.
– Gravity lows L4 and L5. Located near La Palma, there are two remarkable gravity lows, one in the north and the other in the southern region. The southern one exhibits a larger geographic extent and amplitude than the northern one.
– Gravity lows L6 and L7. Two gravity lows, close to each other, are associated with El Hierro, one towards the NE and the other towards the SW.

Other features that were not observed in the Bouguer anomaly unfiltered map appear clearly after filtering. For example in the SE area of the map, nearest to the African continent, a net aspect is seen with two principal orthogonal directions, NE-SW (A1 and A2) and NW-SE (A3 and A4).

Geodynamic discussion of the gravity data

The new gravity data available and its qualitative and quantitative interpretation, together with other geologic and geophysical data, allow us to discuss different aspects of geodynamic interest. The main features observed in the Bouguer anomaly maps are fundamentally linked to three characteristics:
– The change shown in the Bouguer anomaly values, which oscillate between 30 mGals in the eastern edge to 320 mGal in the occidental one.
– The discontinuous way this change takes place, clearly conditioned by the three high gradient zones previously described: LFGZ, TGZ and PHGZ (Figures 8 and 11a).
– The general orientation of the Bouguer anomaly contours. There is a NE-SW trend in the eastern area

(nearest to the Africa continent) which progressively changes to a N-S direction towards the western edge, as we advance to more 'oceanic' areas.

Anomaly values under 40 mGals located in the SE edge of the research area seem to indicate the presence of continental crust. Between the values of 40 mGals and east of Fuerteventura and Lanzarote, where values of 140 mGal are found, the crust has a transitional character. In this area, with values around 140 mGal to the west, change takes place abruptly at the LFGZ.

All of the study area located west of LFGZ has Bouguer anomaly values between 140 and 320 mGal, which typically correspond to oceanic crust (e.g. Talwani et al., 1965). The evolution of the anomalies in this area is not defined in a continuous way. There are areas of small increments, limited by narrow high gradient zones (TGZ and PHGZ).

The Canary Islands are located between two high gradient zones, LFGZ and PHGZ (140–250 mGal) and there is a high gradient zone (TGZ) that divides the archipelago into two areas. The eastern islands (Lanzarote, Fuerteventura, Gran Canaria and part of Tenerife) are located in an area of lower Bouguer anomaly values than the western islands (La Gomera, La Palma, El Hierro, and probably part of Tenerife). From the PHGZ towards the west, high Bouguer anomaly values are higher than 250 mGal, and could correspond to normal oceanic lithosphere not affected by hotspot volcanism.

The closer to the continent, the more pronounced the NE-SW orientation of the gravity anomaly contours becomes, hile in the west, they become more N-S oriented. In this sense, the variation in the high gradient zones orientation confirms an increase of the oceanic influence towards the west, because they are more parallel to the fabric of the oceanic crust.

In the island area, between LFGZ and PHGZ, there are some closed anomalies, several tens of km in diameter. Important differences have been observed between them. In the western islands the anomalies are gravity lows (L1 and L2, Figure 11a), while in the eastern islands they are gravity highs (H1 and H2, Figure 11a). In the central islands (Tenerife and La Gomera) these types of anomalies are absent. The displaced position of the gravity lows of the western islands with respect to the centre of the islands and its high wavelengths suggest the influence of the upper mantle as a source for these anomalies. The gravity highs observed in the eastern islands are eccentric for Fuerteventura and centred on Gran Canaria. In the first case the sources can be related to crustal density changes, flexures in the crust (as it is located near the transitional area) and upper mantle related effects. For Gran Canaria the source of the anomaly can be associated with crustal density variations and mantle related effects. The absence of this kind of anomaly in relation with Tenerife and La Gomera is of special interest.

Smaller gravity anomalies and high gradient zones overlie the main features, indicating the presence of more localized structures near the surface. Some of these gradients have a rectilinear layout which together with other geophysical and geological information, allow us to interpret them as fracture zones: a) The E-W gravimetric gradient (G1, Figure 11b) that separates and displaces in a dextral sense the gravity highs of Lanzarote and Fuerteventura. On both sides of this gradient the crust has different thickness and seismic velocity characteristics (Dañobeitia and Canales, 2000); b) Another gravity gradient is the N-S oriented structure located east of Gran Canaria (G3, Figure 11b) which limits an axis of minima with a N-S orientation (Figure 8). In this area there is a strong structural control to the orientation of submarine canyons that sharply changes its direction to N-S, as can be seen in bathymetric maps (Llanes, 2000; Acosta et al., this issue); c) The channel between Tenerife and Gran Canaria is structured by gravimetric gradients of NW-SE direction (G5, G6 and G7, Figure 11b) and a strong structural control in this direction can be also inferred from the morphology when looking at detailed bathymetry (Acosta et al., this issue). An important NE-SW fracture has been traditionally interpreted in this channel (Dash and Bosshard, 1969; Bosshard and MacFarlane, 1970; Mezcúa et al., 1992), but there is no gravimetric evidence related to it. This is the area with the highest seismic activity within the instrumented period. An earthquake of 5.2 magnitude (09/05/1989) was provisionally interpreted as being related to a strike-slip fault of possible NE-SW direction (Mezcúa et al., 1992). However, there is another solution calculated for the same earthquake (CMT Catalogue, Dziewoinski et al., 1981) as a reverse strike-slip focal mechanism that implies the possible presence of an E-W or a NW-SE fault. There are gravimetric gradients with both directions in the channel, but none with a NE-SW orientation; d) The last gravimetric gradient interpreted as a fault is located between Tenerife and La Gomera with a NW-SE direction (G7). The detailed bathymetry also shows a strong structural control with the same direction (Acosta et al., this issue). In this location, Dash and Bosshard (1969) and Bosshard and MacFarlane

(1970) interpreted a fault running from El Hierro to La Gomera and Tenerife, perpendicular to the one proposed here (G7). However, no gravity evidence to support this has been found on our Bouguer anomaly maps.

The gravity highs located in Fuerteventura and La Gomera (H2 and H9) are interpreted here as being produced by the presence of the Basal Complex that outcrops on both islands. The Basal Complex (Bravo, 1952; Fúster et al., 1968, Ancoechea et al., 2003) is a thick Upper Cretaceous sedimentary sequence overlaid by submarine volcanoes and intruded by an intense NNE-SSW trending sheeted dyke swarm, which formed in association with the location of alkaline plutons. Their outcrops have been found on three islands: Fuerteventura, La Gomera and La Palma. On La Palma, the Basal Complex outcrops in the central area of the island, where a relative gravity high is observed, but it is partially masked by two gravity lows (L4 and L5, Figure 11b). When comparing the described gravity highs related to the Basal Complex with that on Gran Canaria, we found some similarities. Two of the described gravity highs related to it (H6 and H8) have very high relative values of Bouguer anomalies, which could suggest the possible presence of the Basal Complex near the surface (although there is no outcrop on the island). There are also some positive anomalies of short wavelength and high amplitude in the submarine extension of Fuerteventura (H3, H4 and H5) that we interpret as high density intrusive bodies.

Conclusions

From a methodology point of view it is necessary to study Bouguer anomaly maps and not only free air anomaly maps, in order to avoid water influence and to have enough resolution for marine geological interpretation. This is because Bouguer anomalies enhance sub-seafloor density changes.

From a geodynamic point of view it is remarkable that the Canary Islands have complex gravity characteristics. The explanation is the superposition of different effects. The following conclusions can be drawn on the basis of this study:

- Bouguer anomaly values above 140 mGal have been observed for all the islands in the Canary Archipelago, which suggests an oceanic crustal character.
- Two areas are differentiated by their Bouguer anomaly values, one between 140 and 320 mGal that corresponds to oceanic crust, and the other under 140 mGal related with the transition and continental crust. The oceanic crust, located to the west, is separated from the transitional crust by the high gradient zone LFGZ.
- The area of oceanic crust can be divided into three bands limited by two high gradient zones, TGZ and PHGZ.
- The orientation of the high gradient zones LFGZ, TGZ and PHGZ progressively varies from NE-SE, in the eastern edge, to N-S in the western edge. This variation reflects an increasing oceanic influence towards the east.
- Circular anomalies with various tens of km in diameter located in the eastern and western islands may have different origins. The gravity lows observed in western islands may be related to upper mantle effects, while the gravity highs observed in the eastern islands may be related to mantle and crustal effects. In the central islands (Tenerife and La Gomera) these kind of anomalies are absent.
- The gravity highs of Fuerteventura, La Gomera and La Palma are related with the Basal Complex. The gravity highs of Gran Canaria and the areas of 'Amanay Bank' and 'El Banquete Bank', located southwest of Fuerteventura, are produced by high density intrusive bodies.
- There are some high gravity gradients that present a linear layout and their study, together with other geophysical and geological information, allows us to interpret them as fracture areas. These fracture areas have not been correctly interpreted until now or were assigned to a different structural direction.

Acknowledgements

This work was funded by the EEZ project (Oceanographic and Hydrographic Research of the Spanish Economic Exclusive Zone) and the 'Universidad Complutense de Madrid'. We thank the 'Conserjería de Educación de la Comunidad de Madrid y Fondo Social Europeo' for pre-doctoral grants awarded to Pilar Llanes and Juan Álvarez. We also thank the captain, officers and crew of RV Hespérides on cruises EEZ-98, EEZ-99 and EEZ-2000 for their cooperation at sea. Finally, we are grateful for the efforts of re viewers whose comments have helped improved the manuscript.

References

Ablay, G.J. and Kearey, P., 2000. Gravity constraints on the structure and volcanic evolution of Tenerife, Canary Islands, *J. Geophys. Res.* **105** (B3): 5783–5796.

Acosta, J., Palomo, C., Uchupi, E., Muñoz, A., Escartin, J., Herranz, P. and Sanz, J.L., 1997. Morphology and seismic character of north slope of Tenerife, Canary Island. Evidence for episodic massive landslides, *J. Geophys. Res.* **102**: 20325–20342.

Acosta, J., Uchupi, E., Muñoz, A., Herranz, P., Palomo, C., Ballesteros, M. and EEZ Working Group, this issue. Geologic evolution of the Older Canary Islands: Lanzarote, Fuerteventura, Gran Canaria and La Gomera, with a Brief description of the avalanches on the Younger Islands: Tenerife, La Palma and El Hierro, *Mar. Geophys. Res.*.

Ancochea, E., Fúster, J.M., Ibarrola, E., Cendrero, A., Coello, J., Hernan, F., Cantagrel, J.M. and Jamond, C., 1990. Volcanic evolution of the island of Tenerife (Canary Islands) in the light of new K-Ar data, *J. Volcanol. Geotherm. Res.* **44**: 231–249.

Ancochea, E., Brändle, J.L., Huertas, M.J., Cubas, C.R. and Hernán, F., 2003. The felsic dikes of La Gomera (Canary Islands): identification of cone sheet and radial dike swarms, *J. Volcanol. Geotherm. Res.* **120**: 197–206.

Anguita, F. and Hernán, F., 2000. The Canary Islands origin: a unifying model, *J. Volcanol. Geotherm. Res.* **103** (1–4): 1–26.

Araña, V., Camacho, A.G., García, A., Montesinos, F.G., Blanco, I., Vieira, R. and Felpeto, A., 2000. Internal structure of Tenerife (Canary Islands) based on gravity, aeromagnetic and volcanological data, *J. Volcanol. Geotherm. Res.* **103** (1–4): 4364.

Banda, E., Dañobeitia, J.J., Surinach, E. and Ansorge, J., 1981. Features of crustal structure under the Canary Islands, *Earth Planet. Sci. Lett.* **55**: 11–24.

Bendat, J.S. and Piersol, A.G., 1986. Random data, 2nd revised ed., John Wiley and Sons.

Blakely, R.J., *Potential theory in gravity and magnetic applications*, Cambridge University Press, Cambridge, 1–441, 1995.

Bosshard, E. and MacFarlane, D.J., 1970. Crustal structure of the western Canary Islands from seismic refraction and gravity data, *J. Geophys. Res.* **75**: 4901–4918.

Bravo, T., 1952. Aportación al estudio geomorfológico y geológico de la costa de la fosa tectónica del valle de la Orotava, *Bol. R. Soc. Esp. Hist. Nat.*, Tomo L, 1–30.

Camacho, A.G., Montesinos, F.G., Vieira, R. and Arnoso, J., 2001. Modelling of crustal anomalies of Lanzarote (Canary Islands) in light of gravity data, *Geophys. J. Int.* **147**: 403–414.

Canales, J.P. and Dañobeitia, J.J., 1998. The Canary Islands swell: a coherence analysis of bathymetry and gravity, *Geophys. J. Int.* **132**: 479–488.

Carracedo, J.C., Day, S.J., Guillou, H., Rodriguez Badiola, E., Canas, J.A. and Pérez Torrado, F.J., 1998. Hotspot volcanism close to a passive continental margin: the Canary Islands, *Geol. Mag.* **135** (5): 591–604.

Dañobeitia, J.J., Canales, J.P. and Dehghani, G.A., 1994. An estimation of the elastic thickness of the lithosphere in the Canary Archipelago using admittance function, *Geophys. Res. Lett.* **21**: 2649–2652.

Danés, Z.F., 1982. Short note: An analityc method for the determination of distant terrain corrections, *Geophysics* **47** (10): 1453–1455.

Dañobeitia, J.J. and Canales, J.P., 2000. Magmatic underplating in the Canary Archipelago, *J. Volc. Geother. Res.* **103**: 27–41.

Dash, B.P. and Bosshard, E., 1969. Seismic and gravity investigations around the Western Canary Islands, *Earth Planet. Sci. Lett.* **7**: 169–177.

Dziewoinski, A.M., Chou, T.A. and Woodhouse, J.H., 1981. Determination of earthquakes source parameters from waveform data for studies of global and regional seismicity, *J. Geophys. Res.* **86**: 2825–2852.

Hausen H., 1962. New contributions to the geology of Gran Canary (Gran Canaria, Canary Islands), *Soc. Sci. Finnica, Comment. Phys.–Math.* **27** (1): 418.

I.G.N. (Instituto Geográfico Nacional de España), 1996. Banco de datos gravimétricos de España. Ministerio de Fomento. Internal Repport.

Jachens, R.C., Simpson, R.W. and Blakely, R.J., 1989. Isostatic residual gravity and crustal geology of the United States, *Geol. Soc. Am. Bull.* **172**: 405–423.

Jin, Y., McNutt, M.K. and Zhu, Y., 1994. Evidence from gravity and topography data for folding of Tibet, *Nature* **371**: 669–674.

Karner, G.D. and Watts, A.B., 1983. Gravity anomalies and flexure of the lithosphere at Mountain Ranges, *J. Geophys. Res.* **88** (B–12), 10449–10477.

Lin, J., Purdy, G.M., Schouten, H., Sempere, J.C. and Zervas, C., 1990. Evidence from gravity data for focused magmatic accretion along the Mid-Atlantic Ridge, *Nature* **344**: 627–632.

Llanes, P., 2000. *Análisis morfoestructural basado en técnicas geofísicas del margen submarino centro-oriental del archipiélago Canario*, Bachelor Thesis, Universidad Complutense de Madrid, 133 p.

MacFarlane, D.J. and Ridley, W.I., 1968. An interpretation of gravity data for Tenerife, Canary Islands, *Earth Planet. Sci. Lett.* **4**: 481–486.

MacFarlane, D.J. and Ridley, W.I., 1969. An interpretation of gravity data for Lanzarote, Canary Islands, *Earth Planet. Sci. Lett.* **6**: 431–436.

Mezcua, J., Buforn, E., Udías, A. and Rueda, J., 1992. Short note: Seismotectonics of the Canary Islands, *Tectonophysics* **208**: 447–452.

Nettleton, L.L., 1976. *Gravity and magnetics in oil exploration*. New York, Mac Graw-Hill.

Plouff, D., 1976. Gravity and magnetic fields of polygonal prisms and application to magnetic terrain corrections, *Geophysics* **41**: 727–741.

Ranero, C.R., Torné, M. and Banda, E., 1995. Gravity and multichannel seismic reflection constraints on the lithospheric structure of the Canary swell, *Mar. Geophys. Res.* **17**, 519–534.

Russo, R.M. and Speed, R.C., 1994. Spectral analysis of gravity anomalies and the architecture of tectonic wedging, NE Venezuela and Trinidad, *Tectonics* **13** (3): 613–622.

Roest, W.R., Dañobeitia, J.J., Verhoef, J. and Collete, B.J., 1992. Magnetic anomalies in the Canary Basin and the Mesozoic evolution of the Central North Atlantic, *Mar. Geophys. Res.* **14**: 1–24.

Rothe, P. and Schmincke, H.V., 1968. Contrasting origins of the eastern and western islands of the Canarian Archipelago, *Nature* **218**: 1152.

Sandwell, D.T. and Smith, W.H.F., 1997. Marine gravity anomaly form Geosat and ERS 1 satellite altimetry, *J. Geophys. Res.* **102** (B-5), 10039–10054.

Smith, W.H.F. and Sandwell, D.T., 1997. Global seafloor topography from satellite altimetry and ship depth soundings, *Science* **277**: 1957–1962.

Simpson R.W., Jachens, R.C. and Blakely, R.J., 1986. A new isostatic residual gravity map of the Conterminous United States with a discussion on the significance of isostatic residual anomalies, *J. Geophys. Res.* **91** (B-8), 8348–8372.

Stillman, C.J., Fuster, J.M., Bennell-Baker, M.J., Muñoz, M., Smewing, J.D. and Sagredo, J., 1975. Basal complex of Fuerteventura (Canary Islands) is an oceanic intrusive complex with rift-system affinities, *Nature* **257**: 469–470.

Talwani, M., Le Pichon, X. and Ewing, M., 1965. Crustal structure of the mid-ocean ridges 2. Computed from gravity and seismic refraction data, *J. Geophys. Res.* **70**: 341–352.

Talwani, M., 1971. *Gravity in the Sea*, 4 (1), edited by A. Maxwell, 251–297, John Wiley, New York.

Vieira, R., Toro, C. and Araña, V., 1986. Microgravimetric survey in the Caldera of Teide, Tenerife, Canary Islands, *Tectonophysics* **130**: 249–257.

Watts, A.B., 1994. Crustal structure, gravity anomalies and flexure of the lithosphere in the vicinity of the Canary Islands, *Geophys. J. Int.* **119**: 648–666.

Watts, A.B. and Talwani, M., 1975. Gravity Field of the Northwest Pacific Ocean Basin and its Margin: Hawaii and Vicinity, *The Geological Society of America*, Inc., Map and Chart Series MC-9.

Watts, A.B. and Marr, C., 1995. Gravity anomalies and the thermal and mechanical structure of rifted continental margins, in *Rifted Ocean-Continent Boundaries, NATO Science Series II: Mathematics, Physics and Chemistry*, 463, edited by E. Banda et al., Kluwer Academic Publishers, Dordrecht, 65–94.

Watts, A.B., Pierce, C., Collier, J., Dalwood, R., Canales, J.P. and Henstock, T.J., 1997. A seismic study of lithosphere flexure in the vicinity of Tenerife, Canary Islands. *Earth Planet. Sci. Lett.* **146**: 431–447.

Welch, P.D., 1967. The use of Fast Fourier Transform for the estimation of power spectra: a method based on time averaging over short, modified peridograms, *IEEE Transactions on Audio and Electroacoustics*, vol. AU-15, 2.

Wessel, P. and Watts, A.B., 1988. On the Accuracy of Marine Gravity Measurements, *J. Geophys. Res.* **93** (B-1): 393–413.

Wessel P. and Smith, W.H.F., 1995. New Version of the Generic Mapping Tools (GMT), http:/www.agu.org/eos_els, *American Geophysical Union*.

Yale, M.M., Sandwell, D.T. and Smith, W.H.F., 1995. Comparison of along-track resolution of stacked Geosat, ERS 1, and TOPEX satellite altimeters, *J. Geophys. Res.* **100** (B-8), 15117–15127.

Ye, S., Canales, J.P., Rihm, R., Dañobeitia, J.J. and Gallart, J., 1999. A crustal transect through the northern and northeastern part of the volcanic edifice of Gran Canaria, Canary Islands, *J. Geodyn.* **28**: 3–26.

Marine Geophysical Researches (2003) 24: 129–148
DOI 10.1007/s11001-004-5442-y

A magnetic anomaly study offshore the Canary Archipelago

M. Catalán, J. Martín Davila & ZEE Working Group*
Real Instituto y Observatorio de la Armada, San Fernando 11000, Cádiz, Spain

Key words: Canary Islands, geomagnetism, marine magnetic anomalies, underplating

Abstract

A scalar magnetic anomaly map of the regions offshore the Canary Archipelago is presented here. This map is based on measurements taken inside the Exclusive Economic Zone Project framework. This paper contains a description of the data set, whose accuracy and internal consistency are analysed. The magnetic anomalies are described and the main structural trends are highlighted. This analysis has served to detect two possible fractures with a NW–SE component: one separates Fuerteventura from Gran Canaria Island, and the other apparently runs through Fuerteventura. The latter finding agrees with a 1.7 km depth offset, between its northern and southern halves, detected by the algorithm based on the Euler Deconvolution. A similar approach has been performed, estimating the depth to the top of the most significant anomalies. This result shows that the most noteworthy magnetic anomaly source of the archipelago, which lies between Tenerife and Gran Canaria Island, seems to be located at an average depth of 4 km below sealevel. A spectral analysis was performed to estimate the depth extent of the deepest anomalies. It argues the presence of sub-crustal magnetic sources (underplating) in the archipelago. A standard Euler Deconvolution analysis was executed to analyse the spatial distribution of these mantle-like sources. Our results seem to support the existence of magmatic underplating under Gran Canaria, Tenerife, and Fuerteventura, and suggest this possibility for La Palma.

Introduction

This paper focuses on the processing and inversion of magnetic anomaly data from the Canary Archipelago, inside the Exclusive Economic Zone framework. We have produced a scalar magnetic anomaly map, and, by applying some operators (e.g., analytical signal and Euler Deconvolution), we have highlighted structural features and trends around this area, and inferred the depth and location of the causative bodies. These results improve the resolution and expand on earlier work in the area. Additionally, the magnetic data can now be correlated with other data with the same level of accuracy and resolution.

*J.A. Marín[a], L.M. Agudo[a], J. Acosta[b], C. Palomo[b], P. Herranz[b], M. Ballesteros[b], A. Muñoz[b], A. Carbó[c], A. Muñoz[c], F. Perez-Carrillo[d]

[a] Real Instituto y Observatorio de la Armada, San Fernando 11100 (Cádiz), Spain.

[b] Inst. Esp. de Oceanografía. Grupo de Cartografía Multihaz, C/Corazón de María 8, 28002 Madrid, Spain.

[c] Dpto. de Geodinámica, Facultad de Ciencias Geológicas, Univ. Complutense de Madrid. 28040 Madrid, Spain.

[d] Instituto Hidrográfico de la Marina. C/Tolosa Latour S/N. 11071 Cádiz. Spain.

Magnetic geological setting

The Canary Archipelago is located on the border of the northwestern Africa passive continental margin. Its geographical location indicates that the oceanic crust on which these islands were built is Early Jurassic in age (150–170 Ma). To improve our understanding of this zone we have compiled magnetic data from the Geophysical Data System (GEODAS) (Metzger and Campagnoli, 2003) and merged them into a single 0.2° × 0.2° grid covering the following geographical frame: 35°N–25°N and 5°W–25°W (Figure 1).

Figure 1 shows a magnetic anomaly pattern which is clearly unbalanced and is dominated by a spectacular feature, M0–M4 anomalies, previously known as the J anomaly (Hayes and Rabinowitz, 1975). Its shape delineates a N30°E trend, showing a quite remarkable continuity northward. Moving eastward, other M-series spreading anomalies (black solid line in Figure 1) stand out, e.g., M16, M21 and M25, the oldest alignment consistently present in the Eastern Atlantic (Vogt, 1986; Roest et al., 1992a).

The M0–M4 series lose definition and amplitude at the Canary Lineament (near 30°N paral-

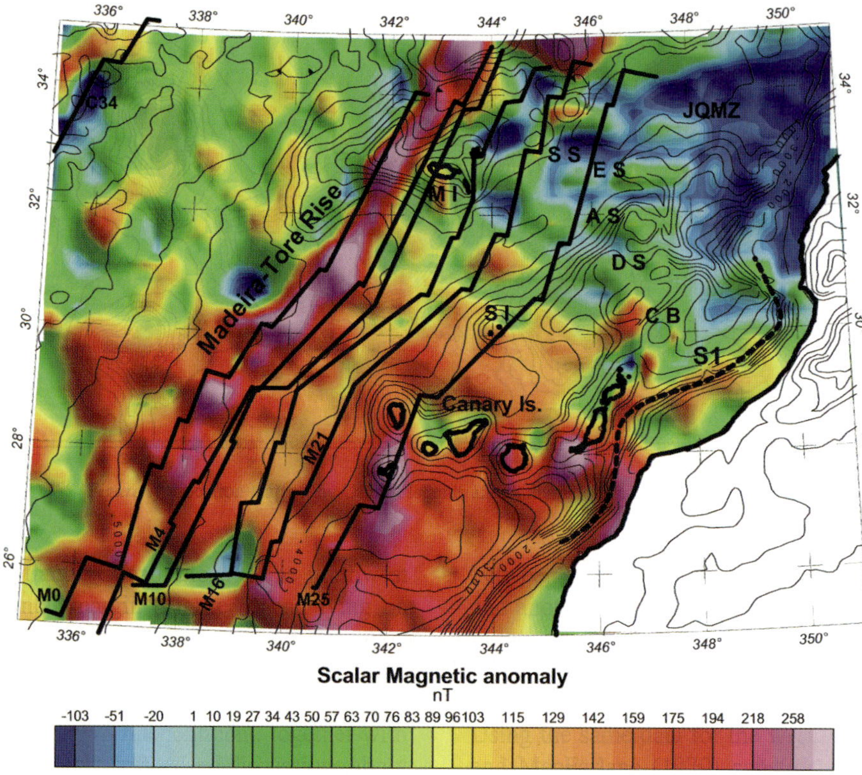

Figure 1. Shaded relief map of scalar field magnetic anomalies, illuminated from the northwest. Figure shows principal features: volcanic islands, linear seafloor spreading magnetic anomalies after Verhoef et al. (1991), in solid lines bathymetry, contour interval: 200 m. CB: Conception Bank; DS: Dacia Seamount; SI: Selvagens islands; MI: Madeira islands; SS: Seine Seamount; ES: Essaouira Seamount; AS: Agadir Seamount, JQMZ: Jurassic Quiet Magnetic Zone. Map projection is Lambert Conic Conformal (scale 1:10.000.000).

lel). A magnetic quiet zone is observed landward of the M-sequence throughout the whole area (Jurassic Magnetic Quiet Zone (JMQZ), 150–170 Ma). Areas of disturbances almost always correspond to the presence of volcanic islands or seamounts, as was inferred from the correlation with bathymetry or with prominent basement peaks (Hayes and Rabinowitz, 1975). By including the Madeira and Canary Archipelagos, Figure 1's geographical frame contains the highest density of islands/seamounts on the African continental margin (Ye et al., 1999).

Eastward of anomaly M25, inside the JMQZ, no seafloor spreading anomalies are visible. Only an elongated anomaly, just off the African margin, can be detected easily. This anomaly, which has a remarkably similar geometry to that of anomaly M25, could be associated with the so-called slope anomaly S1. This anomaly could mark the approximate landward edge of the oceanic crust (Roest et al., 1992a; Roeser et al., 2002).

Previous magnetic investigations

The Canary archipelago has been the subject of many geophysical investigations, but not much magnetic study. In 1993, the Spanish National Geographical Institute (I.G.N.) carried out an aeromagnetic survey composed of seven independent grids flown at different heights and in a north-southward orientation. For further technical details see Socias and Mezcua (1996). As a final result, an aeromagnetic map at 3200 m was published in 1995 (Instituto Geográfico Nacional, 1995).

In April 1995, a marine magnetic survey was carried out in the Tenerife offshore area by the Spanish Oceanographic Institute (I.E.O.). This geophysical campaign, called "Teide-95", included not only magnetic, but also bathymetric and gravity data acquisition. Its main aims involved the mapping of offshore area morphology, trying to determine the volume of pyroclastic rocks and extend the survey as close to the coast as possible.

Ultimately, the goal was to define the relationships between the offshore slope morphology and some onshore collapsed valleys in Tenerife (Teide Group, 1997).

Araña et al. (2000) established a structural model for Tenerife, combining mainly magnetic and gravity data, as well as other volcanological and similar geophysical data. Blanco-Montenegro et al. (2003) performed a similar study focused on magnetic data. They propose a crustal structure model beneath Gran Canaria island and suggested, from spectral analysis, the presence of rocks at mantle-like depths (from the Moho to about 23 km) that could behave as magnetic sources. They discussed the possible relation with magmatic underplating proposed by previous studies (Ye et al., 1999; Dañobeitia and Canales, 2000).

Verhoef et al. (1991) published a magnetic study performed after compilation of magnetic data collected off West Africa over a period of nearly 30 years (1961–1988). It allowed the accurate identification of sea floor spreading anomalies, showing offsets and disruptions in the anomalies caused by fracture zones. It not only focused on the Canary islands but covered a wider area, limited to the north by the Azores Gibraltar Fracture Zone (AGFZ, 37°N) and to the south by the Kane Fracture Zone (KFZ, 21°N). This dataset, in combination with data from the western Atlantic, was used in later work to decipher the African and conjugate margin spreading histories (Roest et al., 1992a).

Since 1998, the Spanish Exclusive Economic Zone Project (ZEEE) has studied the Canary Islands area. This paper presents a magnetic anomaly map derived from the compilation of scalar magnetic measurements. They were obtained during three one-month marine surveys during 1998, 1999 and 2000 carried out in the Canary Islands area, aboard the Spanish Research Vessel "Hespérides". Two additional campaigns, undertaken in October 2001 and 2002, used the Spanish Oceanographic Ship "Vizconde de Eza".

Data compilation and methods

Instrumentation

In order to measure the geomagnetic field at sea, three marine magnetometers were used: Geomet-

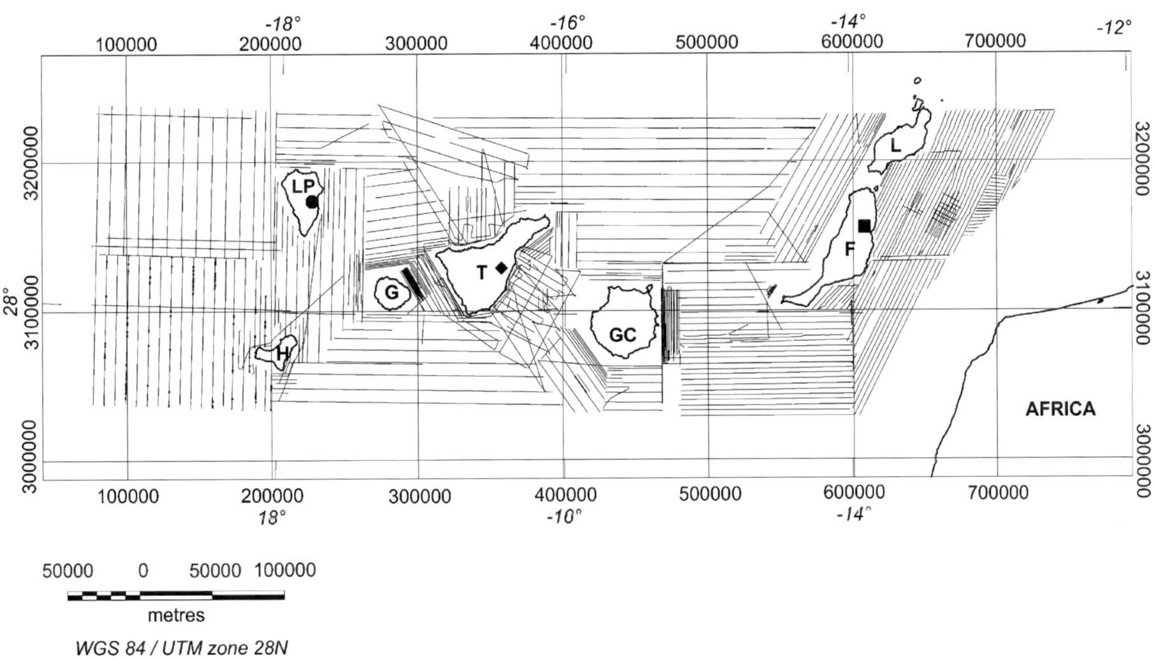

Figure 2. ZEEE track line in the region of the Canary islands (1998–2002). Reference station sites are included, using square and circle symbols respectively for Fuerteventura and La Palma. Güimar observatory (Tenerife) is marked with a diamond symbol. H: Hierro; LP: La Palma; G: Gomera; T: Tenerife; GC: Gran Canaria; F: Fuerteventura; L: Lanzarote.

rics *G-801* Proton Precession Magnetometer, (ROA), Geomag *SMM II* Overhauser effect magnetometer (provided by the University of Cadiz), and a Geometrics *G-876* type, installed on the B.I.O. "Hespérides".

To ascertain the external geomagnetic field component, three reference station sites were used on land (Figure 2). Güimar Observatory (I.G.N.) on Tenerife island provided data during the whole survey period. Additionally, a portable reference station was installed and maintained by ROA staff on La Palma and Fuerteventura. Their periods of activity were adapted to the surveying. The permanent presence of Güimar Observatory, considering its geometric emplacement, allowed us to infer the external field error budget contribution. This difference has been checked and proved not to be greater than 5 nT (in root mean square sense) even during magnetic storm periods. This unexpected result could be justified by the survey shape, which mainly evolved in a Westward progression.

Error estimates and data compilation

In order to evaluate the precision of the surveys and the internal consistency of the whole data set, a crossover analysis was performed. The results, from 639 crossings, show a standard deviation of 12.9 nT and a mean value of less than 0.5 nT. Thus, we can infer that the results have not been influenced by any systematic effect.

Results are gathered in the following histogram (Figure 3). Residuals could be due to several factors:

1. An incomplete cancellation of the diurnal variation.
2. An indirect effect produced by a possible lack of precision in ship's positioning. Considering the magnetic field local gradient, incorrect positioning could generate differences in crossover readings near the coast, where the spatial gradient reaches maximum values.
3. Internal field contribution.
4. Error due to the secular variation models used.
5. Instrumental errors.

The data has been subjected to classical corrections, dating both in time and position using precise satellite navigation (DGPS). Its precision (better than 20 m) induces an expected error contribution for the navigation uncertainty not greater than 1 nT. All tracks were corrected using the Definitive Geomagnetic Reference Field

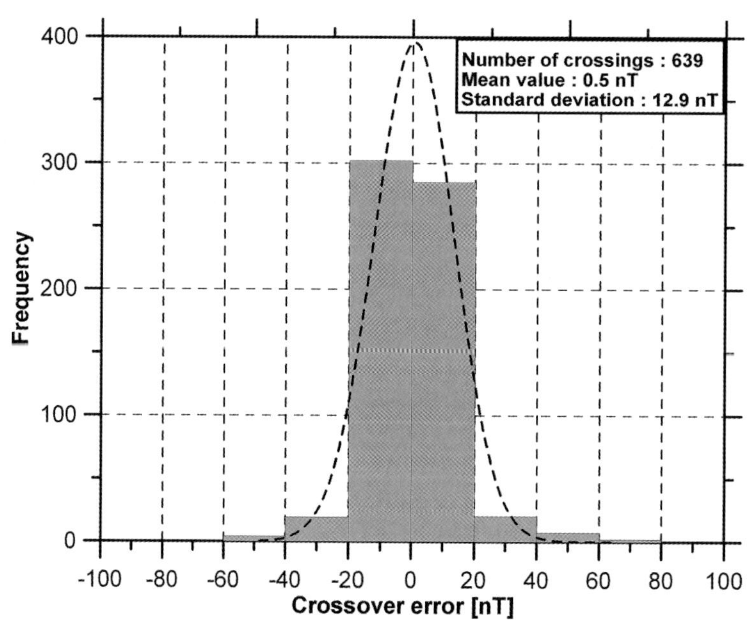

Figure 3. Histogram of crossover errors.

model DGRF-95 for the period 1998–2000 and the IGRF-2000 for 2000–2002. All measurements were referred to 2000.0 epoch.

In order to take into account a possible error contribution due to discrepancies caused by the use of two different regular field models (DGRF-95 and IGRF-2000), we have compared two grids built with the different regular field models used, dated at 2000.0, using the Canary Archipelago's geographical frame. The results conclude that this contribution is smaller than 1 nT and is therefore not significant.

Survey coverage was not uniform. It consisted of irregularly spaced and irregularly oriented ship tracks (Figure 2) with an across-track distance which ranging from 10 km to 500 m, with an approximate 30 m along-track resolution conditioned by local water depth. To convert this into observations over a regular grid, we have applied a Kriging algorithm, which suits our needs quite well when we merge different non-parallel line data. Finally, a 5 km × 5 km grid cell size was used.

To avoid possible edge effects we have digitised the 3200 m aeromagnetic map (Instituto Geográfico Nacional, 1995) shoreline contour anomalies and downward continued these new magnetic anomaly values to sea level. There is no question about instabilities due to this downward continuation because no magnetic sources were cut (Blakely, 1995).

Particular attention must be paid to the aeromagnetic anomaly data used. The aeromagnetic anomaly grid was originally obtained using a IGRF-85 model, and referred to the epoch 1993.8 (Instituto Geográfico Nacional, 1995). In order to avoid the application of the IGRF-85 model on an epoch outside its validation period (1985–1990), we have removed this analytical model, finally extracting the regular field by use of the DGRF-90 evaluated in the flight epoch (1993.8).

As required by a spectrum analysis explained below, we need to fill in those gaps that correspond to the different islands' land coverage. We have digitised seven 5 km × 5 km grids (one grid per island) from the 3200 m aeromagnetic map (Instituto Geográfico Nacional, 1995) and proceeded in the same way as described above to avoid the IGRF-85 model and extract data with the DGRF-90 model. We have upwardly continued our shipborne data until 3200 m and merged them with the aeromagnetic data, obtaining a new grid without any geographical gap.

In order to be certain that no magnetic sources actually exist in the region of continuation, we have upwardly continued our new 3200 m grid up to the highest summit height (the Teide peak: 3800 m, in Tenerife). Not doing so could have introduced some type of instability (Blakely, 1995).

Magnetic anomaly map

Description of the map

As described in the previous paragraph, we were able to set up a 3800 m grid for the whole archipelago, but we decided not to use it, except for two specific instances in order not to lose spatial resolution. In those cases, where we have merged our shipborne dataset with the 3200 m aeromagnetic's sub-grids, it is explicitly indicated in the corresponding figure caption. The magnetic anomaly map (Figures 4 and 5) shows the typical signature associated with volcanic areas, that is, short wavelengths and high amplitudes. It is noteworthy that the greatest anomalies are not represented because they are associated with different calderas located on the islands. Although we lack onshore data, we could discern the main trends. From a detailed analysis of the magnetic map, it is possible to mark out three different magnetic areas.

The first area includes the zones close to the islands (Figure 4: inside white dashed polyline). These zones contain both the greatest amplitudes, as well as the highest frequencies. The aeromagnetic map (Socias and Mézcua, 1996) shows the real onshore situation, where every island's map is more complex than a single dipole pattern.

We could detect dipolar shape anomalies associated with every island, always showing an induced field shape, which meant that remanence was acquired during a period of normal polarity of the Earth's magnetic field (Figure 4). Younger islands (e.g., La Palma and El Hierro) appear to have the most simple characteristics, extending their magnetic influence southward.

In La Palma we mapped two large anomalies just off the island. A negative lobe projects to the north, and a positive one to the south. From the aeromagnetic map we could discern they are not

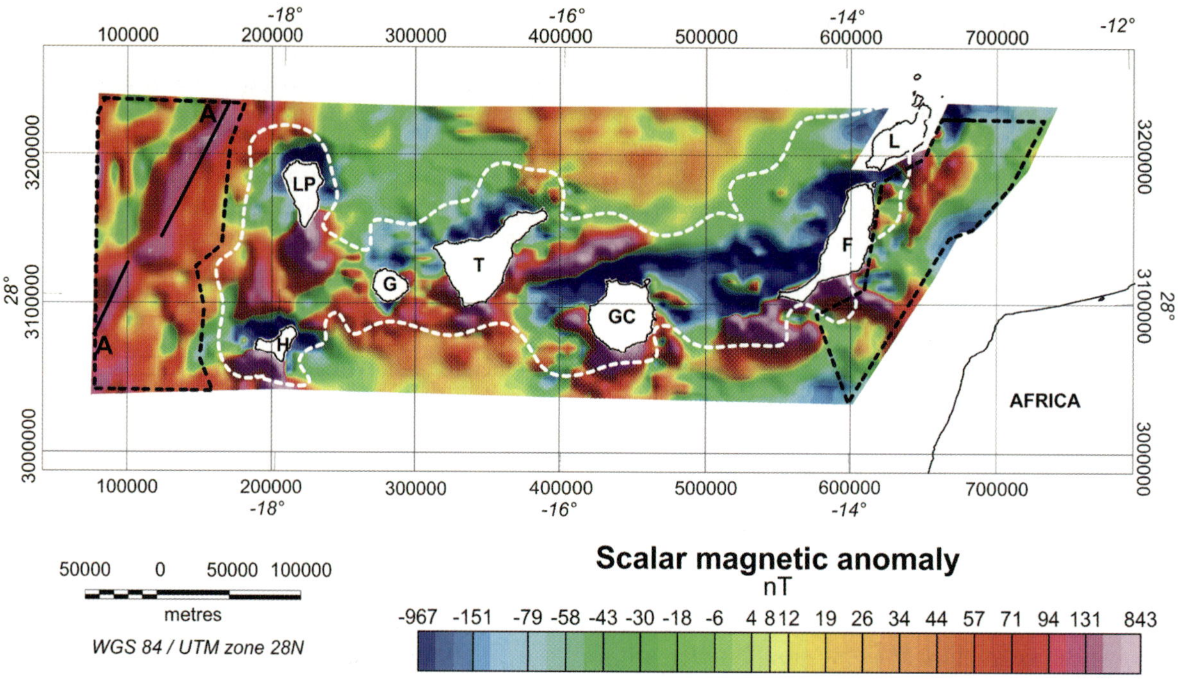

Figure 4. Detail shaded relief scalar field map of the area. Illumination from the northwest. The different magnetic areas are marked (see text for comments).

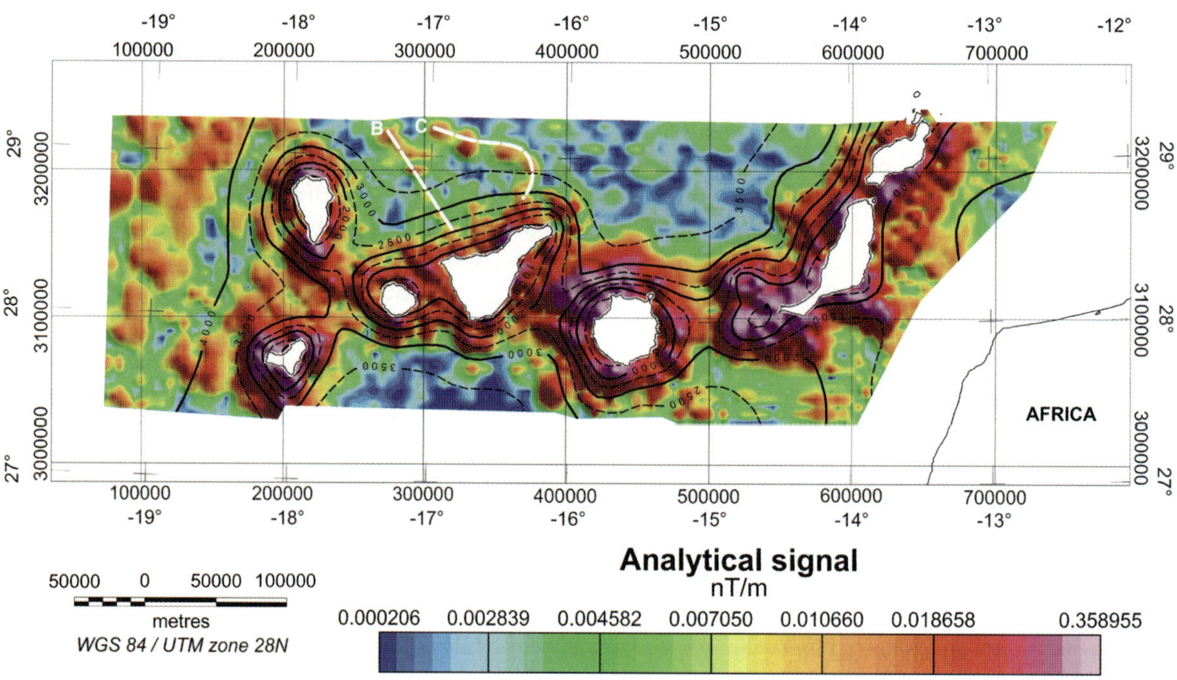

Figure 5. Shaded relief analytical signal map of the area. Illumination from the northwest. For clarification on label meaning see text. In solid and dashed lines bathymetry contour lines every 500 m.

related and belong to different magnetic sources (Instituto Geográfico Nacional, 1995).

La Gomera and Tenerife appear to be influenced by a wide, deep body and so seem to be magnetically connected. The magnetic anomalies get more complex further eastward. The central islands (Tenerife and Gran Canaria) show broader anomalies. A gently anti-clockwise trend rotation from E-W to NE-SW was detected as we moved toward the eastern island group (Fuerteventura and Lanzarote).

It is noteworthy that, between Tenerife and Gran Canaria, a very elongated anomaly is noted. Its pattern seems to be related to an inverse magnetised source. Mézcua et al. (1992) related its origin to an active fault, although this has not been verified by local seismic experiments (Funck and Schmincke, 1998).

The second area (Figure 4: inside black dashed polylines) is situated in the westernmost and easternmost part of the archipelago. The former shows strip-shaped anomalies with a nearly north or N30°E trend. This fact is more clearly demonstrated when we examine the directional derivative map (Figure 6a and b). These trends also could be detected in the eastern part of the map.

In the northern and southern central regions we recognised another magnetic style (Figure 4), and, hence, a third magnetic province (Figure 4: outside black and white dashed polyline boundaries). This third province shows great wavelength anomalies, which include more local peaks with a nearly E-W trend (Figure 6b). Particularly, in the north a great positive terrace could be discerned.

Analytical signal

To make description easier, we combined the scalar anomaly field map with the analytical signal grid. The amplitude of the analytical signal is defined as the square root of the squared sum of the vertical and two orthogonal horizontal derivatives of the magnetic field anomaly (Roest et al., 1992b; Roest and Pilkington, 1993; Salem et al., 2002).

$$|A(x,y)| = \sqrt{\left(\frac{\partial M}{\partial x}\right)^2 + \left(\frac{\partial M}{\partial y}\right)^2 + \left(\frac{\partial M}{\partial z}\right)^2}, \quad (1)$$

where $|A(x,y)|$ is the amplitude of the analytical signal, and M the magnetic anomaly field intensity. Therefore, the resulting 3D analytical signal summarises the net variation of the gradient of the magnetic anomaly field intensity in 3D.

The application of the analytic signal operator in magnetic interpretation was pioneered by Nabighan (1972). It was initially applied on profiles using a 2D approach. Lately, this method has been generalised to 3D. (Roest and Pilkington, 1993; Mohan and Babu, 1995; Hsu et al., 1998; Bastani and Pedersen, 2001). The method is a straightforward way to estimate the depth to sources or to quantify remnant effects (Salem et al., 2002).

One of the most attractive aspects of this 3D operator is the fact that its amplitude produces maxima over magnetic contacts nearly regardless of the direction of magnetisation or its induced and/or remnant character (Roest et al., 1992b; Roest and Pilkington, 1993; Keating and Sailhac, 2004). In this sense, there is no need to assume a purely induced magnetisation effect hypothesis or to discuss possible remnant vector space orientation. This simplistic approach (a purely induced magnetisation effect hypothesis) could lead to severe distortion, mainly in volcanic environments where remnant effects play an important role (Roest et al., 1992b; Roest and Pilkington, 1993; Blakely, 1995).

To analyse the area from the point of view of its magnetisation, we take advantage of the analytical signal signature (Figure 5). The analytical signal map shows three domains: One in green colour, another in yellow to orange, and the last one in red. The red region roughly corresponds to the 0.05–0.03 nT/m range in analytical signal units. The orange to yellow region is 0.02–0.01 nT/m, and the green domain delimits the smallest amplitudes ($<10^{-2}$ nT/m). If we track the bathymetry contour line map, we do not observe a single isobath fitting the green-yellow boundary but rather roughly defined zones that present higher bathymetric gradients. It is noteworthy that this operator (analytical signal) is based on first derivatives of the scalar magnetic anomaly field, highlighting boundaries according with their magnetisation contrast. These analytical signal domains present a reasonable correlation with the Free Air gravity map (Carbó et al., this issue), while this coherency is not noticeable in the Bouguer grid. All

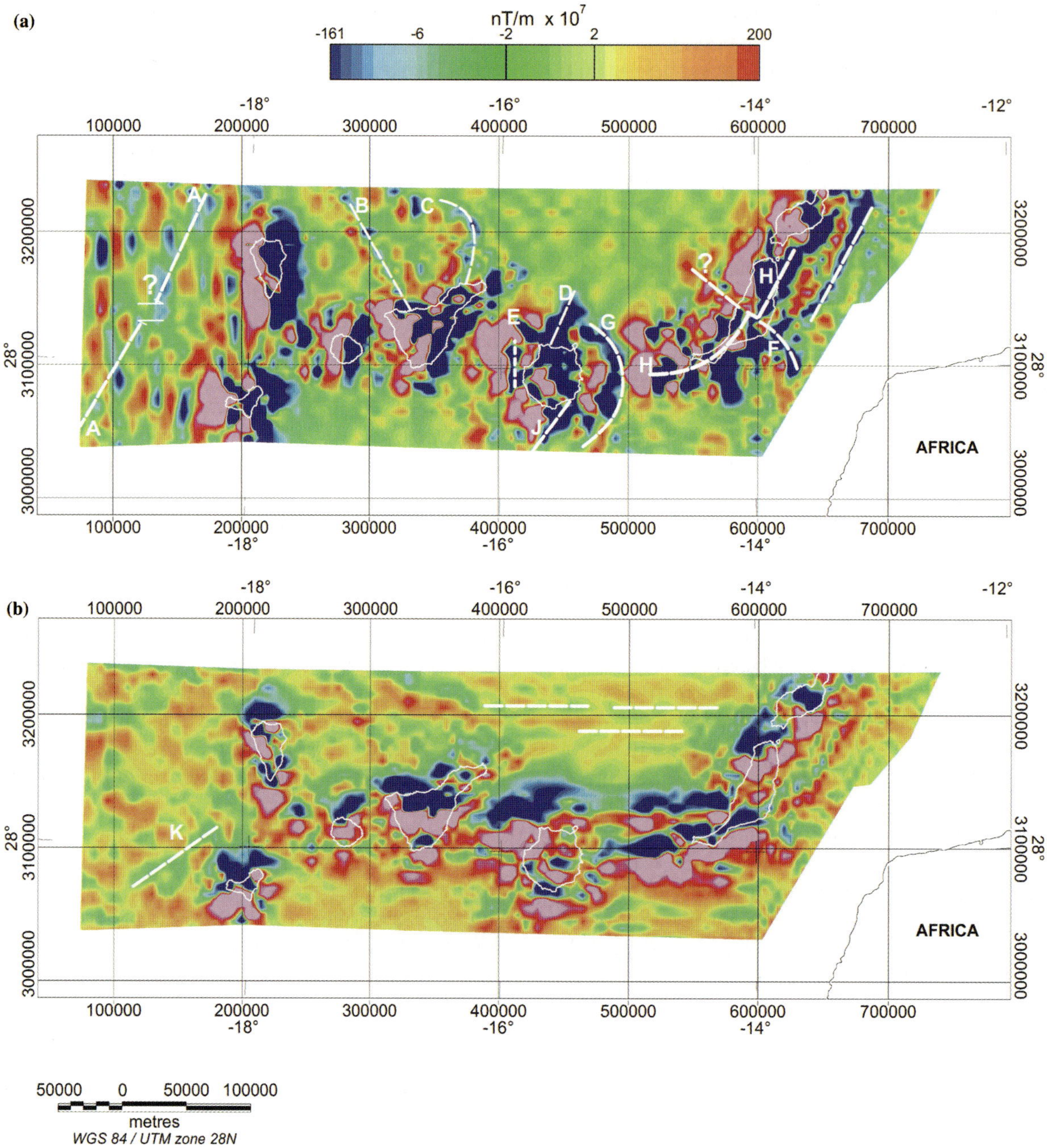

Figure 6. (a) x-Axis analytical signal horizontal derivative. (b) y-Axis analytical signal horizontal derivative. Labels A–H show suggested main magnetic alignments inferred from analytical signal horizontal directional derivatives. In order to properly track onshore features we have merged our shipborne dataset with seven 5 km × 5 km sub-grids from the 3200 m aeromagnetic grid (see section "Error estimate and data compilation" for comments).

these reasons lead us to associate the previously established analytical signal regions with topography; particularly their largest signatures are conditioned by the submarine prolongation of every island.

Depth to the source

Depth to the top: Euler Deconvolution

Magnetic methods are able to image structures at greater depths than other geophysical procedures.

In order to have a rough idea concerning the depth to the top of the different magnetic horizons, we have applied an algorithm based on the Euler Deconvolution.

The Euler Deconvolution (hereafter called "Standard Euler Deconvolution") is applied by selecting a square window of data from a total field grid and its orthogonal derivatives. Euler's equation is solved, by using the least squares method, simultaneously for each grid position within every window. The equation includes a parameter, the degree of homogeneity "N", which is usually referred to as the structural index (S.I.). This index is a measure of the rate of change with distance of a field.

This semi-automated inverse method has proved to be quite useful in several recent studies (i.e. Muszala et al., 1999; Barbosa et al., 2000). For a complete discussion on its mathematical details, refer to Thomson (1982) and Reid et al. (1990).

The "Standard Euler Deconvolution" produces a large set of solutions. Due to its vulnerability to an incorrect structural index value, and/or to the window size where it performs the inversion, solutions may be generated in areas that are free of anomalies or on the edges of anomalies. This is a difficult task to solve, especially if the area, as is the case here, is full of anomalies with quite different wavelengths. To avoid this, we have applied what we will call the "Located Euler Deconvolution".

The new method begins by calculating the analytical signal. It finds its peaks and then uses these locations for Euler Deconvolution (Blakely, 1995). This procedure produces far fewer solutions than the standard one, estimating solutions only over recognised anomalies. The final solution involves only a few depth estimates.

In the first place, in order to get an adequate S.I., we have followed the classical approach. To provide accurate solutions, the square window of data must adequately sample each anomaly on the grid. This occurs when the window size is about half the anomaly width (Durrheim and Cooper, 1998). As most of the anomalies that interest us are 30–40 km in width, a good estimate for the search window size is 20 km × 20 km. We emphasise that for our current purpose (selection of a correct S.I.) the window size is not critical.

Next, we applied an Euler Deconvolution on the grid using three different structural indices (0.5, 1.0 and 2.0). Looking for the best solution cluster (smaller scattering in the Euler Deconvolution solutions) we discarded 2.0 and 1.0 and in the end selected a value equal to 0.5. This S.I. value conditions the type of magnetic source resolved with accuracy. Because it belongs to the magnetic contact type, it has a long wavelength.

Aware that the subjectivity of the assumed "structural index" might limit the effectiveness of this technique, we have tried to compare our results with some possible depth solutions derived from other independent techniques. Ranero et al. (1997) used multichannel seismic reflection, to study three long profiles west of El Hierro and La Palma Islands. Their "B" and "C" profiles fit our needs. They detected a gradual change from smooth (eastern side of each profile) to rough basement topography (central to westernmost part of their referred profiles). Since the area we are interested in falls eastward of M21 (eastern side of their referred profiles), its basement topography is stable, and this situation favours solution quality. Top of basement is situated at 7000 ms (TWT: two-way travel time), about 5.5 km depth below sea level (b.s.l.). Urgeles et al. (1998) in the same zone, estimated, from seismic reflection, a 6900 ms (TWT) basement depth. Our "Located Euler Deconvolution" solution (based on 0.5 S.I.) estimates depth values between 4400 and 6100 m b.s.l. (Figure 7), in reasonable agreement with previously cited seismic values (Urgeles et al., 1998).

The map shows that the most significant sources lie in the 4000–6000 m range. In fact, for the great inverted dipole anomaly just NW of Gran Canaria we have obtained three depth values (b.s.l.): 5400 (label "A"), 4100 (label "B") and 3600 m (label "C"). All of them show a depth uncertainty smaller than 15% (±800 m).

Depth to bottom

The method to estimate the depth extent of magnetic sources has been treated from two points of view: by analysing the shape of isolated magnetic anomalies (e.g., Bhattacharyya and Leu, 1975), to calculate the location of the centroid, and when the depth to the top of the source is also known, then the depth to the bottom of the

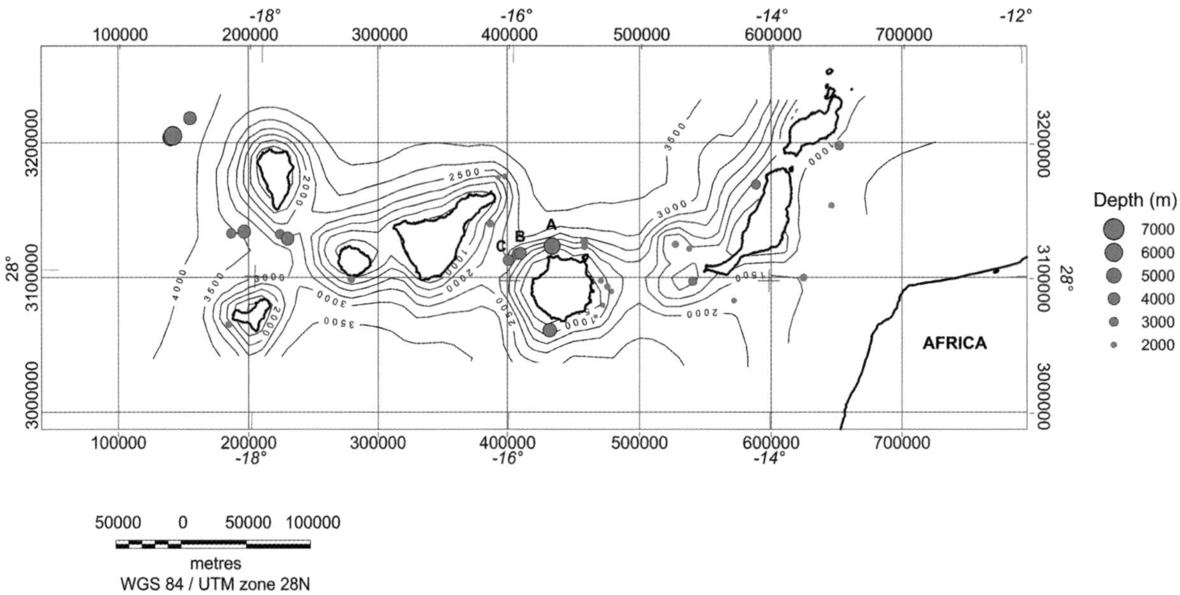

Figure 7. Euler Deconvolution solutions obtained from the scalar field map using the so-called "Located Euler Deconvolution" for SI 0.5. Bathymetry in meters. Contour interval: 500 m.

source could be obtained (Blakely, 1995). Other studies concentrated exclusively on the statistical properties of patterns of magnetic anomalies (e.g., Spector and Grant, 1970). Both methods utilise the relationship between the spectrum of the magnetic anomalies and their depths. Tanaka et al. (1999) used a similar procedure to the method of Spector and Grant (1970). The top bound and the centroid of a magnetic source were calculated from the power spectrum of magnetic anomalies and were used to estimate the basal depth of a magnetic source to magnetic anomalies from East and Southeast Asia. Shuey et al. (1977) have used the shape of radially average spectra to estimate the depth extent of magnetic sources. In particular, the position $k_{\max}$ of the maximum along the $|k|$ axis is related to the depth to the bottom of the layer according to the following equation:

$$k_{\max} = \frac{\log z_b - \log z_t}{z_b - z_t}, \tag{2}$$

where z_b and z_t are the depths (measured from the survey height) to the bottom and to the top of the layer, respectively. Although spectrum analysis provides depth estimates with simple assumptions we must highlight that these methods have fundamental limitations. Following Blakely (1995), we note that the determination of the depth to the bottom of a source (z_b) cannot be made without knowledge of the depth to its top (z_t). Second, this estimate of the depth to the bottom focus on the lowest wavenumber parts of the Fourier domain, which is quite sensitive to poorly known regional fields, and, to be able to discriminate with accuracy the peak in the spectrum, the magnetic survey must have a minimum dimension, L, which fulfils the equation:

$$L \geq \frac{4\pi(z_b - z_t)}{\log z_b - \log z_t}. \tag{3}$$

Blanco-Montenegro et al. (2003) applied the method based on the $k_{\max}$ estimation on a grid centred in Gran Canaria, suggesting the presence of rocks located at mantle-like depths (from the Moho to about 23 km) which could behave as magnetic sources.

We have followed a similar procedure to that of Blanco-Montenegro et al. (2003). In order to be able to resolve the peak in the spectrum, and following Equation (3), we obtained a 347 km size for the window. We have extracted from the 3800 m upward continued grid (section "Error estimates and data compilation") three different 347 km east–west windows for three different geographical frames (Figure 8). The first one centred

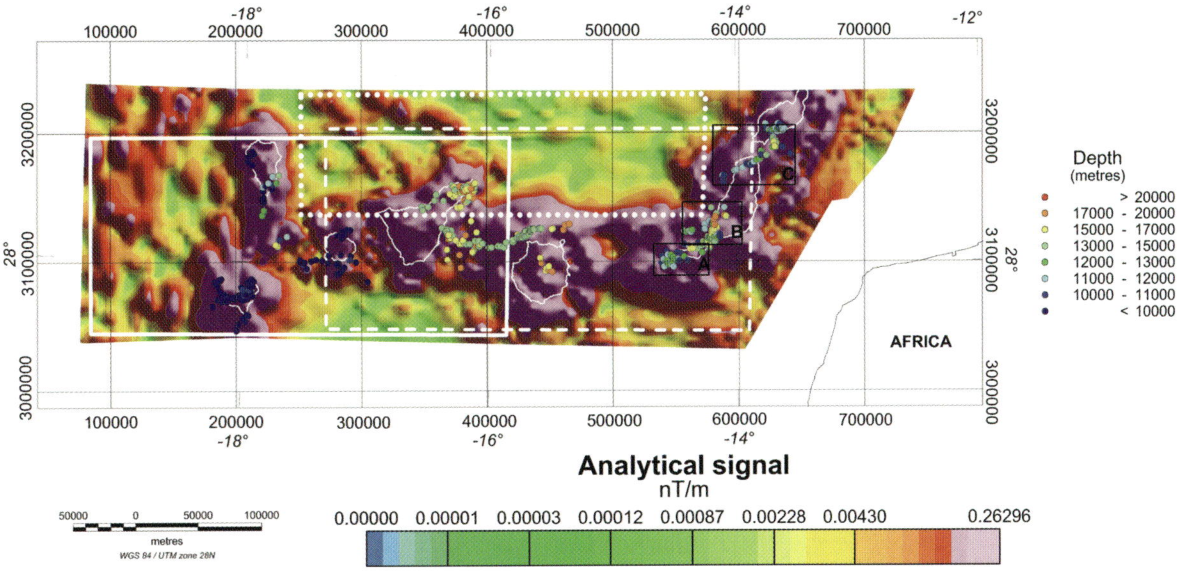

Figure 8. Shaded relief Analytical signal map of the area. Illumination from the northwest. In white (solid, dashed and spotted) lines windows used for the depth to the bottom spectral calculus (see text for comments). In black boxes, areas "A", "B" and "C" (see text for comments). Coloured circles show depth to the top of deepest magnetic sources. We have merged our shipborne dataset with seven 5×5 km sub-grids from the 3200 m aeromagnetic grid (see section "Error estimate and data compilation" for comments).

on Gran Canaria Island (dashed white line in rectangle, Figure 8), occupies the southern-central zone. The second one covers a northward parallel area (dotted white line in rectangle, Figure 8). The third one, located to the west, centred midway between El Hierro and La Palma (solid white line in rectangle, Figure 8). Their locations were not arbitrary. We have tried to isolate three different domains from the point of view of their magnetic character. One mainly concerned with volcanic rocks (Figure 8: Southern area), a second one which covers a nearly off-volcanic domain (almost oceanic) (Figure 8: Northern area), and a last one which merges both previously cited zones (Figure 8: Western area). Their location was constrained by: (a) the 347 km size area inferred from Equation (3) and (b) trying not to include areas with no data.

We have obtained the radial averaged power spectrum of every grid, which is shown in Figure 9. For the first grid (Figure 9: squares) we obtained a $k_{max} = 0.007$ cycles/km, and a $k_{max} = 0.0022$ cycles/km for the second (Figure 9: circles). The third grid (Figure 9: blue triangles) shows a different pattern. We recognised a peak, located in $k_{max} = 0.0021$ cycles/km, but a second one, with smaller amplitude, could be identified in $k_{max} = 0.0092$ cycles/km.

To obtain the z_b we have to resolve Equation (2). Following Blanco-Montenegro et al. (2003), we have assumed that $z_t = 15$ km, which is the average depth to the bottom of the oceanic crust deduced by Ye et al. (1999). The result shows that the depth to the bottom of this magnetic source (first grid) lies at 24 km b.s.l., which provides a 9 km thick magnetic source.

For the second grid we have selected a value (z_t) of 13 km, which, according to Watts et al. (1997), is the average value of the flexed crust and a typical oceanic crust domain in the area. The resolution of Equation (2) due to such a small k_{max} (0.0022 cycles/km) does not offer any numerical solution.

For the third area we lack refraction information from previous studies. We consider indirect information including: (a) The basement morphology shows no moat depression flanking the El Hierro and La Palma Islands group, and no seaward peripheral bulge is noticeable (Urgeles et al., 1998). So there is not expected to be any lithospheric flexure that could affect the Moho depth. (b) Additionally the presence of an oceanic magnetic anomaly in its westernmost margin (Figure 4) forced us to select a z_t value even smaller than 15 km. Again the resolution of Equation (2) due to such a small k_{max} (0.0021 cycles/km) does

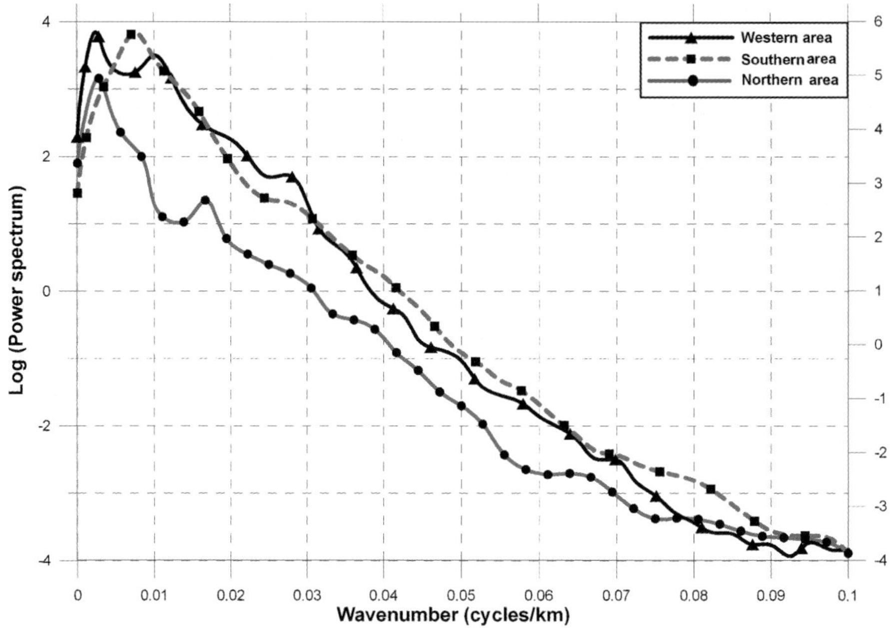

Figure 9. Radial average power spectrum from three different 347 km length windows in the east-west direction. We have merged our shipborne dataset with seven 5 km × 5 km sub-grids from the 3200 m aeromagnetic grid (see section "Error estimate and data compilation" for comments).

not offer any numerical solution. Moreover, after a careful analysis it appears to be independent of the z_t-selected value. Nevertheless as has already been mentioned, the existence of another peak at 0.0092 cycles/km needs a more specific analysis. This feature is discussed below (section "Magmatic underplating").

Alignments and trendings

We have taken advantage of the fact that the magnitude of the analytical signal is to a great extent independent of induced and remnant amplitude ratio and also of the direction of magnetisation. Accordingly, a directional derivative has been applied to two orthogonal directions, the x-axis and y-axis on the analytical signal grid, to enhance magnetic alignments (Figure 6a and b). In order to track features not only offshore but on land also, this analytical signal grid is derived from the scalar magnetic anomalies merged with seven 5 km × 5 km sub-grids from the 3200 m aeromagnetic grid, upwardly continued to 3800 m (see section "Error estimates and data compilation"). This technique highlights the main magnetic alignments around the archipelago.

Discussion of results

Lineations and magnetic trendings

The Canary Archipelago was constructed on old oceanic crust (150–170 Ma). Without accounting for the so-called "slope anomalies" (Roeser et al., 2002), its westernmost margin marks the limit of the oldest magnetic oceanic anomaly (M25) recognised in the Atlantic Ocean, as well as the beginning of the Jurassic magnetic quiet zone (Hayes and Rabinowitz, 1975; Verhoef et al., 1991; Roest et al., 1992a). Roest et al. (1992a) proposed that M25 approached the eastern flank of El Hierro and La Palma group (Figure 1).

In Figure 4, a linear trend (label A–A) appears to be associated with these western anomalies. This linear magnetic anomaly shows, to the southwest, a branch that runs sub-parallel with a nearly 21 km offset. This could be interpreted as a fracture zone (Figure 6a). We suggest, as did Verhoef et al. (1991), that this anomaly (A–A) probably corresponds to M21.

The main fractures in the islands and ocean floor have NE-SW and E-W orientations. Both cases, due to the fact they are parallel or similar to the spreading direction of oceanic basement, are considered results of this tectonic process. These alignments could be identified in the west and north area. Likewise, the E-W trend could be discerned mainly in the north and south central areas, and always far from magnetic volcanic disturbances (Figure 6b). This could support a correlation with the tectonics of Atlantic Ocean opening.

In the eastern part of Lanzarote and Fuerteventura, NE-SW trends could also be discerned. This NE-SW pattern has suggested to different authors that the eastern islands form part of a NE-SW ridge system parallel to the African coast and related with NW-African tectonic events (Roest et al., 1992a).

Mezcua et al. (1992) noted that the eastern islands show almost a single block magnetic pattern (Figure 5). Local magnetic trends, mostly showing arc shapes, are roughly perpendicular to the inferred structural axis pointed out by Carracedo (1980), which initially runs NE-SW, turning in a clockwise direction to the E-W (Figure 6a: label "H–H").

Offshore Gran Canaria to the north and south, there are rough NE-SW magnetic alignments, and on the island's northwestern coast there is a NW-SE alignment. "D" and "E" alignments (Figure 6a) are almost perpendicular to the most remarkable anomaly of the whole survey area. Several authors have interpreted this anomaly as a fault (Anguita and Hernan, 1975; Mezcua et al., 1992; Romero et al., 2000).

The most noteworthy magnetic trend is a long wavelength feature, seen in blue (Figure 6a). This signal shows a reasonable geographical correlation with the region, which includes the strongest analytical signal amplitude (Figure 5). We interpret this wide lineation as an indicator of a region where old oceanic crust was altered by extrusive processes. This trend runs almost continuously through the archipelago and is only locally disrupted in an area between Gran Canaria and Fuerteventura (Figure 6a: label as "G"). As there is no seismic refraction information for this zone, it is difficult to assess its cause. Collier and Watts (2001) use multichannel seismic reflection profiles to determine the seismic stratigraphy of the flexural moat that flanks the Canary Islands. Their Line 15 crosses this area. Nevertheless, the top of the oceanic crust cannot be inferred due to a shallowing bathymetry and a thickening sediment layer. The geometry of the experiment (N-S) is also poor and cannot help in this regard. The Bouguer map (Carbó et al., this issue) reflects a high gradient area. Figure 5 shows a disruption in the most intense part of the analytical signal signal (red colour), but the yellow background seems to be intact. As was previously stated, it is difficult to conclude anything, although a fracture, which does not affect the whole crust, would agree with our data.

The derivative image (Figure 6a) west of El Hierro Island reveals many alignments delineating an almost north–south course, parallel to the isobaths, except the previously cited linear magnetic anomaly (Figures 4 and 6: A–A) which shows a slightly NE course. East of this oceanic magnetic anomaly we were able to observe a roughly N60°E trend (Figure 6b: label as 'K') that marks the limit of a wide area with low scalar magnetic values (Figure 4). Its inclination seems to be greater than average. Urgeles et al. (1998) recognise three seismic units west of El Hierro and La Palma bounded by strong reflectors. These seismostratigraphic units were interpreted as distal products of the volcanic evolution (erosion, mass wasting) of the archipelago. They show that the contours defined by these reflectors have a nearly N50°E trend. In contrast, the present-day seafloor contours run oblique to them, at a 30° northward angle. As was previously stated, Figure 6a shows anomaly A–A (Figure 4), which is interpreted as an oceanic anomaly. This trend shows a nearly anticlockwise 30° oblique strike with respect to the "K" alignment. Considering the strike of this alignment, and that it limits a region of low positive scalar magnetic amplitudes (to the west) and of low negative ones (to the east), we interpreted the "K" trend as a boundary between two domains: first, the transition from a marine basin, whose structure is closely related with Canary Island evolution from the point of view of its seismostratigraphic units, and, second, a typical oceanic region.

We detected a linear segment surrounding the northwestern part of Tenerife (Figure 6a: "B"), which also appears on Figure 5 (same label). These alignments join several short wavelength signals and probably are correlated with some submarine features classified by Romero-Ruiz et al. (2000) as seamounts (their Figure 4). We suggest a probable volcanic origin for them on the basis of their high analytical signal value, in spite of the short wavelength character and a water depth greater than 3500 m. Just east, an arc-shaped set of analytical signal anomalies is well established (Figures 5 and 6: "C"). They were detected in the previously cited work and classified as seamounts (its Figure 4). We suggest a volcanic origin for these anomalies as well.

The inverse magnetised anomaly between tenerife and Gran Canaria

Funck and Schmincke (1998) analysed three seismic profiles around Gran Canaria. Their P104 and P107 profiles show a thin sediment layer on the Gran Canaria western flank (about 0.5 km thick). Considering the water depth (2500 m on average) and depth solutions b.s.l. obtained from Euler Deconvolution (section "Depth to the Top: Euler Deconvolution"): 5400 m (Figure 7: label "A"), 4100 m (Figure 7: label "B") and 3600 m (Figure 7: label "C"), we infer an upper crust location for the top of the great inverted dipole anomaly and that the causative body seems to be tilted to the NE. Nevertheless, the latter finding must be considered with caution because depth differences between the solutions are within the uncertainty associated with the measurement, roughly 1 km.

Several works (Anguita and Hernan, 1975; Mezcua et al., 1992; Romero-Ruiz et al., 2000) concluded there was an African type fault between Tenerife and Gran Canaria due to the relationship with the tectonics of the Atlas range in the African continent. This solution corresponds to a submarine fault some 50 km long with transcurrent (left-lateral) and reverse components.

Blanco-Montenegro et al. (2003) consider this linear magnetic anomaly to be an intrusive body cooled during a period of reversed polarity of the Earth's magnetic field, and suggest the possibility that it could have acted as the magmatic source of the submarine volcanism detected between Tenerife and Gran Canaria. Probably this is the most likely possibility. This magnetic source shows amplitude of 1000 nT peak to peak. After testing some forward models we estimated that an extremely thick sediment layer would be needed (greater than 6 km) so it is improbable that this magnetic anomaly could be attributed to a purely structural feature (a fault) alone. We attribute this magnetic anomaly principally to the effect of remnant magnetisation.

We consider that the fault hypothesis cannot account for this strong anomaly, at least as a major contributor because structural features usually cause small anomalies. Moreover, the Bouguer anomaly map (Carbó et al., this issue) does not show either a significant signal or a high gradient area with this strike. The possibility that the anomaly represents a reverse polarised intrusive body seems more reasonable.

Magmatic underplating

Underplating is defined as magmatic material added to the base of the Earth's crust. This geophysical feature has been reported under Hawaii (Lindwall, 1988) and the Marquesas (Caress et al., 1995) and it is still an open question and a matter of debate for the Canary Islands. Watts et al. (1997) performed a refraction seismic study in Tenerife and found no evidence of underplating beneath the island. Ye et al. (1999) performed a similar refraction seismic study, but on Gran Canaria, covering its northern and northeastern sector up to 60 km away from the coastline. A zone of magmatic underplating, as deep as 26 km, was found beneath the island, suggesting the possibility that if the underplating beneath Tenerife occurs in the same way as in Gran Canaria, that is, restricted to the main volcanic edifice without extending towards the ocean basin, the geometry of the experiment and the quality of the onshore data could have affected its detection. Dañobeitia and Canales (2000) reviewed the previously cited works (Watts et al., 1997; Ye et al., 1999) and re-examined records sections of Banda et al. (1981), reconsidering their interpretation. Dañobeitia and Canales showed a velocity depth section for both eastern islands (Fuerteventura and Lanzarote) as an alternative model. This velocity depth section shows a smooth boundary between the lower crust and a diffuse Moho, slightly deepening towards the centre of Fuerteventura. It also showed a 8–10 km thick layer in the lower crust, with a high average velocity of 7.4 km/s, overlying an anomalous upper mantle velocities of 7.6–7.8 km/s. There was always a fundamental difference with other oceanic islands, like Hawaii and the Marquesas, specifically the limited radial extension of this magmatic material incorporated at the base of the crust, which does not underlie the whole volcanic edifice, but is restricted to specific parts and ends abruptly (Dañobeitia and Canales, 2000). The key question is whether magnetic data can provide some clue concerning this geophysical feature?

Two main problems appear. The first concerns the possibility that at mantle-like depths, temperatures could be low enough to remain below Curie temperature. The second concerns whether it is possible to detect such a potentially deep magnetic source. Concerning the first question, magnetisation measurements as well as petrologic studies of rock samples brought to the surface from the mantle by structural and igneous processes indicate that ferrimagnetic minerals, which are the sources of most magnetic anomalies, generally are not present in the mantle. However, the existence of sub-crustal lithospheric sources of magnetic anomalies cannot be ruled out because the possibility exists that there are localised zones where mantle processes produce ferrimagnetic minerals and that ferrimagnetic elemental iron could be present in the upper mantle (Langel and Hinze, 1998). This feature has been discussed by Blanco-Montenegro et al. (2003), taking into account the pressure dependence of the Curie point, and its relation with crustal and sediment layer thickness (Hayling and Harrison, 1986). In the case of Gran Canaria Island, this sediment layer could be 4 km thick in its northern region. Therefore, Blanco-Montenegro et al. (2003) following the Hayling and Harrison (1986) pure conductive heat flow model, concluded that a 200–300 °C temperature could be reached at the base of the crust. Considering this temperature, Blanco-Montenegro et al. (2003) stated that, if the plutonic complexes located beneath Gran Canaria were of gabbroic composition, they could contribute to the measured magnetic anomaly.

Here we summarise what is known about the Moho depth and sedimentary layer thickness knowledge for the whole archipelago before beginning an interpretation of each of the following magnetic sources analyses. The analyses seek to determine if the sources are sub-crustal, and, hence, plausible candidates for underplating. Recent local seismic studies performed in the Canary archipelago (Banda et al., 1981; Watts et al., 1997; Urgeles et al., 1998; Ye et al., 1999) suggest that a value of 15 km for the Moho depth, and a 5 km sedimentary cover thickness could be considered representative for the central islands (Gran Canaria and Tenerife). A different situation occurs in the eastern islands, where seismic models (Banda et al., 1981; Dañobeitia and Canales, 2000) show a smooth boundary between the lower crust and a diffuse Moho, which deepen slightly towards the centre of Fuerteventura, reaching a maximum depth of 24–25 km. The upper limit for the sediment cover thickness is 5 km.

With respect to the western islands (La Palma and El Hierro), we note that the sedimentary layer on their western submarine flanks reaches nearly 5 km thickness (Ranero et al., 1997; Urgeles et al., 1998). As there is no refraction and reflection seismic data for the islands' immediate surroundings, nothing can be said about the Moho and the actual thickness of the sedimentary layer, except that it seems implausible that the Moho is deeper than 15 km. The reason, according to the Bouguer anomaly map (Carbó et al., this issue), is that the islands are located on the westernmost margin of a transitional crust, and from 342° meridian westward, isolines delineate a strongly north-southward component, showing Bouguer anomalies ranging in accord with oceanic values.

This situation leads us to believe that in the offshore marine area an underplated source in Canary Archipelago could have a magnetic signature and thus appear on the magnetic anomaly map. In the same way, we could expect there to be a similar source attached to a Fuerteventura and Lanzarote islands, possibly at a depth of 13–15 km, with a magnetic signature. However, its sub-crustal origin cannot be asserted "a priori".

Determining whether or not such a potentially deep magnetic source can be detected is key to answering the second question. From the depth to the bottom analysis (section "Depth to bottom") we have obtained three profiles that represent the radial average power spectrum of three different grids (Figure 9). In the southern area, the peak situated at $k_{\max} = 0.007$ cycles/km could be compatible with the presence of a magnetic source with a depth to the bottom of 24 km b.s.l., indicating a nearly 9-km-thick magnetic body. As was stated in section "Depth to Bottom", the northern area spectrum is not compatible with the presence of any sub-crustal source. Since in the Canary Archipelago any magmatic underplating evidence has always been connected with volcanic islands (Ye et al., 1999; Dañobeitia and Canales, 2000), this result is supported by the absence of any island in its interior.

The third grid, in the westernmost area, shows two peaks, one at $k_{max} = 0.0021$ cycles/km, and a second one at $k_{max} = 0.0092$ cycles/km. Although the first is the greatest, in power terms, the latter shows a slightly lower value: 1.39 times in power terms units and must not be discarded. We interpret this anomalous behaviour to the fact that this mathematical procedure is based on a probabilistic behaviour of the distribution of magnetisation, which assumes magnetisation to be isotropic, that is, its statistical behaviour is the same in all horizontal directions (x and y) (Blakely, 1995). From Figure 8, we see that this is not correct for the westernmost window, which shows an unbalanced situation with a lineated anomaly in its western border, and strong magnetic anomalies in its right half associated with volcanic islands (Tenerife, Gomera, Hierro and La Palma). In this sense, we must take into account the presence of the other peak ($k_{max} = 0.0092$ cycles/km). If we solve for a $z_t = 15$ km, Equation (2) provides a depth to the bottom solution at 20 km b.s.l. This value corresponds to a 5 km thick magnetic source, but it is surely influenced by the variability of the Moho depth in this area, where we link a transitional crust domain and an oceanic one. If we solve with $z_t = 13$ km, which could better fit an average Moho depth value in the westernmost frame, the solution turns out to be 22.5 km, which corresponds with an almost 10 km thick magnetic source and shows reasonable agreement with seismic values (Ye et al., 1999).

Another interesting aspect is the location and distribution of all these possible sub-crustal magnetic sources. We have taken advantage of the Euler Deconvolution algorithm, but must take into account some drawbacks: (a) This algorithm incorporates three gradients. This fact makes it sensitive to noise, which could affect the quality of the solution by introducing fluctuations and incoherencies into solutions between adjacent grids. (b) We are interested in a potential deep source that could have a low magnetic signature that must be isolated. (c) Previous geophysical studies of the archipelago have shown the local character of the signal we are seeking. It seems to be associated always with islands (Tenerife, Gran Canaria, and possibly Lanzarote and Fuerteventura), and shows a nearly 10 km vertical extent.

To overcome all these drawbacks we have applied the "Standard Euler Deconvolution" instead of the so-called automatic "Located Euler Deconvolution" procedure. As was stated for drawback (b), the signal we are interested in does not necessarily show a high magnetisation contrast, and this semi-automated method, based on the detection of analytical signal peaks, could be inadequate.

Drawback (c), states that the underplated body is limited in both the vertical and horizontal planes. We have selected a high SI. An index equal to 3 should represent an ideal body, a sphere. An index equal to 2 can describe a body with only one infinite dimension. We believe this is an intermediate case between both and is best seen in an index of 2.5 (Reid et al., 1990).

Another significant question concerns window size. This method employs overlapping windows, resolving Euler's Equation for each. The window of data must adequately sample each anomaly on the grid. This window must be large enough to include the entire anomaly being analysed, but not so large that it could contain multiple anomalies, which could lead to confusing solutions. Additionally, a small window could be more sensitive to noise as the number of data points in the window is decreased.

As the anomaly associated with every island has different sizes, we have selected a variable window size, different for every island. To set a criteria to determine the window size we have used the analytical signal wavelength. We have selected, as a balance, a window size equal to half the characteristic analytical signal wavelength associated with every island. This window size goes from a 15 km × 15 km value for Gomera island to 40 km × 40 km for Gran Canaria and Tenerife islands. With all these conditions, especially the window size criteria, we have highlighted the deepest sources. To reduce topographic influences, as well as to highlight deeper sources, we have upwardly continued the mixed "shipborne plus aeromagnetic" 3800 m grid (see section "Error estimate and data compilation" for details) to 5000 m. Those solutions that show a 15% depth uncertainty or that fall outside their correspondent window cell are discarded.

Gran Canaria: This island shows an homogeneous set of solutions, all of them deeper than 15 km. It is observed that they appear to be located in the northern half of the island (central, NNW and NNE sector). This result shows good agreement with constraints from seismic methods (Ye et al., 1999).

Tenerife: A similar pattern as that found at Gran Canaria was observed. A clear cluster of solutions is displayed in the NE sector of the island (Anaga Peninsula). The data set of solutions obtained on the eastern offshore side of the island shows a certain spatial variability. Nevertheless, most of solutions (72%) fall into the 14–16 km segment, which takes into account the uncertainty associated with this type of depth estimation and fits quite well with the Moho under this island.

No refraction seismic experiments have been performed in this area. Additionally, in roughly the same area, the Bouguer map (Carbó et al., this issue) presents a wide terrace whose boundary is set by the 200 mGal isoline. This area contains a body with a lower than expected density, considering the monotonic transition from a continental domain to an oceanic one observed to have a strong north-south component. This transition basically shows a change in Moho depth, apparently disrupted in this area due to the presence of the slightly low-density body that magnetic solution indicates could have its top at Moho-like depths. We suggest that it is plausible to consider the body to be magmatic underplating.

We observed a set of solutions, marking depths from 13.5 to 14.5 km, that link the island of Gran Canaria and Tenerife. This set of solutions shows a remarkably linear trend and a reasonable correlation with the inverse magnetic anomaly track (Figure 4). Ye et al. (1999) inferred that both ocean islands' uplift and evolved volcanism were caused by the same process of magmatic underplating. The up-welling magma did not make it to the surface but, due to its density and rheology, was trapped below the crust and fractionated into more dense, basic magma added to the base of the crust and lighter, more evolved magma able to continue rising. This explanation agrees with the interpretation of the inverse magnetised body between Tenerife and Gran Canaria. In the section "The inverse magnetised anomaly between Tenerife and Gran Canaria" we conclude that the possibility that this anomaly represents a reverse polarised intrusive body seems more reasonable. With an "Extended Euler Deconvolution", we located its top at an upper crust level. We suggest that the linear magnetic anomaly we reflect in the total field anomaly grid could be a consequence of two contributions, one deeper than the other: one with a top at upper crust like depths, and the other at a sub-crustal level, but both linked in their origin. This possibility offers another clue that cannot be discarded as long as we do not have refraction seismic information for this particular area. There is the possibility that the same plutonic complex links both Tenerife and Gran Canaria.

La Gomera: The entire set of depth solutions falls below 10 km, in fact 88% are less than 7.6 km. There is lack of information concerning Moho depth at La Gomera, nevertheless a 15 km average value for crustal thickness at Tenerife seems to support a crustal origin (possibly upper crust) for causative magnetic bodies.

El Hierro: A similar analysis could be applied on El Hierro. The depth to the top of the causative source ranges from a minimum 6.5 km to 11.2 km depth. Most of them (66%) are grouped around 9 km b.s.l. Its westernmost position makes it difficult to conclude anything about its crustal or upper mantle origin.

La Palma: There are some depths values situated on the NW side of the island. They are concentrated around 6.1 km depth (minimum = 5.1 km; maximum: 7.4 km). On its southern side we found certain variability, in fact there was a 10.2 km average depth value, and 13.1 km and 8.2 km as the maximum and minimum depths to the top of the magnetic source, respectively. We inferred a step of nearly 4 km between north and south solutions. We suggest a crustal origin for the north set of depths, but the south group of solutions deserves a more careful analysis. In spite of the depth variability, we see that solutions are clearly deeper than in the north (almost 4 km). Once again we lack seismic information to locate the Moho. Offshore La Palma we could see another depth solution cluster. They show similar statistics with regard to the depth solution group just mentioned. Moreover, in the scalar field map (see Figure 4) there is a southward positive anomaly. In the same sense, the Bouguer map (Carbó et al., this issue) shows a non-absolute minimum anomaly in the area. Taking both facts into consideration, the possibility of an anomalous mantle seems plausible and coherent.

Fuerteventura and Lanzarote: The depth solution shows variability. This has to do with the estimative nature of the algorithm and that is why we prefer to use statistical terms. In order to simplify the discussion we have drawn three boxes (Figure 8, labelled A, B and C). These groups of boxes have three different average "depth to the top of the deepest magnetic sources" levels. In box "A", from a set of solution of 26 values, we have obtained 11.5 km as a mean depth value, with maximum and minimum depths equal to 12.8 and 10.1 km, respectively. In box "B", from a set of solution of 41 values, we have obtained 13.2 km as the mean depth value, with maximum and minimum depths equal to 16.0 and 10.0 km, respectively. It is worth pointing out that only six values were greater than 11 km. In box "C", from a set of solution of 55 values, we have obtained 11.5 km as a mean depth value, maximum and minimum depths equal to 15.5 and 10.1 km, respectively. Again, it is important to stress that only six values are greater than 11 km.

We could discern a nearly 2 km jump between "A" and "B" frames, and in "C" a similar horizon to "A". Dañobeitia and Canales (2000) reexamined old seismic records sections from Banda et al. (1981) suggesting the presence of underplated material but denote, however, a slight deepening of Moho boundary, as well as for the upper to lower crust limit, towards the centre of Fuerteventura as a consequence of a flexure. Our results agree are consistent with that conclusion.

Figure 8 shows an absence of solutions from the south to the north set. Our depth solutions are located over the higher analytical signal amplitude sources. We consider that these places mark the locations where upwelling magmas interact with an old oceanic crust, disrupting and magnetising it and that this process does not affect the whole volcanic edifice but is instead restricted to specific parts. Additionally, in Figure 6a we observed the presence of a high gradient in the NW-SE direction in this zone (Figure 6a: label "F"). It may be an indicator of a Fault. This aspect could justify the absence of solutions in this area, and the existence of a nearly 1.7 km vertical average offset between "B" and "C" frame solutions.

Conclusions

In this paper we have described the main geomagnetic characteristics of ZEEE in the region of the Canary Islands. In the first part, we concentrated mainly on purely technical aspects including: data acquisition and treatment, instrumentation, and quality of the surveys. The analysis of this data set has led to several conclusions, summarised as follows.

From a spectral analysis we have obtained evidence, which shows the possible existence of magnetic sources. The evidence locates the depth to the bottom at a sub-crustal level. They seem to be located especially in the southern islands area (Figure 8). They show a 24 km b.s.l depth extent, which indicates a 9-km-thick magnetic source.

Additionally, we applied a Located Euler Deconvolution algorithm to the scalar magnetic anomaly grid and obtained three depth values b.s.l. 5400 m (label "A"), 4100 m (label "B"), and 3600 m (label "C"), for the inverse magnetised anomaly between Gran Canaria and Tenerife. These results suggest a top upper crust location for the basement top of the magnetic body.

Using a Standard Euler Deconvolution algorithm we have located the deepest magnetic sources in the archipelago, analysed their spatial distribution and their possible sub-crustal magnetic origin. We have found evidence of these sources, particularly at Gran Canaria and Tenerife. Their distribution seems to be restricted to specific areas, in good agreement with previous seismic studies (Ye et al., 1999; Dañobeitia and

Canales, 2000). Fuerteventura and La Palma show a cluster of solutions at the 11.5–13.2 km range, and 10.2 km b.s.l. depth, respectively. Seismic studies (Dañobeitia and Canales, 2000) show that at Fuerteventura there is a diffuse Moho boundary. At La Palma there is no refraction seismic data to delineate the Moho. In this situation, nothing can be inferred yet, concerning the sub-crustal origin of these deep magnetic sources.

The same analysis (Euler Deconvolution) provided evidence which suggests the possibility that the inverse magnetised body between Tenerife and Gran Canaria could tie together two contributions: one at upper crust levels and the other, deeper, seems to be mantle in its origin. The continuity of the solution data set suggests another possibility, that is, that Tenerife and Gran Canaria could be connected at depth by the same plutonic complex. Concerning the westernmost islands we have found no evidence that supports the presence of magmatic underplating under either La Gomera or El Hierro islands.

Additionally, south of La Palma a clear cluster of solutions appears. Bouguer map (Carbó et al., this issue) reflects non-absolute minimum values at the area. These factors support the existence of an anomalous mantle, which starts in the south of La Palma and extends nearly 27 km southeastwardly.

The Euler Deconvolution analysis shows a gap in the depth solutions at Fuerteventura and Lanzarote. The magnetic alignment map (Figure 6a) shows a high gradient in the NW–SE direction, in the gap area. We consider this to be an indication of a major fracture.

Acknowledgement

We wish to thank the Spanish Navy staff for their invaluable support, as well as the crews of the B.I.O. "Hesperides", the B.O. "Vizconde de Eza" and the "Unidad de Tecnología Marina (U.T.M.)". Also, we are in debt to the National Geographical Institute and particularly to Benito Casas, for kindly providing the scalar field magnetic data from Güimar Observatory. Thanks are also extended to Peter Clift and two anonymous reviewers who critically read an earlier version of the manuscript and made many helpful suggestions which greatly improved its final version. Finally, we would like to express our gratitude to J. A. Peña, R. Alvarez and J. Gallego for their help and technical support. This is a contribution to the Spanish Ministry of Defense ZEE Program.

References

Anguita, F. and Hernan, F., 1975, A propagating fracture model versus a hot spot origin for he Canary Islands, *Earth Planet. Sci. Lett.* **27**, 11–19.

Araña, V., Camacho, A.G., García, A., Montesinos, F.G., Blanco, I., Vieira, R. and Felpeto, A., 2000, Internal structure of Tenerife (Canary Islands) based on gravity, aeromagnetic and volcanological data, *J. Vol. Geoth. Res.* **103**, 43–64.

Banda, E., Dañobeitia, J.J., Suriñach, E. and Ansorge, J., 1981, Features of crustal structure under the Canary Islands, *Earth Planet. Sci. Lett.* **55**, 11–24.

Barbosa, V.C.F., Silva, J.B.C. and Madeiros, W.E., 2000, Making Euler Deconvolution applicable to small ground magnetic surveys, *J. Appl. Geophys.* **43**, 55–68.

Bastani M. and Pedersen, L.B., 2001, Automatic interpretation of magnetic dike parameters using the analytical signal technique, *Geophysics* **66**, 551–561.

Blakely, R.J., 1995, *Potential theory in gravity and magnetic applications*, Cambridge, England: Cambridge University Press. 441 pp.

Bhattacharyya, B.K. and Leu, L.-K., 1975, Analysis of magnetic anomalies over Yellowstone National Park: mapping of Curie point isothermal surface for geothermal reconnaissance, *J. Geophys. Res.* **80**, 4461–4465.

Blanco-Montenegro, I., Torta, J.M., Garcia, A. and Araña, V., 2003, Analysis and modelling of the aeromagnetic anomalies of Gran Canaria (Canary Islands), *Earth Planet. Sci. Lett.* **206**, 601–616.

Carbó, A., Muñoz-Martín, A., Llanes, P., Álvarez, J. and EEZ Working Group: Gravity analysis offshore Canary Islands from a systematic surveying. *Mar. Geophys. Res.* This issue.

Carracedo, J.C., 1980, Geología del Atlas básico de Canarias, 1:200.000, Interinsular Canaria.

Caress, D.W., McNutt, M.K., Detrick, R.S. and Mutter, J.C., 1995, Seismic imaging of hot-spot related crustal underplating beneath the Marquesas Islands, *Nature* **373**, 600–603.

Collier, J.S. and Watts A.B., 2001, Lithospheric response to volcanic loading by the Canary Islands: constraints from seismic reflection data in their flexural moat, *Geophys. J. Int.* **147**, 660–676.

Dañobeitia, J.J. and Canales, J.P., 2000, Magmatic underplating in the Canary Archipelago, *J. Vol. Geoth. Res.* **103**, 27–41.

Durrheim, R.J. and Cooper, G.R.J., 1998, EULDEP: a program for the Euler Deconvolution of magnetic and gravity data, *Comput. Geosci.* **24**(6), 545–550.

Funck, T. and Schmincke, H.-U., 1998, Growth and destruction of Gran Canaria deduced from seismic reflection and bathymetric data, *J. Geophys. Res.* **103**, 15393–15407.

Hayes, D.E. and Rabinowitz, P.D., 1975, Mesozoic magnetic lineations and the magnetic quiet zone off northwest Africa, *Earth Planet. Sci. Lett.* **28**, 105–115.

Hayling, K.L. and Harrison, C.G.A., 1986, Magnetization modelling in the north and equatorial Atlantic ocean using MAGSAT data, *J. Geophys. Res.* **91**, 12423–12443.

Hsu, S.-K., Coppens, D. and Shyu, C.-T., 1998, Depth to magnetic source using the generalized analytical signal, *Geophysics* **63**, 1947–1957.

Instituto Geográfico Nacional, 1995, Mapa aeromagnético del archipiélago Canario (Anomalías de intensidad del campo magnético), 1:500.000, edited by Instituto Geográfico Nacional, NIPO: 162–95–004-X.

Keating, P. and Sailhac, P., 2004, Use of the analytic signal to identify magnetic anomalies due to kimberlite pipes, *Geophysics* **69**, 180–190.

Langel, R.A. and Hinze, W.J., 1998, The Magnetic field of the Earth's Lithosphere: The satellite perspective, Cambridge, 429 pp.

Lindwall, D.A., 1988, A two-dimensional seismic investigation of crustal structure under the Hawaiian Islands near Oahu and Kauai, *J. Geophys. Res.* **92**, 13687–13707.

Mézcua, J., Buforn, E., Udias, A. and Rueda, A., 1992, Seismotectonics of the Canary Islands, *Tectonophysics* **208**, 447–452.

Metzger, D. and Campagnoli, J., 2003, Marine Trackline Geophysics Data on DVD, Version 4.1.18, National Geophysical Data Center.

Mohan, N.L. and Babu, L.A., 1995, An analysis of the 3-D analytic signal, *Geophysics*, **60**, 531–536.

Muszala, S.P., Grindlay, N.R. and Bird, R.T., 1999, Three-dimensional Euler Deconvolution and tectonic interpretation of marine magnetic anomaly data in the Puerto Rico trench, *J. Geophys. Res.* **104**, 29175–29187.

Nabighian, M.N., 1972, The analytic signal of two-dimensional magnetic bodies with polygonal cross-section: Its properties and use for automated anomaly interpretation, *Geophysics* **37**, 507–517.

Ranero, C.R., Banda, E. and Buhl, P., 1997, The crustal structure of the Canary Basin: Accretion processes at slow spreading centers, *J. Geophys. Res.* **102**, 10.185–10.201.

Reid, A.B., Allsop, J.M., Granser, H., Millet, A.J. and Somerton, I.W., 1990, Magnetic interpretation in three dimensions using Euler Deconvolution, *Geophysics* **55**, 80–91.

Roeser, H.A., Steiner, C., Schreckenberger, B. and Block, M., 2002, Structural development of the Jurassic Magnetic Quiet Zone off Morocco and identification of the Middle Jurassic magnetic lineations, *J. Geophys. Res.* **107** (B10), 2207, doi:10.1029/2000JB000094.

Roest, W.R. and Pilkington, M., 1993, Identifying remanent magnetization effects in magnetic data, *Geophysics* **58**, 653–659.

Roest, W.R., Dañobeitia, J.J., Verhoef, J. and Colette, B.J., 1992a, Magnetic anomalies in the Canary basin and the Mesozoic evolution of the central North Atlantic, *Mar. Geophys. Res.* **14**, 1–24.

Roest, W.R., Verhoef, J. and Pilkington, M., 1992b, Magnetic interpretation using the 3-D analytic signal, *Geophysics* **57**, 116–125.

Romero-Ruiz, C., Garcia-Cacho, L., Araña, V., Yanes Luque, A. and Felpeto, A., 2000, Submarine volcanism surrounding Tenerife, Canary Islands: implications for tectonic controls, and oceanic shield forming processes, *J. Vol. Geoth. Res.* **103**, 105–119.

Salem, A., Ravat, D., Gamey, T.J. and Ushijima, K., 2002, Analytical signal approach and its applicability in environmental magnetic investigations, *J. App. Geophys.* **49**, 231–244.

Shuey, R.T., Schellinger, D.K., Tripp, A.C. and Alley, L.B., 1977. Curie depth determination from aeromagnetic spectra. *Geophys. J.R. Astron. Soc.* **50**, 75–101.

Socias, I. and Mézcua, J., 1996, Levantamiento aeromagnético del archipiélago canario, Instituto Geográfico Nacional, Publicación Técnica No. 35, 28 pp.

Spector, A. and Grant, F.S., 1970, Statistical models for interpreting aeromagnetic data. *Geophysics* **35**, 293–302.

Tanaka, A., Okubo, Y. and Matsubayashi, O., 1999, Curie point depth based on spectrum analysis of the magnetic anomaly data in East and Southeast Asia, *Tectonophysics* **306**, 461–470.

Teide Group, 1997, Morphometric interpretation of the northwest and southeast slopes of Tenerife, Canary islands, *J. Geophys. Res.* **102**, 20325–20342.

Thomson, D.T., 1982, EULDPH: A new technique for making computer assisted depth estimates from magnetic data, *Geophysics* **47**, 31–37.

Urgeles, R., Canals, M., Baraza, J. and Alonso, B., 1998, Seismostratigraphy of the western flanks of El Hierro and La Palma (Canary Islands): a record of Canary Islands volcanism, *Mar. Geol.* **146**, 225–241.

Verhoef, J., Colette, B.J., Dañobeitia, J.J., Roeser, H.A. and Roest, W.R., 1991, Magnetic anomalies off West Africa (20–38°N), *Mar. Geophys. Res.* **13**, 81–103.

Vogt, P.R., 1986, Magnetic anomalies and crustal magnetization, in Vogt, P.R. and Tucholke, B.E. (eds.), The Geology of North America, Vol. M, The Western North Atlantic Region. (Geol. Soc. Am.), pp. 229–256.

Watts, A.B., Peirce, C., Collier, J., Dalwood, R., Canales, J.P. and Henstock, T.J., 1997, A seismic study of lithospheric flexure in the vicinity of Tenerife, Canary Islands, *Earth Planet. Sci. Lett.* **146**, 431–447.

Ye, S., Canales, J.P., Rihm, R., Dañobeitia, J.J. and Gallart, J., 1999, A crustal transect through the northern and northeastern part of the volcanic edifice of Gran Canaria, Canary islands, *Geodynamics* **28**, 3–26.

Marine Geophysical Researches (2003) 24: 149–160
DOI 10.1007/s11001-004-5883-3

Paleoearthquake evidence in Tenerife (Canary Islands) and possible seismotectonic sources

L.I. Gonzalez de Vallejo[1], R. Capote[1], L. Cabrera[2], J.M. Insua[1] and J. Acosta[3]
[1]*Dpto. de Geodinámica, Universidad Complutense, 28040 Madrid, Spain. (vallejo@geo.ucm.es)*
[2]*Laboratorio COAC, 38509 Guimar, Tenerife, Spain*
[3]*Instituto Español de Oceanografía, C/ Corazón de María 8, 28002 Madrid, Spain*

Key words: Canary Islands, Paleoliquefaction, Paleosismicity, Seismites, Seismotectonic, Tenerife

Abstract

A series of clastic dikes and tubular vents were identified in southern Tenerife (Canary Islands). These features are the result of seismic liquefaction of a Holocene sand deposit, as the consequence of a high intensity paleoearthquake. The peak ground acceleration (pga) and magnitude of the paleoearthquake generating these liquefaction features were estimated by back calculation analysis. A representative value of 0.30 ± 0.05 g was obtained for the pga. From this, an earthquake intensity of IX was estimated for the liquefaction site. Magnitude bound methods and energy based approaches were used to determine the magnitude of the paleoearthquake, providing a moment magnitude $\mathbf{M} = 6.8$. The zone in which the liquefaction structures are found has undergone tectonic uplift and is affected by two faults. One of these faults was responsible for displacing Holocene materials. Dating of the uplifted sand formation indicates an age of $10,081 \pm 933$ years, the liquefaction features ranging from this age to 3490 ± 473 years BP. This paleoearthquake was of much greater magnitude than those known historically. Faults with neotectonic activity are significant features that should be borne in mind when assessing the seismic hazards of the Canary Islands, presently considered as low and mainly of volcanic origin.

Introduction and regional seismicity

Several structures attributed to liquefaction phenomena of seismic origin have been identified in exposed sand deposits near El Médano, on the south coast of Tenerife, Canary Islands (Figure 1). These findings prompted subsequent tectonic investigations including the geotechnical characterization of soils, geochronological analysis, and the analysis of geophysical, seismicity, and neotectonic data which we report here. In these investigations, we were able to characterize a Holocene sand formation and analyze the liquefaction structures. Possible formation mechanisms and the origin and age of these structures were evaluated. In the same area, we identified two faults that affected the Holocene deposits. Estimates were made of the acceleration and magnitude of the paleoearthquake that produced these structures, and possible seismic sources were characterized

Based on earthquake information, the Canary Islands have been generally thought to experience low to very low seismicity, with earthquakes always associated with volcanic activity. During the historical period (Figure 2A), which starts in the XIV century with the first references to volcanic eruptions, the most intense earthquakes on the archipelago took place in Yaiza (Lanzarote) in 1730 (intensity X), in Fuencaliente and Cumbrevieja (La Palma) in 1677 and 1920, respectively (both VII), in Ingenio (Gran Canaria) in 1913 (VII), and in Fuerteventura in 1915 and 1917 (both VII). On the island of Tenerife, the maximum intensity recorded was VI for the earthquakes on 1910/03/15 in Icod, 1909/01/4 in Puerto de la Cruz, 1909/09/23 in La Orotava, 1909/11/21 in Vilaflor and 1937/06/21 in Garachico. In this historic record, six earthquakes of intensity VI were registered on Tenerife, all in the 20th century and mostly affecting the north side of the island or its capital city Santa Cruz.

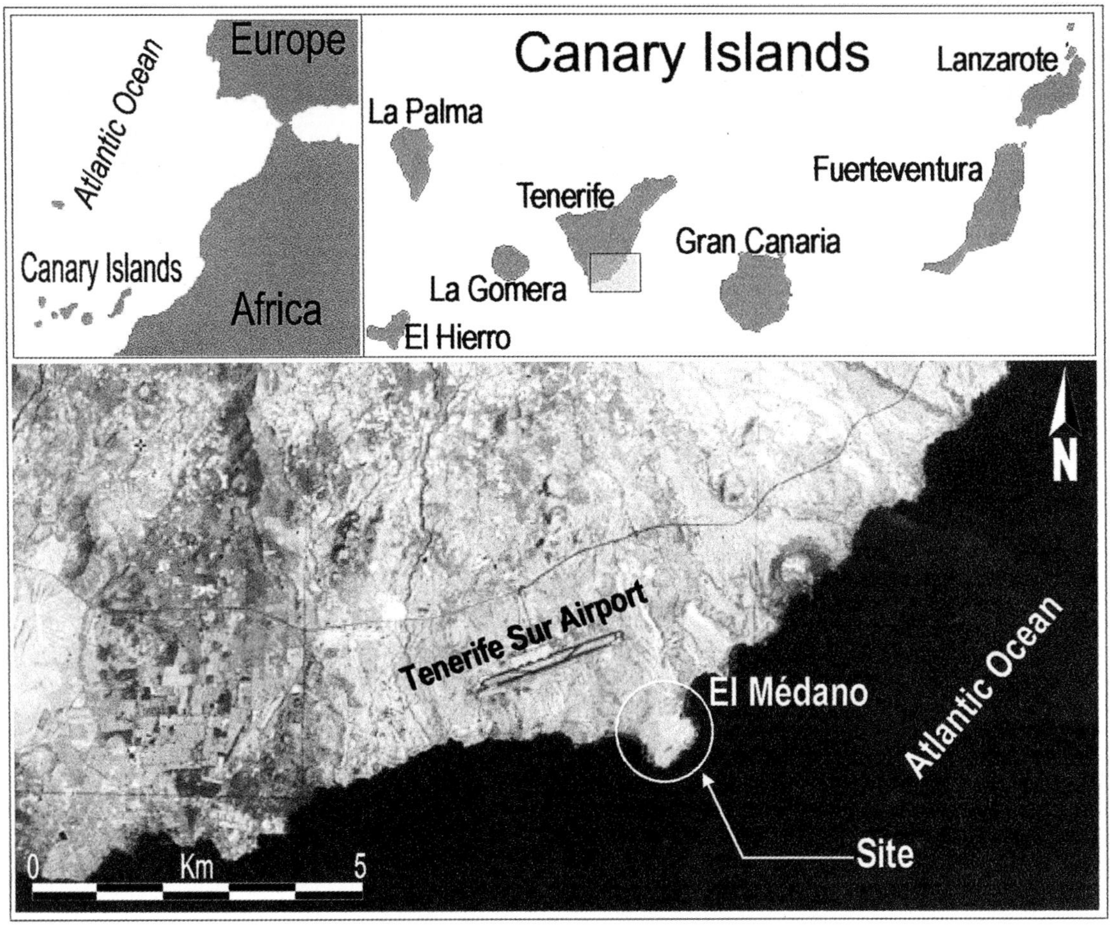

Figure 1. General location of the study area.

This record only reflects earthquakes felt in the most highly populated areas or those associated with volcanic eruptions. Knowledge of events occurring on the islands of El Hierro, La Gomera or the south of Tenerife is practically non-existent.

It was not until 1958 that a seismological station was installed in the Canaries. Two further stations were built in 1975, and over the past few years a more extensive network is being set up, with plans for stations over all the islands (there were seven stations in 2002). This will allow greater precision in locating and characterizing earthquakes. The distribution of epicenters recorded over the instrumented period is shown in Figure 2B.

The largest instrumented earthquake had a moment magnitude **M** = 5.2, its epicenter being in the sea between the islands of Tenerife and Gran Canaria (27°56.8′ N and 16°12.0′ W). Its maximum intensity was even felt on Tenerife. An analysis of this earthquake (Mezcua et al., 1992) has provided some ideas regarding the seismotectonic setting of the Canaries, which could help explain the paleoearthquake that caused the paleoliquefaction discussed here. The distribution of aftershocks recorded by a temporary station set up on the south coast of Tenerife between May 9 and June 17, 1989 indicates concentrated aftershocks along an 80 km long band aligned N33° (Figure 2C). This earthquake corresponds to a fault of around 30 km length. The hypocenter depth of the 5.2 magnitude event has been calculated as 50 km by Mezcua et al. (1992) and as 15 km by Dziewonski et al. (1990).

The analysis of the focal mechanism and that of the aftershocks points to a NNE-SSW alignment and inclination close to the vertical for the fault. This fault could also have been responsible for seismicity of greater magnitude and not related to processes of volcanic activity. Further more, geophysical marine

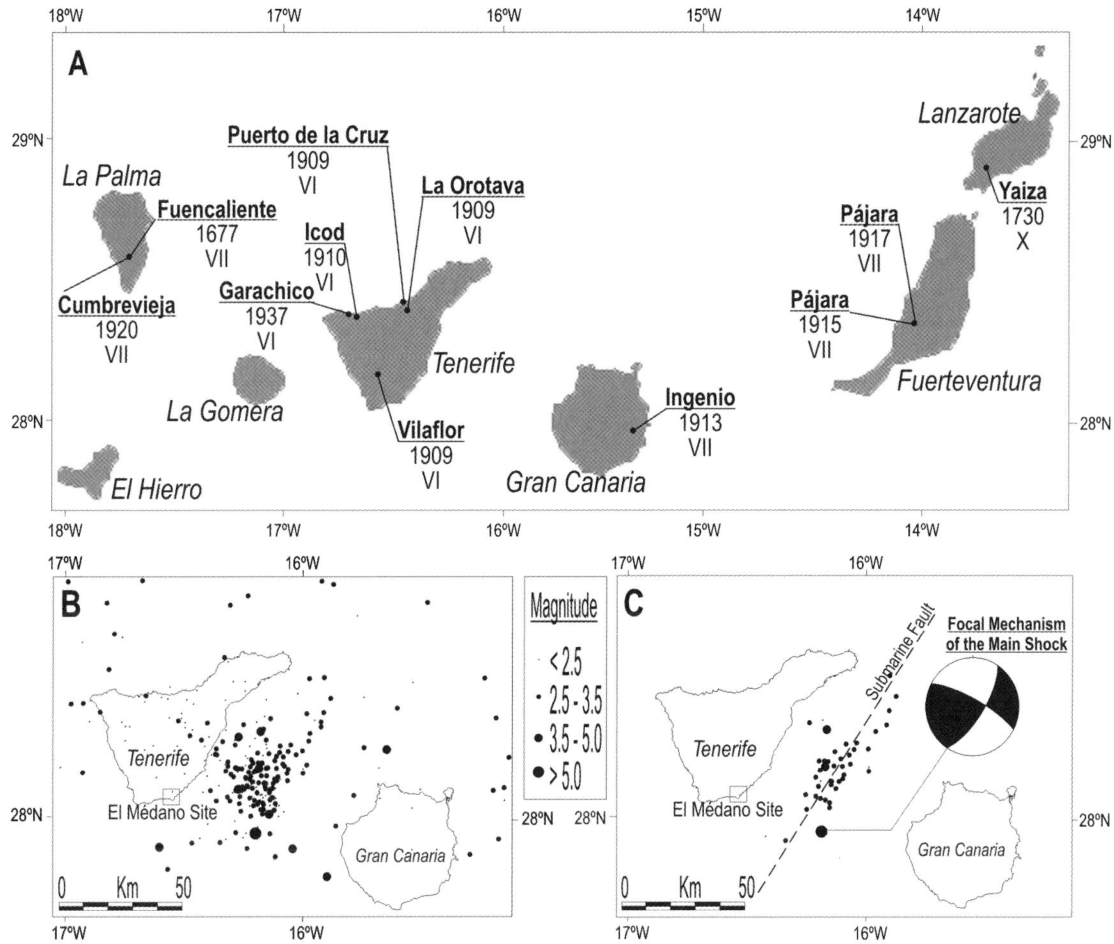

Figure 2. Seismicity of the Canary Islands. A: Historical seismicity until 1975 for earthquakes of intensity $I \geq VI$. B: Earthquake epicenters from 1975 to 2002 between Tenerife and Gran Canaria. C: Epicenters of the 1989 earthquake and its aftershocks. Nodal planes from focal mechanism: A = 33°–71° SE; B = 298°–77° NE.

investigations have revealed the occurrence of significant tectonic events (Llanes et al., this volume) associated with epicenters in the sea.

There is an obvious need for investigations that focus on paleoseismicity and neotectonics in regions for which earthquake information is scarce. This is definitely the case for the Canary Islands, whose instrumental period is shorter than 30 years and historical record is incomplete.

Geology of the Study Area

The area investigated is found in El Médano close to Leocadio Machado Beach (Figure 3). This beach is bounded inshore by a 40–50 meter wide range of coastal dunes orientated in a NE–SW direction. Some small lagoons have formed between the dunes and a coastal platform. This platform overlies a formation of volcanic tuffs of acid composition, and descends from the volcanic central part of the island. Towards the SW, a minor volcanic structure, the Montaña Roja, is composed of pyroclastic basaltic materials that overlie the tuff formation. These materials are overlain by a formation comprising beach sands that rises 2 to 15 m above sea level and shows several liquefaction structures.

The tuffs correspond to a set of pyroclastic units related to a phase of explosive salic eruptions between 0.7 and 0.13 Ma. The material is composed of pumice lapilli, lithic fragments and sanidine crystals. The Montaña Roja volcano lies at the southern margin of the study area (Figure 3). The volcano is a breached cone, open towards the ESE. Its altitude

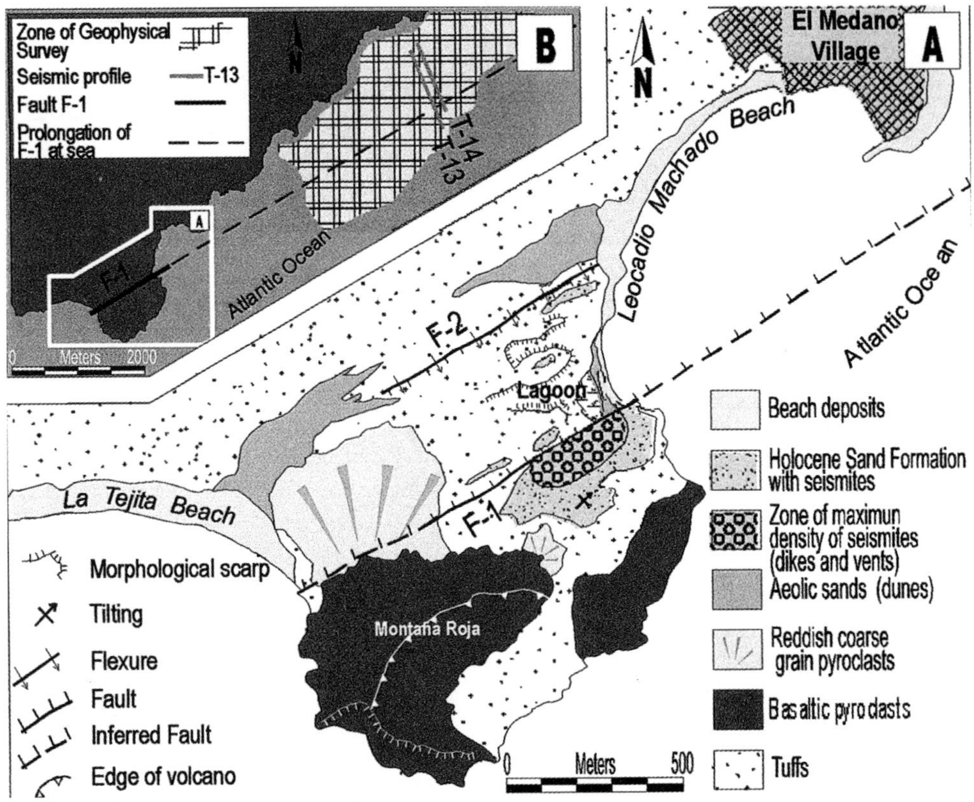

Figure 3. El Médano Site. A: Geological sketch map showing liquefaction features. B: Seismic profiles and prolongation of fault F-1 into the sea.

is 150 m above the lowlands at its base. It is composed of basaltic ash and cinder type pyroclasts and has been dated as being older than 100,000 years. The beach sand formation comprises bioclastic sands or weakly cemented, compact calcarenites and is 2.0 to 2.5 m thick. The coarse sand is made up of shell fragments, lithic grains, and plagioclase and pyroxene crystals. The substrate is composed of massive salic tuffs, within which an alteration level or softer paleosol 0.5 m thick can be observed. At some points, it is easy to distinguish a dense network of plant root structures, indicating the beach is transgressive and lies upon a densely vegetated surface. In turn, the beach material is covered in some areas by a thin pyroclastic level < 1 m thick and by calcareous crusts. The pyroclastic level contains centimetric-size, yellowish, pumice fragments enveloped by a pumice matrix. The white carbonate crust is laminated and 0.1 m thick.

The beach sand formation is slightly inclined at 3.5° towards the NE and is fractured such that faults and a network of joints organized in sets may be observed. The most outstanding tectonic structures are two N55° E trending faults (F1 and F2) running from Leocadio Machado Beach towards the SW (Figure 3). The southern fault F1 is most evident and is marked by an escarpment of 0.7 to 1.2 m height, SE side up, interrupting the beach and chain of coastal dunes, and also bounding the inland lagoon. The scarp disappears towards the SW and the fault's course appears to be covered by recent dunes and a reddish-colored basaltic pyroclastic deposit coating the NW flank of Montaña Roja volcano. The trace of fault F1 can be followed 1.2 km onshore, although it extends under the sea at both its exposed ends. In fact, several seismic reflection profiles were performed in the surroundings of El Médano in the bay's offshore zone. The profiles were obtained using a UNIBOOM system (EG&G) on a catamaran with an 8-element hydrophone, firing 500J as the energy source. The ship was positioned using a Raydist system with two shore-stations working in circular mode. The profiles indicate an acoustic basement and a top series of unconsolidated Quaternary sediments. A NE–SW trending fault cutting the acoustic basement was detected along the trace of fault

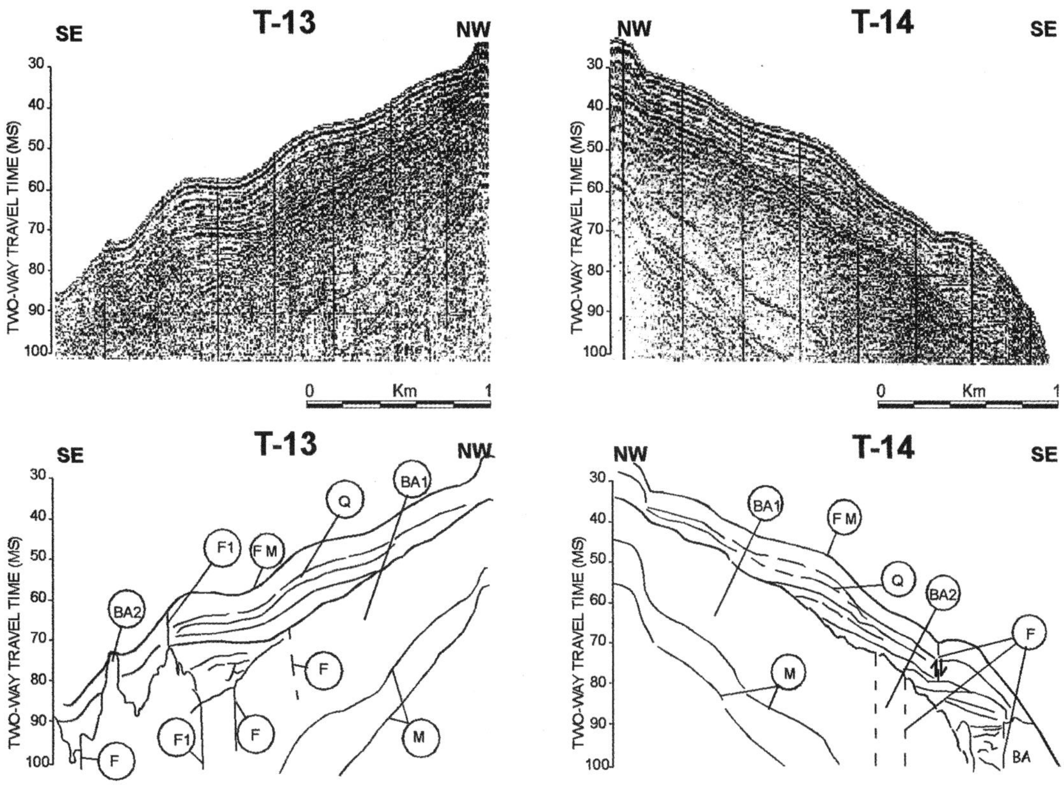

Figure 4. Seismic profiles. Upper panel: Seismic reflection Uniboom profile T-13 and T-14 off the El Médano coast (see Figure 3 for location). Lower panel: Interpretation drawing: FM: Sea floor. Q: Quaternary sediments; BA1: Acoustic basement type 2; F1: Fault F1; F: other fractures; M: Multiple.

F1 (Figure 4). The minimum length of this fault is 5 km.

The northern fault, F2 (Figure 3), is marked by a less pronounced morphological scarp. This fault is best observed at the NE extreme of the mapped area and fades out until it disappears at the SW end. The fault is marked by a slight flexure that produces a scarp 0.5 m high with more erosion of the footwall. A vertical displacement of 0.7 to 1.2 m has been observed in fault F1. The time of the displacements was after formation of the beach, dated as Holocene as described below.

A set of highly continuous fractures interpreted as joints mainly affects the tuff formation. Some of these joints, nevertheless, show an intensely curved trace. Although relatively small in number for the area, their inclinations were always close to 90° vertical. The most common alignments define three sets of joints whose directions in order of highest to lowest frequency are N175°, N56° and N105° (Figure 5).

Description of the liquefaction features

Liquefaction structures were observed in the uplifted beach sand formation (Figures 3 and 6). This formation extends over an area of around 90,000 m^2, but could have reach 650,000 m^2 in the past 50 years. Changes produced in coastal dynamics and anthropogenic effects have substantially modified the zone over the last decades, with the almost complete disappearance of the dunes and acceleration of erosive processes. Artificial removal of a large proportion of the uplifted beach sands has led to the current appearance of the study area.

The section observed in the site was as follows from top to bottom:

- An upper layer H1 located at the top of the deposit composed of coarse to intermediate, highly compact, partially-cemented sands. Its thickness is approximately 1 m. The surface is intensely eroded and shows wind erosion structures indicating its

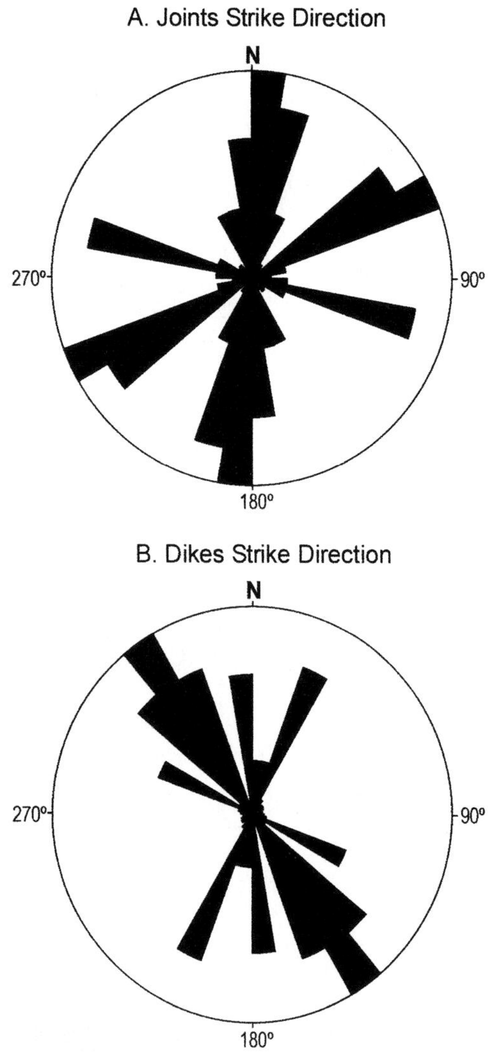

Figure 5. Rose diagrams displaying the two dimensional orientation distribution of data in the form of a circular histogram. A: Joint strike direction; B: Dike strike direction.

thickness was greater, possibly attaining at least 2 m. This layer shows evidence of liquefaction structures, sand dikes and vents described below. Above this layer, small calcareous crusts of 1–10 cm thickness and altered tuffs appear.

- The lower layer H2, beneath the previous layer, is comprised of medium to coarse sands, somewhat finer and less compact than in layer H1. It shows lamination and cross-stratification. This layer is partially crossed by vents but no dikes are observed. Its thickness ranges from 0.5 to 1 m.
- The substrate is composed of two layers of tuffs, T1 at the top and T2 at bottom. T1 comprises reddish tuffs weathered to form a 0.5 m thick paleosol. T2 is formed by yellowish, massive, highly compact pumice tuffs.

Within layers H1 and H2 there are numerous liquefaction structures formed by vents and clastic dikes or tabular structures (Figures 7 and 8). These outcrop both at the surface or in natural exposures. The dikes are composed of sands of the same composition as in layer H2, with strikes of 145°, 25°, 5° and 110° (Figure 5). The 145° striking dikes, besides being the most frequent, lie almost perpendicular to the direction of the topographical slope, while those trending 110° show similar directions to the 105° striking joints. In the mapped area, the length of the dikes reaches 25–30 m, although they are not easy to observe because of the intense erosion and dune deposits that partly cover them. Dikes are commonly 4 to 8 cm thick, although in some cases, thicknesses of up to 20 cm have been measured. Many of these dikes have a central opening or double rim 0.5 to 1.0 cm wide (Figure 9). The main system (145°) is the most continuous and presents the greatest thicknesses. The dikes show lateral terminations in the shape of thin filaments up to 1 cm long. Sometimes they cut into each other and also cut the tubular structures.

The formation mechanism of the clastic dikes seems to be related to lateral spreading and hydraulic fracturing mechanisms (Obermeier, 1990). The orientation of the main system (145°) perpendicular to the slope and its greatest thickness and continuity in relation to the other systems could be explained by a mechanism of lateral spreading. Lateral spreading reflects translational movement downslope and separation between individual blocks where shaking has been especially strong (Obermeier et al., 1993). Movement occurs where there is only minor resistance to lateral translation of the cap sitting on liquefied sediment. Besides lateral spreading, the geometry of the dikes (145° and other directions), their orientation, injected material, apical terminations and central apertures all point to a hydraulic fracturing mechanism.

The sand formation also shows numerous tubular structures in the shape of vents with diameters of 8 to 20 cm, whose greatest density coincides with a zone close and parallel to the scarp that marks the surface evidence for fault F1 (Figure 3). These tubes have a very compact peripheral ring with secondary infilling materials of loose sand inside them. Owing to the greater compactness of the ring, erosion has preserved the structures and these may be seen in the outcrops. These structures are present from the lower layer H2 upwards and cross the upper layer H1 (Figure 7). In

Figure 6. General view of the seismites zone showing an abundance of tubular vents. For scale purposes the height of the paleoseismic features is between 20 and 30 cm.

Figure 7. Longitudinal section of a tubular vent affecting layers H1 and H2. For scale purposes the hammer is 30 cm long.

the zone of greatest density, 3 to 5 tubes occur per m^2; the average being around 2 tubes per m^2.

Origin of the liquefaction features

To establish the origin of the liquefaction structures, possible causes of both seismic and aseismic nature were analyzed. The following causes were considered:

Volcanic activity. This can generate structures that give rise to vents or tubular conduits, injection of materials, fractures, infills, alterations, etc., as a consequence of the ejection of fluids, gases and materials. The last volcanic episode registered in the area took place over 100,000 years ago, while the age of the beach deposits is of the order of 10,000 years. The dikes and vents only affect the paleobeach deposit and not the tuffs of the substrate, ruling out a possible direct volcanic origin, although hydrothermal processes are being investigated, as a potential secondary process related to the tubular vents.

Biological origin. Some marine and coastal organisms can produce channels and orifices in beach deposits. However, the size, geometry and arrangement of the structures analyzed preclude this possibility (Calvo, pers. comm., 2002).

Pressure waves and tsunamis. Tsunamis produced by large offshore displacements of the sea bottom

Figure 8. Clastic dikes in the area of greatest density of liquefaction features. For scale purposes the dikes' height is between 5 and 10 cm, and their length is tens of meters.

Figure 9. Clastic dikes showing a central aperture and large vent. For scale purposes the hammer's length is 30 cm.

are highly improbable in this zone. The most recent landslide, in the valley of Guimar (35 km east of El Medano), occurred around 0.8 Ma (Masson and Watts, 1995). Although tsunamis or submarine slope failures of seismic origin cannot be ruled out, deposits associated with these have not been found in the study area. Nevertheless, this hypothesis is unlikely given the distribution and orientation of the liquefaction structures.

Artesian pressures. The morphological and hydrogeological conditions of the zone exclude this possibility.

The evidence supporting a seismic origin includes:

Upward directed hydraulic forces would be exerted

rapidly or almost instantly. The formation of dikes with the injection of sands transported from a source area by high-energy upward directed hydraulic forces, the apertures or double rims in the dikes and their apical terminations indicate mechanisms of hydraulic fracturing at high pressures. The geological evidence shows that virtually all the dikes must have a seismic liquefaction origin, and could almost certainly have formed solely in response to hydraulic fracturing (Obermeier, 1996). The direction of the tabular dikes is not random as would be expected if the dikes had originated by non-seismic mechanisms. The granulometric characteristics of the sand layers and their high uniformity, geological age and origin, geomorphological conditions and depth of the water table (discussed below), along with the resistance of the sands in which the structures developed are indicative of a high susceptibility of the deposits to liquefaction.

On the basis of the above evidence we can rule out a possible non-seismic origin and propose a seismic origin. This idea is also consistent with the seismotectonic characteristics of the region and the proximity of the previously described faults. The principal mechanism giving rise to the structures analyzed, also known as seismites, was liquefaction of the lower layer H2 that was composed of saturated sands. Due to the effect of intense interstitial pressures, water and sand were transported and expelled towards the surface via the vents, forming sand blows or craters by sand explosion. Hydraulic fracturing or lateral spreading led to rupture of the upper layer H1 of compact sands, giving rise to sand dike injection. The possibility of more than one earthquake occurring in the zone should not be precluded because of the presence of dikes cutting the vents, possibly implying more than one phase of liquefaction.

The beach sands formation has been dated by thermoluminescence as Holocene (Millán et al., 2002 as 10,081 ± 933 years BP). The calcareous crusts that cover some of the liquefaction structures have been dated by the same technique at 3490 ± 473 years BP. According to these data, liquefaction and the seismic phenomenon that produced it took place prior to 3490 years ago but after 10 ka.

Estimates of acceleration and magnitude of the paleoearthquake

The force of a seismic event and the magnitude of a paleoearthquake can be estimated by several methods. These have been reviewed in detail by Obermeier et al. (2001). The following methods are applicable to the present case:

(a) The cyclic stress method based on estimates of the lower-bound peak ground acceleration at individual sites of liquefaction.
(b) The Isihara method, which uses dike height at the site of hydraulic fracturing to estimate the actual value of peak ground acceleration at the site.
(c) The magnitude bound method, which uses the furthest distance from the seismic source to the liquefaction zone.
(d) Energy based solutions.

Methods (a) and (b) serve to calculate the peak acceleration needed for liquefaction to start at a particular site. The cyclic stress method is based on the method of Seed and Idriss (1971) and subsequent updates by Seed et al. (1985) and Youd and Noble (1997). Its application requires the interpretation of the soil profile at the time of liquefaction. To this end, we took into account the current conditions of the sand deposit and the aging processes to which the soil has been subjected from the time of liquefaction to the present.

The most common aging processes are (Olson et al., 2001): destruction of pre-earthquake soil structure and aging effects during liquefaction; post-liquefaction consolidation and densification, and post-liquefaction aging. The main outcome of liquefaction is increased granular packing, which may compact the sediment by some 27% (Owen, 1987). Following deposition, natural and man-made deposits develop a structure resulting from post-depositional mechanical readjustment and possible weak chemical bonding at particle contacts. This process is referred to as aging. The development of soil structure results in the improvement of soil properties such as shear strength, modulus, and penetration resistance (Schemertmann, 1991).

The present uplifted sand beach deposits show evidence of these aging processes. The main factors that have contributed to the compaction and partial cementation of the upper H1 layer are the uplift of the deposits by tectonics and the resultant downdropping of the water table, the geochemical conditions of the environment that favored an input of calcium carbonates and aluminum silicate compounds, and the extremely arid climatic conditions. Sands that could reflect conditions predating the aging processes were identified. These sands occur in the vicinity of the site closest to the coast and show the typical site granulometry and composition. Prior to liquefaction, the soil profile may have been as follows:

- An upper layer H1 formed of coarse to medium sands, dense to very dense with less than 2% fines, apparent natural density 1.7 g/cm^3 and N_{SPT} (number of blows of the standard penetration test or SPT, ASTM – D1586) equal to or more than 30 blows. Estimated mean layer thickness was 2 m. The water table would have lain towards the base of this layer, and would have been subjected to variations in the water table. Its behavior could correspond to that of a hard, semi-confining and nonliquefiable layer.
- A lower layer H2, comprising medium to coarse sands, relative density intermediate, containing less than 2% highly uniform fines, apparent natural density 1.5 g/cm^3 and N_{SPT} between 15 to 20 blows. The water table would lie above this layer and it would therefore be saturated. Mean estimated layer thickness is 1 m. This layer would have acted as a source zone for liquefaction.
- Layer T1 containing red tuffs weathered to 0.5 m thick paleosols.
- Layer T2, substrate composed of massive, very compact tuffs.

The cyclic stress method was applied following these hypotheses to estimate the peak ground acceleration (pga) necessary for the soil to undergo liquefaction. The results obtained give an acceleration of 0.22 g for $(N_1)_{60} = 15$ and 0.30 g for $(N_1)_{60} = 20$. According to Youd and Noble (1997), this would correspond to a 50% probability of liquefaction.

The Ishihara (1985) method considers that the maximum height of liquefaction dikes is controlled by two factors: the thickness of the liquefied sediment and the pga. This method is valid for seismic structures produced by hydraulic fracturing. It is applicable where the cap thickness is reasonably uniform and when source sands range from very loose to moderately compact, at least for earthquakes of moment magnitude $\mathbf{M} \sim 7.5$ or larger, (Obermeier, 1998). It was considered that the hard, semi-confining, nonliquefiable layer was 2 m in thickness and a thickness of 1 m was assumed for the liquefiable source. For these conditions, the resulting pga was 0.35 g according to the Ishihara method. Bearing in mind that the cyclic stress method represents the minimum acceleration value, and Ishihara method considers average conditions, a representative value of 0.30 g was selected from the possible range between 0.22 and 0.35 g.

From the accelerations calculated, intensities at the site can be estimated using one of the available empirical expressions. The equation used in the Spanish Seismic Code is: $I = [3.2233 + \log_{10}(a/g)]/0.30103$, where I are intensities, a is the horizontal pga (cm/s^2) and g is in % gravity. Hence an acceleration of 0.30 g gives a predicted intensity of IX.

The magnitude of paleoearthquakes, in terms of the moment magnitude M, can be calculated using the magnitude bound method and energy based approaches. The magnitude bound method estimates the magnitude of a paleoearthquake using relations between earthquake magnitude and the distance from the tectonic source to the farthest site of liquefaction. It is based on worldwide historical earthquakes (Ambraseys, 1988) and the data described by Obermeier et al. (1993) and Pond (1996). This method requires the identification of the seismic source. In the present case, the closest seismic sources are found between Tenerife and Gran Canaria over a line of epicenters in the ocean, at an approximate distance of 35 km from the El Médano site (Figure 2C). This source is associated with a NE–SW trending fault that runs parallel to the eastern coast of Tenerife, and was inferred from gravimetric data by Bossard and McFarlane (1970). In 1989, this fault produced the greatest earthquake instrumentally recorded on the archipelago ($\mathbf{M} = 5.2$). The length attributed to this fault is 30 km (Mezcua et al., 1992), yet it extends to over 80 km. A further possible seismic source is fault F1 located at the site. Its prolongation beneath the ocean was established by reflection seismic profiles indicating a minimum length of 5 km. However, the instrument record makes no reference to earthquakes with epicenters close to this fault, so that we only consider the submarine fault 35 km from the site as a seismic source when calculating the earthquake's magnitude, which yielded a magnitude **M** in the range 6.4 to 6.8. Given that 6.4 is the lower limit of the data considered by Ambraseys (1988), we took a $\mathbf{M} = 6.8$ as being representative. Wells and Coppersmith's (1994) relationship between fault length and magnitude also gives a $\mathbf{M} = 6.8$.

The second method used to estimate magnitude is based on so-called energy-based approaches that relate magnitude to energy release (Davis and Berrill, 1982) and subsequent reviews by Berrill and Davis (1985) and Trufinac (1995). This method relates magnitude to distance from the epicenter, to the liquefaction site and the $(N_1)_{60}$ (number of blows of the SPT for a pressure of 10 t/m^2 and an effective energy of 60%). For $(N_1)_{60} = 20$, the results obtained indicate a magnitude $\mathbf{M} = 6.8$ according to the method of Berrill and

Davis (1985) and of 7.2 according to that of Trufinac (1995).

In summary, the magnitudes estimated are in the range 6.4 to 7.2; a value of $\mathbf{M} = 6.8$ being considered the most representative. These estimates were based on the assumption that the seismic source was the submarine fault. If fault F1 as being closer to the site was the source of seismicity then models predict lower magnitudes but similar accelerations.

Conclusions

Several liquefaction structures were identified in El Médano, in southern Tenerife.These structures were clastic dikes and tubular vents; their origin being attributed to the liquefaction of sands by an earthquake of great intensity.

The mechanisms that gave rise to the clastic dikes were hydraulic fracturing and lateral spreading of a layer of compact sands in response to high pore pressures of seismic origin. These pressures, in turn, led to the movement and injection of sands across the compact sands level. The vents are the result of high upward hydraulic pressures causing the ejection of water and sand through these conduits to the surface, possibly forming sand blows and explosion craters.

The peak ground acceleration needed to produce liquefaction and the sand dikes was estimated at 0.22 to 0.35 g. An acceleration of 0.30 g, considered to be the most characteristic, would correspond to an intensity of IX at the site of liquefaction. The magnitude of the earthquake causing liquefaction was calculated to be in the range 6.4 to 7.2 with a value of $\mathbf{M} = 6.8$ taken to be representative. This result was obtained assuming that a submarine fault was the seismic source.

The liquefaction structures developed over a tectonically uplifted beach of sand deposits dated as $10,081 \pm 933$ years BP. Over these sands and liquefaction structures, fine calcareous crust levels dated as 3490 ± 473 years BP were observed. The paleoearthquake responsible for liquefaction occurred during the Holocene; its age lying between these two dates. Nevertheless, tectonic and geomorphological data from field observations suggest an age closer to the younger constraint.

Two faults F1 and F2 aligned in a direction N55° close to the liquefaction site were identified. Fault F1 cuts the uplifted beach sand formation. Through seismic reflection profiles, its extension under the sea was identified, and a minimum length of 5 km was established. Both faults limit a small graben, which gives rise to a depression in whose approximate center there are several lagoons.

Possible seismic sources near the site of liquefaction were considered. The main source is inferred to have been a submarine NNE–SSW trending fault some 35 km from the site between the islands of Tenerife and Gran Canaria. Its movement takes the form of a sinistral thrust. This fault shows associated seismicity. Another proposed source is fault F1, which affects the sand formation where the paleoliquefaction is found. No historical epicenters related to this fault have been recorded.

The tectonic structures affecting materials of recent age and the seismicity associated with these structures demonstrates existing seismotectonic relationships and confirms the paleoseismic activity identified in southern Tenerife. The paleoearthquake investigated here is the largest of those registered on the Canary Islands.

The presence of active faults affecting materials of very recent age and their association with a paleoearthquake of high intensity in the south of Tenerife are key factors that need to be borne in mind when evaluating seismic hazards on the Canaries, a region, which up until now, had been considered to be of low seismic activity.

Acknowledgements

We are grateful to Dr. A. Millán of the Universidad Autónoma de Madrid for carrying out the thermoluminescence procedures. Dr. S. Obermeier from the US Geological Survey (Emeritus), Professor F. Ricci Lucchi of the Universita di Bologna and Professor J. P. Calvo of the University Complutense of Madrid are thanked for their helpful comments and advice. Thanks are also due to Dr. M. A. Rodriguez Pascua and P. Llanes for providing useful information. Finally, we are grateful for the efforts of reviewers whose comments have helped improved the manuscript.

References

Ambraseys, N. N. 1988. Engineering seismology. *Earthq. Eng. Struct., D.*, **17**(1), 1–105.

Berril, J. B. and Davis, R. O. 1985. Energy dissipation and seismic liquefaction of sands revised model. *Soils and Found.*, **25**(2), 106–118.

Bosshard, E. and MacFarlane, D. J. 1970. Crustal structure of the westerm Canary Islands fron seismic refraction and gravity data *J. Geophys. Res.*, **75**, 4901–4918.

Davis, R. O. and Berril, J. B. 1982 Energy dissipation and seismic liquefaction of sands. *Earthq. Eng. Struct. D.*, **10**, 59–68.

Dziewonski, A. M., Ekström, G., Woodhouse, J. H. and Zwart, G., 1990, Centroid-moment tensor solutions for April–June 1989, *Phys. Earth Planet. Int.* **60**, 243–253.

Ishihara, K. 1985. Stability of material deposit during earthquakes. *Proc. 11 th. Int. Conf.Soil. Mech. and Found. Eng.* San Francisco, Vol I. 321–376. A. A. Balkema, Rotterdam.

Llanes, P., Muñoz, A. Muñoz-Martín, A., Acosta, J., Herranz, P., Carbó, A., and Palomo, D. Morfological and structural analysis in the Anaga offshore massif, Canary Islands: fractures and debris avalanches relationships. *Mar. Geophys. Res.*, (in this volume).

Mezcua, J., Burforn, E., Udías, A. and Rueda, J., 1992. Seismotectonic of the Canary Islands. *Tectonophysics*, **208**, 447–452.

Millan, A., Benitez, P. and Calderón, T. 2002. Datación absoluta por termoluminiscencia de muestras de paleoplayas de Tenerife. Lab. Datación y Radioquímica. Universidad Autónoma de Madrid. España. (Unpublish).

Obermeier, S. F. 1996. Use of liquefaction -induced features for paleosismic analysis. *Eng. Geol.*, **44**, 1–76.

Obermeier, S. F. 1998. Overview of liquefaction evidence for strong earthquakes of Holocene and latest Pleistocene ages in the states of Indiana and Illinois, USA. *Eng. Geol.*, **50**, 227–254.

Obermeier, S. F., Martin, J. R., Frankel, T. L., Munson, P. J., Munson, C. A. and Pond, E. C. 1993. *Liquefaction evidence for one or more strong Holocene earthquakes in the Wabash Valley of southerm Indiana and Illinois. U.S. Geol. Survey Prof. Paper*, **1536**, 27 pp.

Obermeier, S. F., Pond, E. C. and Olson, S. C. 2001. *Paleoliquefaction studies in continental settings: geological and geotecnical features in interpretations and back-analysis. U.S. Geol. Survey. Openfile Report*, **01–29**. 75 pp.

Olson, S. M., Obermeier, S. F. and Stark, T. D. 2001. Interpretation of penetration resistence for back analysis at sites of previous liquefaction. *Seism. Res. Lett.*, **72**(1), 46–59.

Owen, H. G. 1987. *Deformation processes in unconsolideted sands. In: Deformation of Sediments and Sedimentary Rocks.* Jones, E. M. and Preston, M. F. (Ed.), Geol. Soc. of London, Publ. **29**, 11–24.

Pond, E. C. 1996. *Seismic parameters for the central United States based on paleoliquefaction evidence in the Wabash Valley*. Ph D. Thesis. Virginia Polytech. Inst. Blacksburg, Virginia, 583 pp.

Schemertmann, J. H. 1991. The mechanical aging of soils. *J. Geotech. Eng-ASCE*, **117**(1), 1288–1330.

Seed, H. B. and Idriss, I. M. 1971. Simplified procedure for evaluating soil liquefaction potential. *J. Soil Mech. Found. Div. ASCE*, **97**(1), 1249–1273.

Seed, H. B., Tokimatsu, K., Harder, L. F. and Chung, R. M. 1985. Influence of SPT procedures in soil liquefaction resistence evaluations. *J. Soil Mech. Found. Div. ASCE*, **111**, 1425–1445.

Trifunac, M. D. 1995. Empirical criteria for liquefaction in sands via standard penetration test and seismic wave energy. *Soil Dyn. Earthq. Eng.*, **14**, 419–426.

Wells, D. L. and Coppersmith, K. J. 1994. New empirical relationships among magnitude, rupture length, rupture area, and surface dispklacement. *B. Seismol. Soc. Am.*, **84**, 974–1002.

Youd, T. L. and Noble, K. 1997. Liquefaction criteria based on statistical and probabilistic analysis. *Proc. NCEER Workshop on Evaluation of Liquefaction Resistence of Soils*. Youd, T. L. and Idriss, I. M. (Eds.), Tech. Rep.NCEER-97-0022, State University of New York at Buffalo, New York, 210–216.

Marine Geophysical Researches (2003) 24: 161–169
DOI 10.1007/s11001-004-1100-7

Age and composition of the Amanay Seamount, Canary Islands

E. Ancochea & M.J. Huertas
Departamento Petrología y Geoquímica, Facultad. Ciencias Geológicas, Instituto Geología Económica Consejo Superior Investigaciones Científicas – Universidad Complutense, 28040 Madrid, Spain
**Corresponding Author (E-mail: huertas@geo.ucm.es)*

Key words: Amanay Seamount, Eastern Canarian Volcanic Ridge, Oceanic Island Basalts, K-Ar geochronology, Canary Islands

Abstract

A number of samples have been dredged from the upper parts of Amanay and El Banquete Seamounts, yet volcanic materials have been collected only on Amanay Seamount. Based on textural features and the presence or absence of kaersutite, two main types of olivine pyroxene basaltic rocks have been identified. The rocks are basanites with high enrichment in the most incompatible elements, similar to that displayed by Ocean Island Basalts. Samples from Amanay Seamount formed due to a low degree of melting of an enriched mantle, very similar to that which probably caused the Miocene volcanic activity of Fuerteventura. The age of Amanay volcanic rocks, 15.3 ± 0.4 and 13.1 ± 0.3 Ma, is similar to those of the older volcanic units exposed in the nearby islands (Gran Canaria, Fuerteventura and Lanzarote). This proves the formation of a separate submarine volcanic edifice coeval with the other edifices of the Eastern Canarian Volcanic Ridge. Volcanic activity on the submarine edifice is thought to have ceased at about 13 Ma, simultaneous with the adjacent main volcanic construction.

Introduction

The islands of Lanzarote and Fuerteventura represent the emergent crest of the Eastern Canarian Volcanic Ridge that extends north-northeast below sea level subparallel to the African coastline and to the ocean-floor spreading fabric. To the North, the Ridge is connected to Conception Bank (Figure 1A), a submarine volcanic complex situated above 30°-31° N latitude (Luyendyk and Bunce, 1973; Uchupi et al.,1976; Weigel et al., 1978; Dañobeitia, 1988; Dañobeitia and Collette, 1989).

In the southern part of Fuerteventura two circular shaped seamounts appear west of the Jandía Peninsula. The seamounts were interpreted as corresponding to two volcanic edifices the 'Amanay Submarine Edifice', situated northwest of Jandía, and "El Banquete Submarine Edifice", which occurs southwest of Jandía (Ancochea et al., 1996). The latter edifice is connected to the southern end of Fuerteventura by a flat and shallow platform (Figure 1B and C).

Dredging of Amanay and El Banquete Seamounts was carried out under the scope of the Oceanographic and Hydrographic Research of the Spanish Economic Exclusive Zone (ZEE Project, August 2001) by R/V Vizconde De Eza dealing with the study of the ocean floor around the Canary Islands. The aim of the dredging operations was to collect samples in order to determine both the composition and age of the rocks forming the seamounts.

Amanay and El Banquete

Amanay and El Banquete Seamounts (Figures 1 and 2) lie to the east of Gran Canaria and close to Fuerteventura. In view of their proximity, they were expected to show common geological characteristics. Both seamounts are similar in size, with a basal diameter of 28–29 km, exceeding 2000 m in height, and are separated by a narrow channel exceeding 1000 m in depth.

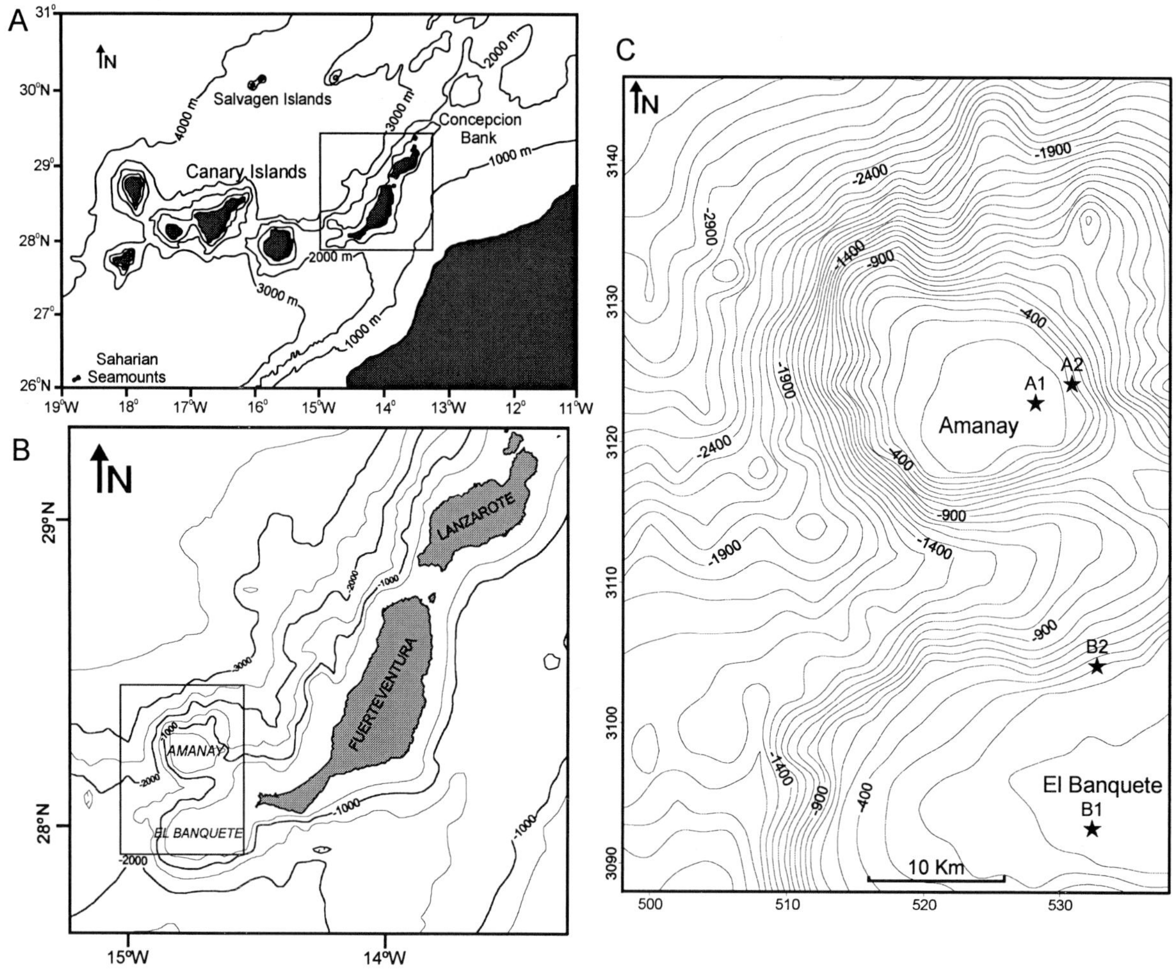

Figure 1. A: Bathymetric map of the Canary Islands and the eastern branch of the Canary Ridge modified after Dañobeitia and Collette (1989). B: Location map of Amanay Seamount using Smith and Sandwell (1997) database and Generic Mapping Tools (Wessel and Smith, 1995). C: Bathymetry map of Amanay Seamount. Contour interval is 100 m. Bathymetry data comes from 'Spanish Economic Exclusive Zone Project'.

Gran Canaria

The subaerial volcanic history of Gran Canaria began about 15 Ma and was divided into three main phases (Schmincke, 1982). The first phase, the subaerial Miocene phase, started with rapid emission of tholeiitic to weakly alkalic basalts (~13 Ma), followed by rhyolitic/trachytic and phonolitic/trachytic lava flows and ignimbrites. After a period of quiescence (from 8.5 to 5 Ma) the Pliocene phase consisted of nephelinites, alkali basalts, basanites, tephrites and phonolites, with peak activity at 4 Ma. Basanites, tephrites, nephelinites and melilitites make up the Quaternary–Holocene phase, which affected only the northern part of the island.

Fuerteventura

Two major structural and petrological units are distinguished in Fuerteventura: the Basal Complex and the Subaerial Volcanic Series. As defined by Bravo (1964), Fúster et al. (1968) and Stillman et al. (1975) the Basal Complex is a thick, Cretaceous sedimentary sequence overlain by submarine volcanic rocks and intruded by an intense NNE-SSW trending sheeted-dyke swarm, which was formed in association with the emplacement of alkaline plutons. The Basal Complex represents the submarine growth stage of the volcanic complex and the subvolcanic roots (plutons and dykes) of their successive subaerial episodes (Ancochea et al., 1996).

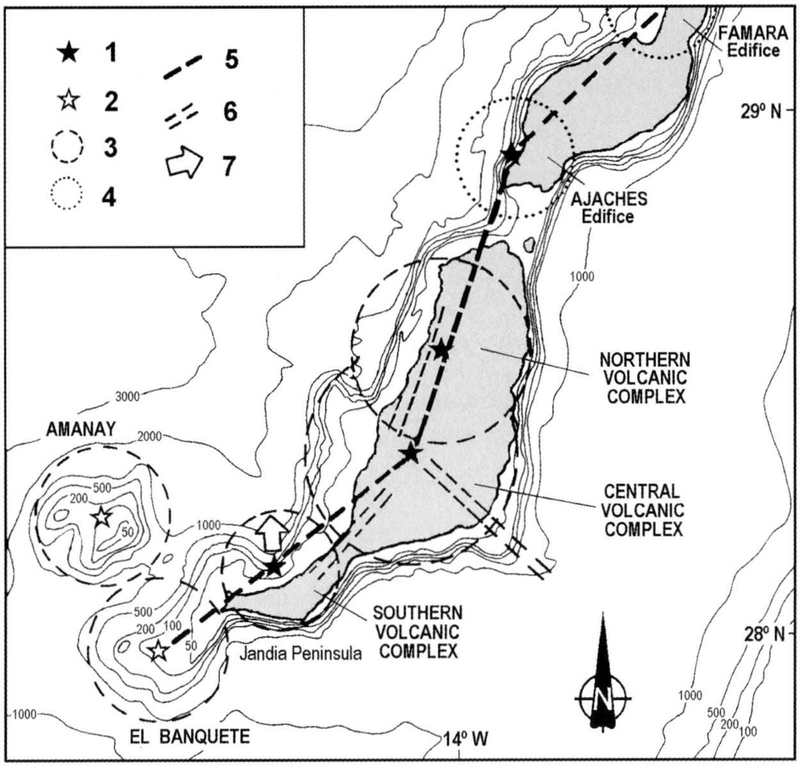

Figure 2. Location of the Volcanic Complexes on Fuerteventura and Lanzarote. 1 : subaerial emission centers; 2 : submarine emission centers; 3 : approximated areas occupied by volcanic complexes in Fuerteventura zone; 4 : approximated areas occupied by volcanic complexes in Lanzarote zone; 5 : alignement of the volcanoes; 6 : main trends of dyke intrusions; 7 : possible landslides. (Ancochea et al., 1996).

Table 1. Age and composition of the Subaerial Volcanic Series of Fuerteventura.

Units			Age (Ma)	Composition
Old Basaltic series (Miocene)	Southern Volcanic Complex (SVC or Jandía)	Lower	20.7–19.3	Mildly alkaline
		Middle	17.2–15.4	Mildly alkaline
		Upper	15.2–14.2	Middle alkaline
		Late Formations	?	Ultra-alkaline
	Central Volcanic Complex (CVC)	Lower	>22.5	Mildly alkaline
		Middle	22.5–20	Middle alkaline
		Upper	17.5–14.5	Middle alkaline
		Late Formations	?–13.2	High alkaline – ultra-alkaline
	Northern Volcanic Complex (NVC)	Lower	>22–15.3	Middle alkaline-Mildly alkaline
		Upper	14.3–12.8	Alkaline
Recent series (Pliocene –Quaternary)		Series II, III and IV	5–0	Middle alkaline Ultra-alkaline

Within the Subaerial Volcanic Series, two episodes are distinguishable. The earliest and most important is the Miocene Old Basaltic Series. Ancochea et al. (1991), Cubas et al. (1992), Hernán et al. (1993) and Ancochea et al. (1993, 1996) have proposed that the Miocene Old Basaltic Series corresponds to three major volcanoes or volcanic complexes: the Southern Volcanic Complexes (SVC or Jandía), Central Volcanic Complexes (CVC) and Northern Volcanic Complexes (NVC) (Figure 2). Each complex has a separate history, in some cases longer than 10 m.y., with several periods of activity alternating with quiescence accompanied by erosion (Table 1). After a period lasting about 7 m.y. (Coello et al., 1992) and characterized by lack of activity and erosion, new basaltic materials were erupted during emplacement of the Pliocene–Quaternary-Recent series. The importance of this stage is lesser in terms of volume, although it continued until present times.

The geochemical character of the volcanic rocks on Fuerteventura is always alkalic, but alkalinity varies from mildly alkaline (normative hyperstene and olivine but without Ca-poor pyroxene), middle alkaline (0 < normative nepheline Ne<5), high alkaline (5<Ne<10) and ultra-alkaline (Ne>10).

Methods

Rocks were dredged in four sites, two from the Amanay Seamount (A1 and A2) and two from El Banquete Seamount (B1 and B2), one of each pair at shallow depth and the other one at a greater depth along the flank (Figure 1 and Table 2).

Volcanic samples have been dredged only in one site (A1). Fragments of volcanic rocks in this site were abundant and appear to be of uniform petrologic type in hand-specimen. They show minor differences in vesiculation, varying from almost 0% (non-vesicular) to more than 10%.

Ten samples from site A1 were selected for detailed petrographic and mineralogical study. Two of the samples (number 24 and 32) were considered to represent the main petrographic types and were later analysed. The analysed samples were treated with deionised water to remove seawater. Major and trace element concentrations were determined by ACTLABS laboratories in Ontario (Canada) by X-ray fluorescence spectrometry and by inductively coupled plasma-mass spectrometry (ICP-MS). Samples 24 and 32 were dated by the K/Ar method by Teledyne Isotopes Laboratory (USA). The material used for dating was represented by 'whole rock' samples of one or two grams, with particle size varying from 0.3 to 1 mm. Argon was extracted by fusion after degassing at moderate temperature in high vacuum and the ^{38}Ar tracer was added to the analysis using a continuous pipetting system. The analytical errors were calculated according to the method of Dalrymple and Lamphere (1969). Converted ages were calculated using the following constants: $^{40}K/K = 1.167 \times 10^{-2}$ *atoms %*; $\lambda\varepsilon = 0.581 \times 10^{-10}\ yr^{-1}$; $\lambda_\beta = 4.962 \times 10^{-10}\ yr^{-1}$; $^{40}Ar/^{36}Ar$ *atmosphere* = *295.5*. All errors are given at the 2 'σ' level.

Chemical analyses of minerals was carried out with a JEOL electron microprobe (JXA-8900 M) equipped with four spectrometers at the Universidad Complutense de Madrid. The operating parameters were: 20 s total counting time, 15 kV accelerating voltage, 20 nA beam current, and 1–5 μm beam diameter. Calibration was against standard minerals provided by Department of Mineral Sciences of the Smithsonian Institution (Jarosewich et al., 1980).

Results

Sedimentary samples

Sampling site A-1: Medium to coarse sandstones. Well-rounded grains of volcanic rocks and skeletal fragments with isopach sparite cement, interpreted as originally aragonitic in nature (calcite now). In spite of cementation, porosity locally reaches 20%. Fossil content is very high, including bioclasts of Lithothamnium algae (largely branching), echinoids and bivalves. Bivalve shells are moderately penetrated by microborings. The sand is interpreted as having been deposited in agitated, clean, shallow-marine waters (upper shoreface).

Sampling Sites A-2, B-1 and B-2: Bioclastic grainstone with Lithothamnium, gastropods, echinoids, bivalves with superficial microboring, benthic and encrusting *Gypsina*-like foraminifers. There are some remains of a peloidal matrix. Most grains are coated by isopach sparite to microsparite cement, interpreted as originally aragonitic. There is some intergranular porosity. The environment of deposition was shallow-marine, clean, and characterized by agitated waters (upper shoreface or carbonate platform).

As these rocks were sampled in deep waters, post-sedimentary downslope transport is inferred, although

Table 2. Location of the sampling points

Seamount	Sampling Points	Coordinates (GPS differential)	Water Depth (m)
Amanay	A1	28° 13.843′ N 14° 42.773′ W	27
	A2	28° 14.574′ N 14° 41.134′ W	290
El Banquete	B1	27° 57.442′ N 14° 40.256′ W	47
	B2	28° 03.707′ N 14° 40.039′ W	312

Table 3. K-Ar radiometric age.

Sample	^{40}Ar* (scc/gr $\times 10^{-5}$)	% ^{40}Ar*	% K	Age (Ma)
24	0.056	79.8	1.10	13.1 ± 0.3
32	0.073	78.6	1.23	15.3 ± 0.4

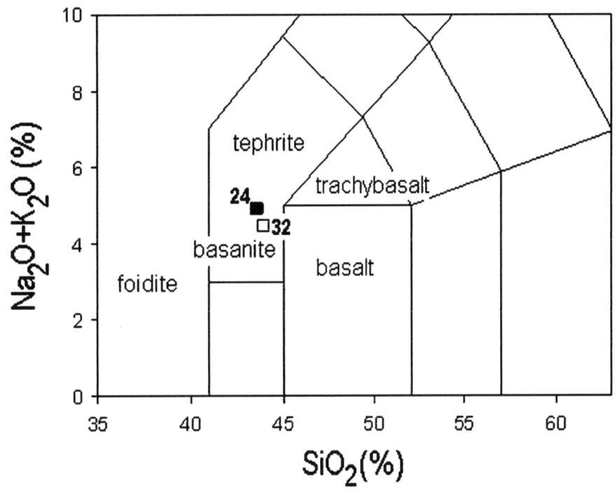

Figure 3. Classification of rocks dredged of Amanay Seamount (24 and 32) using the total alkalis versus silica (TAS).

it is possible that subsidence also contributed to moving these rocks to deeper waters.

Volcanic samples (only site A1)

Petrographic analyses of the rock samples indicate that they are olivine pyroxene basalts, generally non-vesicular or sparsely vesicular.

Two main populations of basaltic rocks can be identified. The most frequent type, represented by sample 24, is a hypocrystalline porphyritic rock in which the groundmass shows variable crystallinity. The main phenocrysts are clinopyroxene and olivine, but minor amounts of amphibole are also present. The basalts display some evidence of alteration to secondary minerals, and locally a few small vesicles (<0.3 mm) filled with calcite or zeolites were observed. Fine-grained granular polycrystalline aggregates (up to 1 cm across) are commonly found; they consist of diopside with variable enstatite contents (44–47%) and rare plagioclase; frequently anhedral grains of pargasitic hornblende surround them. The other group of basalts (represented by sample 32) show holocrystalline texture with clinopyroxene and olivine phenocrysts set in a fine-grained matrix. Amphibole is absent in these rocks which, in addition, do not show glass, vesicular texture, microcrystalline aggregates or any significant alteration. All these features are clearly different from the basaltic group described above.

In both groups, the clinopyroxene is diopside, which forms subhedral to anhedral zoned and twinned phenocrysts with variable size up to 3 mm across and appears to be fresh. The diopside exhibits irregular concentric zonation patterns with a range of enstatite content from core to rim (32.5–46.2%) and cores with variable aegirine contents (3–14%). The high Ca+Na content determined in diopside is characteristic of the clinopyroxene of alkaline lavas (Leterrier et al., 1982). Olivine is present as subhedral microphenocrysts and is less abundant than clinopyroxene. The olivine phenocrysts display complete alteration to iron oxides and hydroxides, as well as resorption rims.

The amphibole shows some variation in abundance within the samples and varies in size from 1 mm to 0.1 mm. It occurs both as anhedral phenocrysts and as microcrysts. Larger phenocrysts frequently show resorbed rims with alteration to iron oxides and sphene. In all samples the amphibole is kaersutite and shows an increase in the $Mg/(Mg+Fe^2)$ ratio towards the cores.

The groundmass in both basalt groups typically consists of fresh idiomorphs laths of labradorite to andesine plagioclase (An_{54} to An_{43}), clinopyroxene (diopside) prisms and oxide minerals (magnetite with exsolutions of ilmenite, Cr-bearing magnetite and ilmenite.)

Two K/Ar radiometric determinations (one from each petrographic type) have been carried out from unaltered or slightly altered samples (Table 3). The

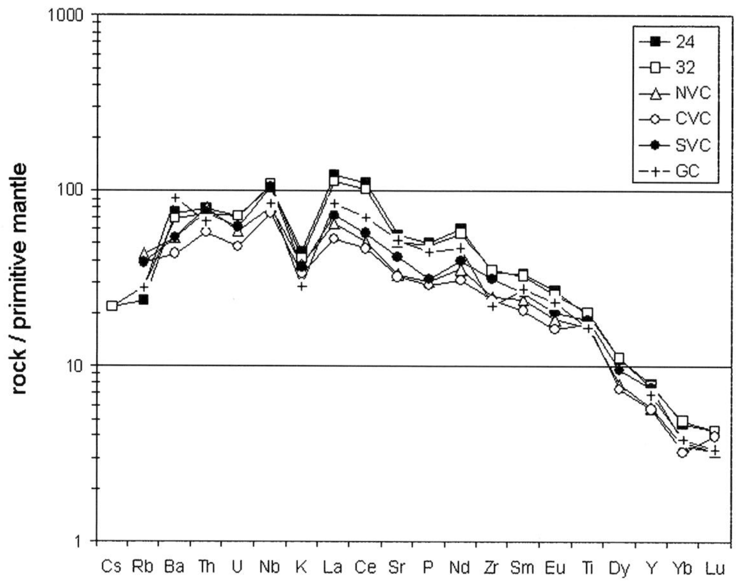

Figure 4. Multi-element diagram for representative samples of rocks dredged of Amanay Seamount (24 and 32). North Volcanic Complex *(NVC)*, Central Volcanic Complex *(CVC)* and South Volcanic Complex *(SVC)* of Fuerteventura Island; Miocene basalts of Gran Canaria Island (*GC*). Trace element concentrations normalized to the composition of upper mantle (Thompson, 1982).

Table 4. Major and trace element data of Amanay rocks (oxides in wt%, trace elements in ppm)

	24	32		24	32		24	32
SiO_2	41.47	41.73	Co	41	39	La	84	77.4
Al_2O_3	12.51	12.99	Cr	130	94	Ce	196	181
Fe_2O_3	13.58	13.84	Cs	0.5	0.5	Pr	19.9	18.4
MnO	0.19	0.2	Cu	95	91	Nd	82.1	77.6
MgO	7.09	7.03	Ga	22	22	Sm	15	14.5
CaO	11.05	10.75	Nb	75	77	Eu	4.52	4.3
Na_2O	3.30	3.04	Ni	99	91	Gd	12.1	11.5
K_2O	1.35	1.24	Sc	25	24	Tb	1.6	1.6
TiO_2	4.24	4.38	V	352	372	Dy	8.1	8.2
P_2O_5	1.09	1.06	Y	36	35	Ho	1.4	1.3
LOI	2.93	2.68	Zn	120	140	Er	3.4	3.3
Total	*98.81*	*98.93*	Zr	384	395	Tm	0.4	0.41
Ba	519	488	Ta	4.9	4.9	Yb	2.3	2.4
Rb	15	18	Th	6.7	6.3	Lu	0.32	0.32
Sr	1171	1071	U	1.5	1.5	Hf	8.3	8.6

percentage of $^{40}Ar^*$ (>20%) is quite high for rocks with such a low K content. The chronologic data correspond to the middle Miocene (16.4–11.2 Ma).

The composition of both basalt types is very similar, with MgO contents around 7% (Table 4) and both plot in the basanitic field of the Total Alkali-Silica diagram (Le Maitre et al., 1989) (Figure 3). Although the rock looks unaltered, a partial alteration of its composition could have resulted from its location in a marine environment Ti, Zr, Y and Nb contents, which are thought to be relatively immobile in aqueous fluids, confirm its alkaline character (Winchester and Floyd, 1976; Pearce, 1982). One can come to a similar conclusion from the high Ti/V ratio (73 and 75) and low TiO_2/P_2O_5 ratio, which are typical of alkaline rocks (Winchester and Floyd, 1976; Shervais,

1982). The samples show depletion in heavy rare earth elements (HREE) relative to the light and intermediate rare earth elements ($(La/Yb)_N$ = 21.5 and 23.9, $(Sm/Yb)_N$ = 6.95 and 6.52).

The compositions of the samples normalized to a primordial mantle (Thompson, 1982; Figure 4) reflect enrichment in the more incompatible elements, especially Nb, Ta, La and Ce; K is depleted relative to Th, Nb, Ta, La and Ce. The pattern is similar to that displayed by Oceanic Island Basalts and Canary Island basalts, showing a strongly negative potassium anomaly and high LREE contents. The low Ba/Nb, Rb/Nb, K/Nb, Th/Nb and Ba /La ratios are similar to those of HIMU basalts (Weaver, 1991).

Comparison with adjacent volcanic edifices

The volcanic rocks from Amanay show some compositional similarities with other Canary Island volcanic rocks, but they display some significant differences. The Amanay rocks show higher alkali, Zr, Y and REE contents at similar MgO contents compared to the basalts from the nearby islands (Gran Canaria, Fuerteventura) (Figure 5). The most similar rocks are those from Jandía (SVC), which is the closest volcanic edifice. Both low HREE contents and the high $(Gd/Yb)_N$ ratios of the Amanay samples suggest low degrees of partial melting, a common feature for Canary Island basalts. $(Gd/Yb)_N$ ratios are higher (3.8–4.1) than those typical of the Miocene basalts of Gran Canaria (3.1–3.4) but similar to the determined $(Gd/Yb)_N$ ratios from basalt of La Palma (3.4–4.5) and Fuerteventura (3.4–5.1, average: 4.1). This could indicate that mantle melting may have started at greater depths than in Gran Canaria, but at the same depth as in La Palma and Fuerteventura (Abratis et al., 2001; Ancochea et al., 1993). High $(La/Yb)_N$ ratios (21.5-23.9) suggest lower degrees of melting than beneath Gran Canaria (10–19), NVC (14–21) and CVC (9–18) of Fuerteventura. The degree of melting of Amanay is within the range of the Jandía edifice (14–29). Zr/Nb, Y/Nb and Ba/Y ratios (5, 0.45 and 14 respectively) are in the same range of those determined in rocks from Fuerteventura, but quite different to the basalts from Gran Canaria (Zr/Nb>6, Y/Nb>0.5 and Ba/Y<10), which suggests a more enriched mantle source for the rocks of Amanay and Fuerteventura.

Ages determined from the volcanic samples of Amanay (13.1 ± 0.3 to 15.3 ± 0.4 Ma) indicate that the rocks formed during a time interval in which Gran

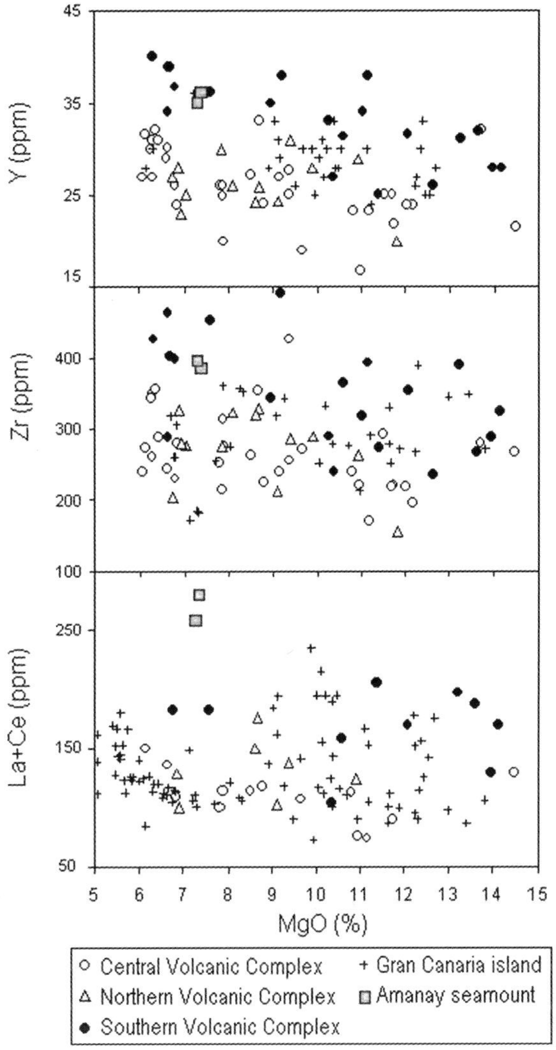

Figure 5. Zr, Y and La+Ce contents versus MgO of Amanay, Gran Canaria and Fuerteventura Miocene basalts.

Canaria (Schmincke, 1982), Fuerteventura and the southern part of Lanzarote (Coello et al., 1992) experienced volcanic activity (Figure 6, Table 1). Regarding Gran Canaria, the age is coincident with the first stage of building of the island, with major activity at 13 Ma. In this period the volcanic activity is important in the three volcanic complexes of Fuerteventura where the youngest volcanic units were being formed (Ancochea et al., 1996). The oldest age determined in the Amanay basalts is coeval with the lower unit of the NVC, and the youngest age is coeval with the upper unit of the NVC. Regarding the CVC, the upper unit of the edifice was being formed at 15 Ma, whilst the volcanic activity (Late Formations, Ancochea et al., 1996) was

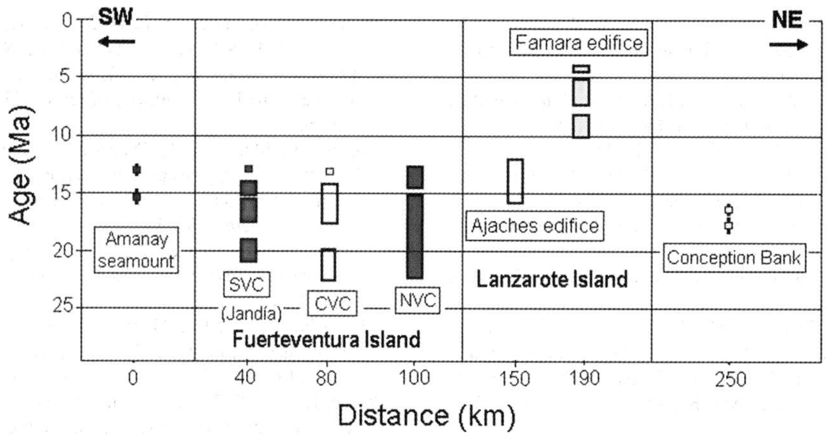

Figure 6. K/Ar radiometric ages from the Amanay rocks and their chronological relationship with the volcanic edifices of the Eastern Ridge of the Canary Islands. X axis: approximate distance (km) to Amanay Seamount.

residual at 13 Ma. Correlation with the SVC shows that the 15.3 ± 0.4 age is located at the boundary between the Middle and the Upper SVC whilst the 13.1 ± 0.3 Ma age is probably contemporaneous with the late formations. The two determined ages are also coeval with the formation of the Ajaches edifice (15.5 ± 0.3 to 12.3 ± 0.3 Ma) in southern Lanzarote. Ages determined in the Famara edifice (northern Lanzarote) are younger (Coello et al., 1992). Finally, two samples dredged from the Conception Bank (Geldmacher et al., 2001) have yielded ages of 16.64 ± 0.2 Ma and 17.50 ± 0.6 Ma, somewhat older than the Amanay rocks, but possibly coeval with earlier stages of the building of Amanay Seamount.

Conclusions

The samples dredged from the Amanay Seamount support the volcanic origin of this edifice. As previously postulated by Ancochea et al. (1993, 1996), the Amanay Seamount is a submarine edifice linked to a row of volcanic edifices extended along a rift subparallel to the African coast-line. The volcanic rocks of Amanay are basanites composed of olivine and pyroxene phenocrysts, erupted at 15.3 and 13.1 Ma.

Two petrographic types are distinguished. The main difference between the two types is the presence or absence of phenocrysts of kaersutite and fine-grained granular polycrystalline aggregates of diopside, enstatite, pargasitic hornblende and plagioclase.

The chemical composition of the two basalts collected in Amanay is quite similar. They show OIB and HIMU patterns with clear, strong LREE enrichments. The composition is also similar to the old basaltic rocks of Fuerteventura. The rocks from Amanay derive from a mantle source more enriched and melted at greater depth than that seen in Gran Canaria, but similar to that found in Fuerteventura.

The volcanic rocks of Amanay formed throughout a period (15.3 ± 0.4 and 13.1 ± 0.3 Ma) of remarkable volcanic activity in the Eastern part of the Canarian Archipelago. During this period, the three volcanic complexes of Fuerteventura and the Southern Edifice of Lanzarote (Los Ajaches) were active. The younger age determined (13.1 Ma) from the Amanay rocks is similar to that shown by the later units belonging to the Fuerteventura and Los Ajaches edifices.

Acknowledgements

This work was supported by the Projects: DGICYT PB1998-0759, funded by the Spanish Ministry of Science and Technology and UCM – PR1/03-11584. We wish to acknowledge Dr C. Dabrio and Dr J.P. Calvo for a critical reading of the manuscript. The authors wish to thank the officers and crew of the R/V Vizconde de Eza for their aid in dredging the seamounts volcanoes.

References

Abratis, M., Schmincke, H.U. and Hansteen, T.H., 2002, Composition and evolution of submarine volcanic rocks from the central and western Canary Islands, *Int. J. Earth Sci. (Geol Rundsch)* **91**, 562–582.

Ancochea, E., Cubas, C.R., Hernán, F. and Brändle, J.L., 1991, Edificios volcánicos en la Serie I de Fuerteventura: rasgos generales del Edificio Central, *Geogaceta*, **9**, 60–62.

Ancochea, E., Brändle, J.L., Cubas, C.R., Hernán, F. and Huertas, M.J. (1993). La Serie I de la isla de Fuerteventura, *Memor. R. Acad. Cienc. Exac. Fís. Nat. de Madrid*, **27**, 151 pp.

Ancochea, E., Brändle, J.L., Cubas, C.R., Hernán, F. and Huertas, M.J., 1996, Volcanic Complexes in the Eastern Ridge of the Canary Islands: The Miocene activity of the island of Fuerteventura, *J. Vol. Geoth. Res.*, **70**, 183–204.

Bravo, T., 1964, Geología General de las Islas Canarias, T.II. Ed. Goya, Santa Cruz de Tenerife, 592 pp.

Coello, J., Cantagrel, J.M., Ibarrola, E., Jamond, C., Hernán, F., Fúster, J.M., Ancochea, E., Casquet, C., Diaz de Terán, J.R. and Cendrero, A., 1992, Evolution of the Eastern Volcanic Rigde of the Canary Islands Based on New K-Ar Data, *J. Vol. Geoth. Res.*, **53**, 251–274

Cubas, C.R., Hernán, F. Ancochea, E. and Brändle, J.L., 1992. El Edificio Sur (Jandía) de la Serie I de Fuerteventura: rasgos generales, *Geogaceta*, **11**, 79–81.

Dalrymple, G.B. and Lanphere, M.A., 1969, Potassium-Argon dating; principles, techniques and applications to geochronology, W.H.Freeman and Co. San Francisco, 258 pp.

Dañobeitia, J.J., 1988, Reconocimiento geofísico de estructuras submarinas situadas al norte y sur del archipiélago Canario, *Rev. Soc. Geol. España*, **1**, 143–155.

Dañobeitia, J.J. and Collette, B.J., 1989, Estudio mediante sísmica de reflexión de un grupo de estructuras submarinas situadas al Norte y Sur del archipiélago Canario, *Acta Geol. Hisp.*, **24**, 147–163.

Fúster, J.M., Cendrero, A. Gastesi, P., Ibarrola, E. and López Ruiz, J., 1968, Geology and volcanology of Canary Islands, Fuerteventura, Inst. Lucas Mallada. C.S.I.C. Madrid, 243 pp.

Geldmacher, J., Hoernle, K., Bogaard, P.v.d., Zankl, G. and Garbe-Schönberg, D., 2001, Earlier history of the $\geq$ 70-Ma-old Canary hotspot based on the temporal and geochemical evolution of the Selvagen Archipelago and neighboring seamounts in the eastern North Atlantic, *J. Vol. Geoth. Res.*, **111**, 55–87.

Hernán, F., Ancochea, E., Brändle, J.L. and Cubas, C.R., 1993, Características generales en el Edificio Norte de la Serie I de Fuerteventura, *Geogaceta*, **13**, 62–64.

Jarosewich, E.J., Nelen, J.A. and Norberg, J.A., 1980, Reference samples for electron microprobe analysis, *Geostandards Newsletters*, **4**, 43–47.

Le Maitre, R.W., Bateman, P., Dudek, A., Keller, J., Lameyre, J., Le Bas, M.J., Sabine, P.A., Schmid, R., Sorensen, H., Streckeisen, A., Woolley, A.R. and Zanettin, B., 1989, A classification of igneous rocks and glossary of terms, Ed. Blackwell, Oxford, 148 pp.

Leterrier, J., Maury, R.C., Thonon, P., Girard, D. and Marchal, M., 1982, Clinopyroxene composition as a method of identification of the magmatic affinities of palaeo-volcanic series, *Earth Planet. Sci. Letts.*, **59**, 139–154.

Luyendyk, B. and Bunce, E.T., 1973. Geophysical study of the northwest African Margin of Morocco, *Deep Sea Res.*, **20**, 537–549.

Schmincke, H.U., 1982, Volcanic and chemical evolution of the Canary Islands, In V.Von Rad et al., (Editors), Evolution of the Passive Margin of NW Africa. Springer, Heidelberg, 273–306.

Pearce, J.A., 1982, Trace element characteristics of lavas from destructive plate boundaries. In R.S. Thorpe (Editor), Andesites: orogenic andesites and related rocks. Chichester: Wiley, 525–548.

Shervais, J.W., 1982, Ti-V plots and the petrogenesis of modern and ophiolitic lavas, *Earth Planet. Sci. Letts.*, **59**, 101–118.

Smith, W.H.F. and Sandwell, D.T., 1997, Global seafloor topography from satellite altimetry and ship depth soundings, *Science*, **277**, 1957–1962.

Stillman, C.J., Fúster, J.M., Bennell Baker, M.J., Muñoz, M., Smewing, J.D and Sagredo, J., 1975, Basal complex of Fuerteventura is an oceanic intrusive complex with rift-system affinities, *Nature*, **257**, 469–470.

Uchupi, E., Emery, K.O., Bowin, C.O and Phillips, J.D., 1976, Continental margin of Western Africa: Senegal to Portugal, *Am. Assoc. Petrol. Geol. Bull.*, **60**, 809–878.

Thompson, R.N., 1982, British Tertiary volcanic province. Scott, *J. Geol.*, **18**, 49–107

Weaver, B.L., 1991, The origin of ocean island basalt end-member composition: trace element and isotopic constraints, *Earth Planet.Sci. Letts.*, **104**, 381–397.

Weigel, W., Goldflam, P. and Hinz, K., 1978, The Crustal structure of the Conception Bank. Mar, *J. Geophys. Res.*, **3**, 381–392.

Wessel P. and Smith W.H.F., 1995, New Version of the Generic Mapping Tools (GMT), http:/www.agu.org/eos_els, American Geophysical Union.

Winchester, J.A. and Floyd, P.A., 1976, Geochemical magma type discrimination; application to altered and metamorphosed basic igneous rocks, *Earth Planet. Sci. Letts.* **28**, 459–469.